ARCTIC EXCEPTIONALISM

ARCTIC EXCEPTIONALISM

Cooperation in a Contested World

Barry Scott Zellen

LYNNE RIENNER PUBLISHERS
BOULDER
LONDON

Published in the United States of America in 2024 by
Lynne Rienner Publishers, Inc.
1800 30th Street, Boulder, Colorado 80301
www.rienner.com

and in the United Kingdom by
Lynne Rienner Publishers, Inc.
1 Bedford Row, London WC1R 4BU
www.eurospanbookstore.com/rienner

Library of Congress Cataloging-in-Publication Data
Names: Zellen, Barry Scott, 1963– author.
Title: Arctic exceptionalism : cooperation in a contested world / Barry Scott Zellen.
Description: Boulder, Colorado : Lynne Rienner Publishers, Inc., [2024] | Includes bibliographical references and index. | Summary: "Explains the exceptional stability of the Arctic region from the colonial era to the present—even during periods of global conflict—and considers the potential impact of today's escalating state rivalries and renewed nationalisms"— Provided by publisher.
Identifiers: LCCN 2024010529 (print) | LCCN 2024010530 (ebook) | ISBN 9781962551106 (hardcover) | ISBN 9781962551274 (ebook)
Subjects: LCSH: Geopolitics—Arctic regions. | Arctic regions—Foreign relations. | Arctic regions—International status.
Classification: LCC G606 .Z45 2024 (print) | LCC G606 (ebook) | DDC 327.011/3—dc23/eng/20240508
LC record available at https://lccn.loc.gov/2024010529
LC ebook record available at https://lccn.loc.gov/2024010530

British Cataloguing in Publication Data
A Cataloguing in Publication record for this book is available from the British Library.

Printed and bound in the United States of America

♾ The paper used in this publication meets the requirements of the American National Standard for Permanence of Paper for Printed Library Materials Z39.48-1992.

5 4 3 2 1

For Tuvia, whose boundless optimism
renews my hope for our world

and for my parents, Bill and Paula, for their
boundless patience and encouragement

Contents

1

Arctic Exceptionalism

Arctic Exceptionalism: Cooperation in a Contested World is an examination of the collaborative foundations of Arctic diplomacy and security that have for centuries aligned the interests of the Arctic states, Indigenous peoples, and nonstate actors at the top of the world, even during periods of regional and global conflict and upheavals to the international order.

These foundations are being tested by the rise of new Arctic stakeholders such as China and other non-Arctic states with emerging economic, military, and diplomatic interests in the region as it opens up to increasing maritime commerce, resource development, and strategic mobility. Beijing, as part of its Polar Silk Road initiative, came under criticism from the United States and US allies for making opaque investments—particularly before Arctic stakeholders had become more familiar with the mechanisms of what came to be described as "debt-trap diplomacy." Nationalism and its impact on Arctic diplomacy are intensifying, as became abundantly clear during the 2019 Arctic Council Ministerial meeting in Rovaniemi, Finland, when Secretary of State Mike Pompeo publicly scolded China as an Arctic interloper, creating diplomatic fireworks at the otherwise collegial gathering. For the first time since its formation in 1996, the council broke with its tradition of producing a consensus statement at the meeting's end—not for Pompeo's undiplomatic dustup with China but rather because of the US pivot away from the climate change consensus that had hitherto united all Arctic stakeholders.

In these long months since Russia's February 2022 invasion of Ukraine, this trend has only accelerated, beginning in March 2022 with the unprecedented boycott of council activities by seven of the eight Arctic Council members under Russia's second rotating term as council chair, in protest to the invasion. This pause on Arctic Council activity was remarkable for the unity of the seven boycotting Arctic states, which in the past had experienced their own significant disagreements (primarily between the coastal states abutting the Arctic basin versus the inland and sub-Arctic states). Notably, five of the boycotting states were NATO members, and the other two, Finland and Sweden, were in the process of joining the alliance (with both now fully accessioned)—which will alter the diplomatic dynamics of the Arctic Council. Some fear it will risk permanently exiling Russia from the council's circle of consensus.

However, the Arctic Council has long prided itself on its collaboration across vast gaps in demography, geography, and economy, with its innovative inclusion of six Indigenous Permanent Participants who enjoy unfettered access to the eight Arctic member states, integrating state and tribal interests in a distinctive and exemplary manner. Yet the united stance of the seven democratic council member states against fellow member and then council chair Russia came without consultation with the Arctic Council's Indigenous stakeholders, who were caught by surprise as much by the boycott as by their exclusion from its discussion, a breach therefore of not only the Arctic Council's interstate harmony but its multilevel state-tribe harmony as well.

The Permanent Participants have largely given their ex post facto approval of the boycott, with one notable exception being the Russian Association of Indigenous People of the Arctic (RAIPON), which is state controlled at present with much of its former leadership in exile. The others gave their approval under immense pressure at a time of global consensus against what is perceived as Russia's naked aggression toward an independent neighboring state. Candid observations by Arctic Athabaskan Council (AAC) chief Bill Erasmus and Inuit Circumpolar Council–Alaska president James Stotts about Indigenous exclusion, as well as the importance of Indigenous engagement to inclusive governance in the Arctic, cannot be overlooked. The long track record of inclusivity across the centuries-old East-West fault line and the millennia-old tribe-state fault line is at great risk. Many fear it will not recover and that the Arctic's long cooperative tradition known as Arctic exceptionalism will be forever altered. But as this book chronicles, the roots of Arctic exceptionalism are deep and have endured numerous tests and challenges before.

Structural Foundations of Arctic Exceptionalism

The many complex governing and administrative structures—constitutional rights, legislative mandates, and judicial decisions—that have emerged in the modern Arctic, and their alignment with policies and principles embraced by the member state and Indigenous stakeholders on the Arctic Council, have helped to enshrine Arctic exceptionalism. This has established Arctic exceptionalism not as a normative aspiration but as an enduring dimension of Arctic international relations. It survived not only the bipolar global conflict of the Cold War but also post–Cold War efforts to protect and restore the fragile Arctic environment and post-thaw efforts to combat the unprecedented threats of climate change to the stability of the Arctic system.

How the intensification of state rivalry and renewed nationalism in the Arctic are affecting Arctic exceptionalism, and they in turn are affected by it, will be the focus of this book. Rooted in history and international relations (IR) theory, readers will see how realism in a world of anarchy is systemically impacted by the region's unique extremes, fostering alignments of interests among a diverse coalition of states, Indigenous peoples, and organizations who jointly govern the region and share a common experience of seeking order and survival in the remote, harsh, and ever-challenging Arctic. This was as true during World War II and the alignment of the Western Allies with Stalin's Russia as it was during more peaceful times. The Arctic, with its remote geography, harsh climate, and historic state weakness, has functioned better as an incubator of cooperation than of conflict. In today's contested world, with Europe aflame, it can and should continue to do so.

Indeed, the story of Arctic exceptionalism began long before the Arctic Council's formation a quarter century ago. I trace the roots of the region's continued, underlying commitment to consensus back to the centuries-long experience of collaboratively managing Arctic lands and resources between tribal peoples and the states that would come to assert sovereignty over their homelands. First the chartered companies of the colonial-era fur trade, with minimal numbers of settlers, integrated the vast subarctic and much of the Arctic region into the global political economy—leaving Indigenous polities largely intact (relative to other colonized regions). Then, as the newly formed modern states that now govern the Arctic expanded north in the nineteenth century, they by and large adopted this collaborative approach of asserting sovereignty by partnership with native proxies. This resulted in today's complex Arctic institutional environment defined by a patchwork of co-management systems enshrined by treaty, legislation, and constitutional mandate. The seeds were planted for an enduring, multigenerational commitment to work together despite obvious asymmetries in power, wealth, and demography.

This collaborative governing framework took international diplomatic form with the creation of the Arctic Council; the same players who learned to work together on domestic issues extended their cooperation into the international realm. New interests are managed largely by welcoming observer technical expertise into Arctic Council working groups while limiting their formal decisionmaking influence, which carries forth the spirit of collaboration from earlier eras. What defines this book, and my research on the Arctic, is this synthesis of deep history with IR theory. Thus we do not start with contemporary structures but rather come to understand their emergence over time, and we see in the collaborative sentiment expressed by so many Arctic stakeholders of great variety (large, small, weak, and strong states; stateless tribal peoples; cross-border Indigenous nations; multinational corporations and newly empowered native corporations, among others) not an idealist aspiration but rather the pragmatism born of realism, a balancing of interests amid anarchical pressures that are enhanced by geographical extremes.

So, when we look to today's diplomatic and strategic challenges, we see each rival state leveraging the perception of an intensifying Arctic race for domestic audiences and stakeholders, even though the region in fact has remained relatively stable. Its borders were mutually respected as the Great Game played out largely in headlines, a staged show to maximize budgets and modernize infrastructure, not unlike the earlier balance of power era when small wars were waged and alliances rebalanced to prevent a recurrence of great war.

While I accept and defend the premise of Arctic exceptionalism, I recontextualize it here for the real world of geopolitics and military conflict and find persuasive evidence for its continuation amid recent challenges and tensions. Embedded firmly in a realist framework, I present a sober look at the domestic dynamics and components of the Arctic order, each seeking to maximize its own gains, and how these competing interests align at the top of the world to establish a peaceful and stable region in an otherwise anarchic world where peace is in every stakeholder's interest and war itself is, on a large scale, a logistical impossibility.

Historic hot wars in the Arctic, such as the Japanese lightning conquest of the outer Aleutians, and earlier the Confederacy's final naval assault upon Yankee commerce in the Arctic Ocean in the summer of

1865, and before that the Battle of Hudson's Bay that brought the Seven Years' War into the Arctic region with the bombardment of the Hudson's Bay Company post at York Factory, were intense but brief battles that were part of wider conflicts whose centers of gravity lay far from the region and had modest local impacts at most. The Arctic, for a variety of reasons that will be explored in the book in detail, is inhospitable to many of the realist pillars of world order, including war, and this contributes to the region's tendency toward cooperative outcomes. But the Arctic is not entirely immune from the power-political pathologies of international relations, or completely insulated from their ravages.

This argument is important as it grounds Arctic exceptionalism in both realism and history. Much of the literature on Arctic IR overstates the emergence of a new Arctic cold war or great game and the competitive dynamics of Arctic international relations, overlooking the remarkable capacity of the region to resist the perils of international anarchy and its divisiveness. Its distinct geopolitics, born of isolation, remoteness, and cold, may be transitioning from what Mackinder called "Lenaland" into a more Spykmanian "Rimland," but this transition is not instant, nor does it completely offset the region's underlying harshness.

Because of the Arctic's unique political geography, the regional and domestic forces that shape Arctic diplomacy remain intact, even as new Arctic stakeholders arise on the world stage. Lacking their own territory, these emerging stakeholders (including rising imperial powers like China and long-declined imperial powers like Japan) are really just interlopers who may increasingly pass through and interact with the Arctic states and their structures, but they will always remain subordinate to the Arctic states and their empowered Indigenous peoples who jointly govern the region. All eight Arctic states, including Russia, have shown a remarkably durable commitment to collaboration, even as regional crises erupted around the world, up until the present collapse in circumpolar unity precipitated by Russia's invasion of Ukraine. Indeed, should Russia and its fellow Arctic states go to war—whether in the Baltics, Scandinavia, or the high North Atlantic—as a result of a collapse in international order, the Arctic will face a challenge unseen since World War II. Today's crisis thus threatens Arctic exceptionalism itself. Though not necessarily likely, scenarios of interstate war in the Arctic are no longer viewed to be entirely implausible, and these will be considered in the pages ahead, as will other "internal" scenarios of disruption to the Arctic system including the potential for a secessionary cascade starting with Greenland and expanding across the Inuit homeland of Arctic North America.

Arctic Exceptionalism Takes Root

As Canadian Arctic scholars Whitney Lackenbauer and Ryan Dean explain in their thoughtful and comprehensive overview, "Arctic Exceptionalisms" in *The Arctic and World Order*, "In its conventional application since the 1990s, the idea of 'Arctic exceptionalism' anticipates and promotes the building of a peaceable regime across the circumpolar north," where "either different norms or rules are or should be followed in the Arctic region, or that the region is exempt from 'normal' drivers of international affairs."[1] But at the same time, they note,

> Critics argue that Arctic exceptionalism (in its conventional conceptualization) perpetuates naïve, utopian faith in regional cooperation that cannot override global strategic competition, while simultaneously advancing the view that Arctic states must undertake extraordinary responses to protect their sovereignty and provide security in the Arctic because the region is exceptionally vulnerable. . . . Accordingly, while Arctic exceptionalism was originally used to advance the cause of peace across the region, our analysis illustrates how Arctic exceptionalist logic is also used to support narratives that portend future conflict and thus call for extraordinary action to defend the Arctic as a region apart.[2]

The authors trace the concept back to Gail Osherenko and Oran Young's *The Age of the Arctic: Hot Conflicts and Cold Realities*, where they observe that "'Arctic exceptionalism' had already emerged 'as a powerful force in the world' by 1989 when the Cold War was thawing,"[3] as reflected by the last Soviet premier, Mikhail Gorbachev, in his famed 1987 Murmansk Speech in which he called upon the example of Arctic exceptionalism to guide the world out of the Cold War. While the Cold War ended on its own accord with the collapse of the Berlin Wall, and not the Ice Curtain as Gorbachev had hoped, Lackenbauer and Dean point out that "Young and Osherenko observe that the Murmansk Speech encouraged the Arctic states, which had 'developed policies regarding their own part of the Arctic with little regard for other parts of the Arctic region,' to conceptualize a common region where they had 'much in common with each other.'"[4] Indeed, Lackenbauer and Dean further note, "The prospect of de-militarizing the Arctic agenda opened space to consider political, economic, and environmental issues previously subordinated to military security interests. In Canada, Prime Minister Brian Mulroney's Conservative government

(1984–93) shifted from a strong sovereignty and military emphasis in the mid-1980s to propose an Arctic Council of circumpolar cooperation that would foster peace and normalize political engagement on issues of common concern."[5]

The highly innovative Arctic Council, conceived near the end of the Cold War, would transform Arctic international relations, and would be defined as much by fostering a united, circumpolar Arctic international cooperation inclusive of the Soviet Union (and, after that state collapsed, post-Soviet Russia) as it was by fostering inclusive cooperation between the Arctic states and the region's Indigenous peoples, and it is this distinctive synthesis of tribe and state that reflects my own understanding of Arctic exceptionalism, not just after the Cold War but across the ages, dating much further back in time than the late Cold War thaw of the Ice Curtain that divided West from East. As I will demonstrate in the pages ahead, Arctic exceptionalism dates back at least three centuries and can be rooted even further back in time to prehistory and deep geological time where its underlying geographical context first emerged. Central to Arctic exceptionalism's meaning and relevance for North America is Franklyn Griffiths's prescription (as cited by Lackenbauer and Dean) for the newly proposed Arctic Council to exhibit for "civilized behaviour in relations between Arctic states, between these states and their aboriginal peoples, and in the way southern majorities treat their vulnerable northern environment," with relations "between these states and their aboriginal peoples" serving as a bidirectional causal bridge affecting the former ("relations between Arctic states") and the latter ("the way southern majorities treat their vulnerable northern environment," with environment being not just the geographical environment but also the sociocultural environment of the Indigenous Arctic majority). While this central proposition and its resultant collaboration has long been the hallmark of Arctic North America, it is less the case in many parts of the Eurasian Arctic, and so it is not yet a universal axiom of the whole Arctic, despite its importance as an aspirational value for the Arctic. Recognizing this limit, I focus primarily on Arctic North America in the pages below, and view Arctic exceptionalism primarily through a North American lens, even as I consider regions of the Arctic outside of North America, as well as non-Arctic states and interests, whose interests would continue to at times collide, straining Arctic exceptionalism time and again, but without dooming it.

Indeed, Oran Young (as cited by Lackenbauer and Dean, who are in turn citing Clive Thomas):

> conceptualized "the Arctic as a testing ground," where novel approaches to managing political issues and developing regional governance could yield important lessons and insights for other parts of the world. This concerned "indigenous peoples, the resolution of conflicts between the values of development and environmental protection, and international cooperation on such topics as fishing rights, animal migration and the preservation of cross-border ecosystems in general." While the region had distinctive hallmarks that allowed it to serve as a "testing ground," its "exceptionalism" had to be tempered for regional dynamics or experiments to offer broader lessons.[6]

But it's certainly no coincidence that these issues for which the Arctic could serve as a testing ground and model for the world were also the new, salient post–Cold War Arctic security concepts often described broadly as "soft security," including human security, in particular relating to Indigenous peoples, as well as environmental, ecological, and climate security. Nor is it a coincidence that during the next great era of international bifurcation to follow the Cold War, the two-decade-long global war on terror, Arctic exceptionalism proved to be not only a test bed for new relations between Arctic states and tribal peoples but a model that at least indirectly informed and inspired the coalition of states leading the war on terror that sought to understand the complex, localized conflicts along the periphery of the Westphalian world, and to introduce new institutions of governance to better integrate what was described by Thomas P. M. Barnett as "the Functioning Core" and the "Non-Integrating Gap." Arctic exceptionalism thus becomes a model for much of the world to emulate, relevant to all of those regions at the edge of the Westphalian world system where tribal peoples and remote geography combine to insulate their tribal homelands from the reach of the modern state. Arctic exceptionalism's relevance to the generational conflict that followed the Cold War—even if not fully appreciated by the stakeholders of that conflict who did not make a conscious connection between their state-building efforts in the remote periphery of the world system and the experiment already underway in the Arctic—cannot be overstated. This book is, in a nutshell, my effort to connect these dots.

As Lackenbauer and Dean observe,

> For most commentators, however, the idea of "Arctic exceptionalism" became inextricably linked to the twin assumptions that the region was a cohesive and cooperative space insulated from geopolitical tensions elsewhere, and that it was "exceptional" when compared to other regions. Heather Exner-Pirot and Robert Murray define the concept as

> "the successful effort" both "to maintain cooperation in the region despite internal competition for resources and territory," and "to compartmentalize Arctic relations from external geopolitical tensions." They argue that the Arctic regional order is exceptional insofar as Arctic states and those states with involvement in the area have worked "to negotiate an order and balance of power predicated on norms such as cooperation and multilateralism." . . . While "the Arctic is not immune from the possibility of war and conflict," [it requires] "conscious steps to maintain a strong institutional framework that protects Arctic internationalism."[7]

Such "conscious steps" have been the norm for much of the post–Cold War era. However, with the rise of China and its increasing assertion of interests in the Arctic region, combined (and for the most part aligned) with the resurgence of Russia in recent years, there have been new tensions and retreats from what had only recently emerged as a new and welcome circumpolar commitment to Arctic exceptionalism.

But even amid these retreats, there have been renewed intra-alliance and intra-bloc efforts to foster regional unity to offset the erosion of universal Arctic unity. Arctic exceptionalism thus does not transcend geopolitics, but rather geopolitics and its underlying geographical systems provide the context for and impose limits upon the extent of Arctic exceptionalism. The amazing thing is how malleable and durable Arctic exceptionalism has proven, even in the post–Cold War period. And if we extend Arctic exceptionalism back further in time, it's all the more impressive for its staying power, enduring tectonic changes in global geopolitics, the rise and fall of numerous powers, the recurrence of imperial expansion and war, and even now, the dynamic consequences of climate change. Arctic exceptionalism is thus not an exception from geopolitics but the product of geopolitics: a fusion of realism in international politics and geography, which constrains realism by buffering international anarchy and limiting the reach of the state in the Arctic.

With such deep and enduring geopolitical roots, we thus find the condition of Arctic exceptionalism predates its conceptualization. This interpretation at first seems to disagree fundamentally with what Lackenbauer and Dean describe as "conventional Arctic exceptionalist thinking," as exemplified by the prolific research of University of Lapland scholar Lassi Heininen, who explains, "'The globalized Arctic is an exceptional political space in world politics and international relations, based on intensive international, functional cooperation and high geopolitical stability. . . . This stability does not result from either the classical approach of Great-Game geopolitics or the Hobbesian zero-sum approach. It results from applying a critical and constructivist approach to geopolitics.'"[8]

But as my earlier work on realism in IR theory[9] has observed, realism is very much misunderstood to be antithetical to constructivism and critical theory, when in fact, realism may be more fairly viewed as a two-millennia-long tradition of constructivism in action, what I call "constructive-realism," an ideational response to the ubiquity of anarchy in world politics that dates at least as far back to the ancient Greek polis. Thus Arctic exceptionalism emerges from the very same "Great-Game" and "Hobbesian" dynamics that define international politics worldwide, with both imperial expansion into the Arctic and the arrival of the Westphalian state (with its "Hobbesian zero-sum approach") driving the emergence and endurance of Arctic exceptionalism. States aspire to order everywhere but find the world greatly lacking and insecurity the regrettable result. In the Arctic, however, geography, climate, and the human response to the region's remote austerity have more kindly cooperated, enabling a convergence of interests that is distinct from what we find elsewhere in the world—positioning the Arctic as not only a testing ground for new international relations but a model for application, with appropriate refinement and customization, to other parts of the world, particularly the "Non-Integrating Gap," where NATO was embroiled for two decades of asymmetrical warfare.

Much of what we think we understand about realism in world politics breaks down increasingly with distance from the Westphalian core, as states themselves increasingly diverge from the Westphalian model with this distance. Thus my view of Arctic exceptionalism, as consistent with and the by-product of Arctic geopolitics and realism in international relations, aligns with the view of Juha Käpylä and Harri Mikkola, also cited by Lackenbauer and Dean, "that the geographical and political distance between the Arctic and the southern metropoles that governed it facilitated the characterization of 'a unique region detached, and encapsulated, from global political dynamics, and thus characterized primarily as an apolitical space of regional governance, functional cooperation, and peaceful co-existence.'"[10] In fact, what we witness in the Arctic is less a separation from world politics than a reflection of the interplay of world politics with Arctic geography, yielding distinct regional dynamics no more detached from international relations than other regions, of which there are many beyond the Westphalian core.

Classical geopolitical theorist Halford J. Mackinder recognized this when he organized world politics into the geographical regions of

Heartland, Inner Crescent, and Outer Crescent, along with his lesser known Lenaland, a distinctly Arctic subregion of the world geopolitical system. And similarly, Michael Bravo's "tidy explanation" (as described by Lackenbauer and Dean) of Arctic exceptionalism "as a regional security complex with its own, independent, political calculus that is poorly explained by conventional realist theories of international relations"[11] suggests that Arctic exceptionalism is itself an exception from realist principles, when in fact, as I have argued here, it is more aptly understood as a reflection of realism as filtered through an Arctic lens. Lackenbauer and Dean point out how such efforts to define Arctic exceptionalism as separate from realism and its many power-political features have rendered the concept vulnerable by divorcing it unnecessarily from military history and the reality of political power, which is as real in the Arctic as it is at the Westphalian core, albeit in less centralized form and often more salient at the tribal than the nation-state level. As they note,

> By marginalizing traditional military and security issues, the Arctic exceptionalism embedded in these articulations of an Arctic security complex also creates vulnerability in suggesting that the reintroduction of defence considerations inherently undermines them. Furthermore, by prescribing that the logic of exceptionalism points to a certain type of regime predicated on liberal institutionalism, we might overlook different ways that other commentators—rooted in other schools of thought—also identify "exceptional" characteristics to justify or explain national behaviour and regional dynamics.[12]

In this volume, rooted in a realist understanding of world politics and a classical (albeit updated for our warming world) approach to geopolitics, I find a very close and intimate connection between the Arctic's exceptional qualities and its foundation in imperial expansion and state consolidation, shaped by such forces but reflecting local, regional, national, international, and transnational responses to them. Thus as Lackenbauer and Dean point out, those who worry about the dangers of new conflicts in the Arctic arising from the collision of interests by Arctic and non-Arctic states and non-, sub-, and trans-state actors are no less a part of the Arctic exceptionalism family. Their views and concerns, and cautious approaches to the Arctic future, are with equal merit and essential for us to understand the limits of Arctic exceptionalism in a less unified world. These "exceptional dangers" described by Lackenbauer and Dean, so acutely perceived and in many cases proselytized by what they equally elegantly call the "purveyors of polar peril," have indeed "projected a logic of 'Arctic exceptionalism' rather different from that advanced by the liberal internationalist school," one that is "inherently predicated on a form of exceptionalism positing that the Arctic Ocean was different than every other ocean—a narrative that inherently questioned Arctic state rights and control under established rules."[13] In short, Arctic exceptionalism has its darker side, but one no less exceptional, even if ultimately less cooperative in nature. And this too is rooted in Arctic exceptionalism's intimate connection to the realist world.

Notes

1. Lackenbauer and Dean, "Arctic Exceptionalisms," in *The Arctic and World Order*, edited by Spohr and Hamilton, 327.
2. Ibid.
3. Ibid.
4. Ibid., 328, citing Gail Osherenko and Oran R. Young, *The Age of the Arctic: Hot Conflicts and Cold Realities* (Cambridge: Cambridge University Press, 1989), 12.
5. Ibid., 328, citing Franklyn Griffiths, "Let's Invite Yeltsin to Join Our Club," *Toronto Star*, November 6, 1991.
6. Ibid., 329, citing Thomas, "Intergovernmental Relations in Alaska," 17.
7. Ibid., 329, citing Exner-Pirot and Murray, "Regional Order,"47–48.
8. Ibid., 330, citing Heininen, "Special Features," in *Global Arctic Handbook*, edited by Finger and Heininen, 216–217.
9. See Zellen, *Realist Tradition.*
10. Lackenbauer and Dean, "Arctic Exceptionalisms," 330, citing Juha Käpylä and Harri Mikkola, "On Arctic Exceptionalism," FIIA Working Paper 85, Finnish Institute of International Affairs, April 2015, chap. 4.4.
11. Ibid., 331, citing Michael Bravo, "The Postcolonial Arctic," 2015, https://core.ac.uk/download/pdf/35279618.pdf.
12. Ibid.
13. Ibid., 332–333, attributing the "purveyors of polar peril" phrase to Franklyn Griffiths, and citing Griffiths, Huebert, and Lackenbauer, *Canada and the Changing Arctic.*

2

Strategic Expansion to the Arctic Region

International relations, and in particular its realist tradition, has for much of the post-Westphalian period been state-centric, since the nation-state became the dominant unit in the European, and soon global, international order. As a consequence, modern international relations (IR) theory has followed from the European experience, positing the state as the foundational unit of world order. But this has not always been the case, even in Europe—and much less so elsewhere in the world. In fact, the current global asymmetry, while a break from the post-Westphalian tradition as understood through a Western lens, is not as discontinuous with history as is seen from other vantage points, whether from the tribal perspective along the modern state's fringes or from the Southern Hemisphere where the modern state was imposed from the north, with traditional boundaries between languages and peoples often ignored, sometimes intentionally as a result of colonial divide-and-conquer policies. The tribe-state fault line is thus important for an accurate understanding of world order, but it is only one fault line in the complex substate topology of world politics. The real challenge for us is understanding how to assemble the substate components into states and what sorts of sovereign units will make for a more durable world order. As with the debate over Iraq's future (and indeed, its existence during the darkest days of the war) and with the current debate over Syria's future, the answer is far from obvious. Indeed, it requires an understanding of the quantum level of world order, knowledge of the very foundational fabric of world politics.

The Three Pillars of Order

Most modern IR theory posits the state as the foundation of world order, with an international "system" in which states struggle to survive. During the Cold War, the "system" came to surpass the subsystemic components to some, known as the neorealists or structural realists, who became wedded to the very mechanisms of international order that they theorized about. As a consequence, even smaller units such as the solitary and foundational individual, the archetypical individual at the starting point of Kenneth Waltz's *Man, the State, and War*, and small aggregations of individuals, like the clan, sect, or tribe, were all but ignored. But it was these subsystemic components that would come to define the conflicts of both the immediate post–Cold War period and the global war on terror (also known as the Long War). And one can expect that they will continue to be enormously influential, especially with the proliferation of social networking sites and platforms around which smaller groups can coalesce, fostering their unity. Waltz became famous for his three images, associated with three levels of analysis (system, state, and individual coalesce, defining the architecture of world order and consigning all substate forces to his second (state) or first (individual) images. But the substate dimensions of the world order are in fact more durable, and while not always sovereign or independent, sometimes they are enough so that in times of conflict, they emerge as driving causal catalysts.

At this level, causality can be said to be rooted in what I have described as the "fourth image."[1] It need not always be tribal. There is also the more localized clan, the extended family network so powerful in prestate Indigenous societies (and central to current postwar plans for the restoration of order in Gaza), and the ethereal sect (as we saw during Christianity's fractious origins and like we see today in the many Islamist and jihadist movements and groups organized along sectarian lines). Clan, tribe, and sect all may be thought of as members of the fourth image, substate but causally distinct and politically and militarily efficacious. But they are also different, especially the sect, since this is a spiritual or theological distinction, albeit one that can be mapped to tribal or clan identity in many cases. In the process of exploring the underlying subcomponents of world order, I have developed a new taxonomy that is distinct from but ontologically linked to the levels-of-analysis concept (famously described by J. David Singer as the "level-of-analysis problem in international relations" in his 1961 *World Politics* article of the same title),[2] whether Singer's two-level framework or Waltz's more famous three images, which Singer himself found to be illuminating but unnecessarily restricted to a discussion of causality of war.[3] Singer noted that "during the debate between 'realism' and 'idealism' the analytical implications of the various levels of analysis received only the scantiest attention; rather the emphasis seems to have been at the two extremes of pragmatic policy and speculative metaphysics."[4] Just as Singer and Waltz found some methodological clarity in their discussions of the levels of analysis, we may turn to them to help organize our thinking on this quantum level of world politics operating in the subsystemic, and often in the substate, realm.

Singer ties the levels-of-analysis problem back to the fundamental issue of ambiguity in assessing causality in international politics, like the quantum theorists who probed the seemingly chaotic and at times baffling world within the atom, where causal certainty, which seemed to be resolved so neatly in the Newtonian world, could not be found. The problem is not so much the boundary between order and chaos as argued by the complexity and chaos theorists but rather between what is knowable (and thus predictable) and what is inherently not (and thus forever uncertain but not necessarily chaotic). This riddle of inherent ambiguity, with every bit as much salience as found in quantum theory, is what ultimately confronts the theorist of international politics. Perhaps it was best understood by the classical realists and their idealist counterparts who took a leap of faith and tied their theory to a fundamental view of human nature, rooting international causality in the human heart, the ultimate substate level of analysis, and endeavoring to nurture forth a more ordered world.

The substate realm is not new to theorists of international relations. The intensity of focus on nonstate actors since 9/11 suggests an entirely new level of analysis, but it is only the realization of this substate level's importance that is newly rediscovered. Since 1959, it has been known by many as Waltz's first image (the system), or as part of Singer's subsystemic but imperfectly named "national state" level of analysis. In his comparison of his two primary levels of analysis, the international system and the national state, reminiscent of Waltz's concluding thoughts in *Man, the State, and War* and of his later elaboration in *Theory of International Politics*, Singer writes, "In terms of description, we find that the systemic level produces a more comprehensive and total picture of international relations than does the national or subsystemic level. On the other hand, the atomized and less coherent image produced by the lower level of analysis is somewhat balanced by its richer detail, greater depth, and more intensive portrayal."[5] Intriguingly, Waltz's later work laid the foundation for neorealism, embraced the third image (the international system) for its causal primacy, and postulated a systemic ordering effect induced by the ubiquity of international anarchy. This mirrored the dual image Singer had presented nearly two decades earlier, which similarly saluted the totality of the picture provided by the systemic level while noting the granularity of the subsystemic level (which was not without importance, even in its lack of comprehensiveness).

If we step back to the foundational "trinity of war" so many attribute to Clausewitz, or further back to the three pillars of order enunciated by Machiavelli, or even further back to the three pillars of order that brought stability to Plato's just city in his *Republic*, we can see elements of the levels-of-analysis problem. Clausewitz's work, especially the trinity of war, connected the subsystemic dynamics (domestic political as well as human emotional) with the systemic (the surface boundary of the battlespace, in which two nations, two armies, and two commanders come face to face, with eddies of chaos and an infusion of friction and fog, to disrupt any expectation of a linear outcome, as Beyerchen has noted eloquently in his work).[6] Clausewitz thus grasped not only the ambiguity of the levels of analysis but the supreme importance of charting their boundaries and knowing their limits, something Machiavelli likely comprehended as well, even if his theoretical presentation was, owing to its early-modern emergence, less sophisticated and, therefore, less nonlinear.

But the endurance of the levels-of-analysis problem is notable, as is its connection to the famous trinity of war, or what I call the trinity of order that defined the Western polity and that has endured the passage of two millennia and the evolution from the polis to the modern state (and its fragmentation in some quarters). If we look to Waltz's famous three images, we can see in them an echo of Clausewitz's trinitarian structure. Indeed, Waltz's three images may be understood as a reformulation of the Clausewitzian trinity that conveys their interactive complexity (as noted by Beyerchen in his discussion of the complex, nonlinear movement of the pendulum caught between the magnetic fields of three magnets).[7] The key connecting thread is the inherent ambiguity of causality, which must not be overlooked.

Even as neorealism identified a greater causal probability if not outright causal certainty in the third image, it never slayed this Jabberwock, leaving us with the unresolved riddle of ambiguity that faced Clausewitz as well as the quantum theorists who, through their perseverance and open minds, helped give birth to the nuclear age. As Waltz tried to narrow the effects of ambiguity, emphasizing one level of analysis above the others (in contrast to Singer's more balanced approach or even Waltz's own earlier approach), he imagined a system that was, at heart, unbalanced, and that proved unstable. This instability becomes evident if we imagine the Waltzian images as counterparts to Clausewitz's trinity of war. Waltz eloquently articulated a trinitarian balance inherent in the three images, with man, the state, and war (individual, state, and systemic levels of analysis) largely corresponding to Clausewitz's trinity, with the people (the aggregation of the many individuals who define the first image and who in their collectivity become the *demos* or popular will, the embodiment of the nation but separate from the institutional structures of the state), the government (the structures of governance that defined the state or the sovereign), and the armed forces (the agent of war that operated at the boundary between competing states, sometimes marching across these borders, where they might interact with the raw popular will). The interaction of these three pillars takes place both within the domestic political space and in the more anarchic international arena, and sometimes in the enemy's domestic political space; the result can be a complex series of outcomes, from the conventional clash of battle to popular insurgency, with countless variables to consider.

The core pillars of order, the balanced relationship of the people, the state, and the armed forces, are found in all manner of states and even many substate polities, from the sectarian and tribal militia–dominated neighborhoods of insurgent-plagued Iraq, to the tribal militia–dominated enclaves of Afghanistan, not to mention other ethnically complex weak states such as Burma, where numerous ethnic and tribal militias limit the reach of the national government. Clausewitz's trinity, in essence, defines the limits of sovereignty: where people, state or substate governing entities, and armed force find a balance, an organic equilibrium is achieved that defines a stable sovereign boundary inside of which there is domestic order of some sort.

As Clausewitz observed in the wake of Napoleon, the trinity of war was both remarkably complex and at the same time essential to the maintenance of order; imbalances in any one of the three core pillars could prove catastrophic. Thus the apparent order enjoyed at such high cost (and grave risk to humanity's very survival) during the Cold War evaporated with a suddenness that few theorists of our time had predicted, but Clausewitz may well have understood it intuitively, knowing that a grand convergence of interconnected causal factors at various levels of analysis had taken place—just as it did in his own time. The order that the neorealists imagined was, in many ways, a period of profound disorder, and the chaos many saw at the Cold War's end may in fact be a grand restoration of order as sovereignty itself finds a more organic and hence sustainable equilibrium. Indeed, the Cold War's brief bipolar order was in large measure a synthetic construct and was thus destined to be ephemeral. The world that is now reemerging and has been emergent since the Soviet collapse has the potential to be more organically sound, structurally balanced, and, if fortune smiles upon our generation, durable. Such balance, as I contend, can be found in the Arctic, where geopolitics aligns with domestic and international structures to achieve a state-tribe equilibrium that is distinctive and exemplary for the world. During the protracted asymmetrical conflicts that followed the 9/11 attacks, elements of this balance were applied with much success, though a direct link between Arctic exceptionalism and the global war on terror, or Long War, was seldom identified. Instead, we can draw lessons from the war on terror for the Arctic and, in retrospect, identify the parallels between the tribal zones of conflict that came to define the generational war that the Long War came to be (so long a Long War that critics would come to refer to it as a "forever war") and the tribal polities of the Arctic.

Fur Empires: Colonial Foundations of the Modern Arctic

A long history of collaboration between the Indigenous peoples and sovereign states of Arctic North America has made the Arctic one of the world's most stable and cooperative regions. This long history of economic integration and globalization, dating back to the days of the Hudson's Bay Company (HBC, chartered in 1670) and Russian-American Company (RAC, 1799), has established an enduring foundation for the region's continued stability, from the very first trading posts to today's mix of multinational and native-owned corporations born of the Arctic land claims experience.[8] Institutions in the region have evolved for both domestic self-governance and diplomacy at the international level through the Arctic Council. This widely recognized spirit of international collaboration known as Arctic exceptionalism has come under new pressures from increased great power competition in the Arctic. Numerous states with interests and/or aspirations in the Arctic have articulated revised diplomatic and strategic policies, in notable contrast to prior decades of explicit mutual commitment to collaboration.

An imperial crossroads of economic importance for centuries, Arctic North America remains largely underdeveloped and underpopulated compared to other such crossroads of empire around the world. It is a vast, single region spanning five littoral countries abutting the Arctic basin and three nonlittoral countries, two (Finland and Sweden) with Baltic coasts and one (Iceland) surrounded by the North Atlantic. It encompasses numerous and diverse regional jurisdictions with significant Indigenous populations, who are often local demographic majorities, and when not, usually major pluralities. The transnational nature of the Arctic is rooted in the region's distinct history of global trade—particularly in furs, minerals, whales, and other marine mammal products—with agents of trade from whaling fleets to crown-chartered companies playing an outsized role, acting as proxies for distant sovereigns and in many cases acting as quasi-sovereigns. Combined with the endurance of significant regional Indigenous majorities and pluralities, this long history of economic integration and globalization continues to cast its influence into our era and, increasingly, non-Arctic states, catalyzing dynamic interactions with Indigenous peoples, polities, and corporations across the Arctic.[9]

Imperial Russia's northeastward expansion and conquest of Siberia were driven by the expanding fur empire that integrated its remote Arctic territories with its expanding imperial polity. In the European Arctic, it was a similar thirst for furs that drove not just Russia's expansion but also maritime commercial interests such as those of Britain's Muscovy Company and its predecessor, the Company of Merchant Adventurers to New Lands, which in the mid-sixteenth century cast its yearning along Russia's Arctic coast for a Northeast Passage to the Orient, pushing east to the Barents and White Seas and beyond past Nova Zemlya and the Kara Sea, finding lucrative fur, whale, and fisheries stocks along the way.[10] A similar quest by explorers and their state and corporate sponsors for a Northwest Passage beyond Greenland's western shores and through the frozen archipelago north of Canada's mainland had a similar effect. Combined with the expansion of fur trading networks throughout North America's remote far northwest, the convergence of continued exploration and corporate exploitation of natural resources fueled the integration of the Arctic's remoter territories into the globalizing economies of the late middle ages and early modern world—centuries before a genuine sovereign integration would take root in these frozen lands.[11]

Indeed, while post-Westphalian Europe was dominated by its core nation-states, along the remote periphery of the Arctic the chartered companies that were granted monopoly trading licenses by these very same sovereigns at Europe's core—entities like the HBC, the RAC, and the Royal Greenland Trading Department, all state-sponsored corporate enterprises—were empowered and would effectively govern much of the Arctic region.[12] Such a colonial division was not unique to this corner of the world, but what set it apart was its sustained effort to protect its wildlife resource base first and foremost, which necessarily entailed holding back on colonial settlement. This aligned corporate interests with native interests in a joint effort to sustain the commercial viability of wildlife resources, planting the seeds for the co-management processes and native corporations of the post–land claims and increasingly home-ruled Arctic of today. Only in Russia and Fennoscandia, where settler populations in many areas have for generations outnumbered natives, has the alignment of state and tribal interests not yet achieved parity with Alaska, Canada, and Greenland.[13]

Much of the literature on Arctic land claims views the relatively recent wave of successful negotiations across the Inuit homeland in North America since 1971 as the beginning of an Arctic modernization and globalization process, but history shows a longer integration of traditional northern economies and market economies with roots far outside the region.[14] The formative legacy of this history of economic collaboration has a tendency to be understated, with "Oil Age Eskimos" perceived to be a new phenomenon and Inuit economic interactions with the global oil industry a fundamentally new era. Such a perception has been reinforced by the successful efforts of native rights activists and jurists like celebrated retired BC Supreme Court justice Thomas R. Berger, who galvanized interest in and support for traditional native values through public hearings and thought-provoking reports and books based on those hearings that warned of a sudden, historically transformative moment. In the 1970s, Berger's Mackenzie Valley Pipeline Inquiry froze economic development in the Western Arctic until settled land claims could provide greater native participation in that development. He did it again during the Alaska Native Review Commission a decade later, and again two decades after that with his Conciliator's Report on Nunavut Implementation, and then again a decade later when he helped to save the Peel River watershed from development efforts that bypassed the co-management processes established by the Yukon First Nations Umbrella Final Agreement.[15]

A closer look at the Western Arctic region reveals the depth of these historical processes. The Western Arctic is bisected by a boundary that once separated two global fur empires, Russia's and Britain's, at the zenith of their territorial breadth—until the consequential sale of Russian

America to a reunified United States soon after its victory over the rebellious southern Confederacy. Hoping American possession would serve as an effective buffer to contain British North America, and thereby protect Russia's northeastern flank, Russia's logic held true until the embers of the War of 1812 cooled, along with the strategic tensions between the United States and Britain—paving the way for America's own imperial rise and eventual rivalry with Russia during the Cold War.

The region's relative underdevelopment and continued Indigenous demographic majority—resulting from the fortuitously and relatively light influx of settlers despite the predominant role of the world's first multinational corporations (MNCs), the crown-chartered companies of the colonial era—have contributed to their recent and historic re-empowerment. As the Western Arctic evolved from the age of empire to the post–land claims settlement era, this re-empowerment has positioned the Indigenous peoples of the region to be masters of their own fate. Although this native re-empowerment is the result of newly established institutions created to counterbalance the economic power of sovereign states and MNCs engaged in the region, their experience participating in global market–driven economic activities dates back more than 150 years, and in some cases three centuries or more.

While Russia and Britain asserted quasi-sovereign control over the region for centuries through crown-chartered trading companies, defining an international boundary that continues to subdivide the Inupiat homeland, these competing empires built upon the international fur trade were powered by a vast network of Indigenous hunters and trappers. Some, like the Unangan/Aleuts, were virtual slaves, displaced from their homelands by the conquering Russians, while others, like the trappers of Rupert's Land, maintained their cultural and much of their political autonomy even as they became integrated with the globalized British economy.

Both fur empires, in spite of their vast differences in governance and respect for Indigenous traditions, remained united in their mutual decision to only lightly settle the region, protecting the fur-bearing ecosystems upon which they depended by holding back the pace of colonial settlement.[16] Bockstoce has observed that the northern fur trade was already part of a well-established inter-Indigenous/international trading network linking Alaska natives to Chukchi traders across the Bering Sea, who in turn traded with both China and Russia, when Russia expanded across the Bering, displacing those preexisting networks by force.[17] The existence of such a prior network connecting northern furs to Eurasian markets, and its continuity under direct Russian control after its colonial expansion to Alaska and beyond, to the Russian River settlement in California where further expansion was contained by the northernmost reach of Spain's American Empire, suggests an enduring globalized political economy linking the self-governing era of precontact Indigenous polities to the colonial era, when the earliest MNCs first reached into the Arctic.

But even a light demographic intrusion of the sort initiated by the early Russian fur traders and the whalers who followed could prove calamitous, particularly as waves of new diseases carried by traders, whalers, or miners devastated long-isolated native populations. Indeed, epidemics decimated local populations of Mackenzie Inuit just like they decimated Inuit communities elsewhere international trade brought them together with Europeans during the early years of the twentieth century, exposing the region to a high risk of demographic upheaval. Ironically, it was the very porousness of northern borders and fluidity of cross-border migration by Inuit from Alaska, drawn in part by economic opportunities presented by the integration of their homeland with these expanding global trade networks, that preserved Inuit demographic predominance in the Canadian Western Arctic after the decimation of the Mackenzie Inuit. Similarly fluid cross-border migration by nonnative settlers during the sequential gold rushes forever transformed the demography of the Klondike and interior Alaska, catalyzing the expansion of the Canadian state into its far northwest to better police and administer this hitherto lightly governed region. This challenges many of our preconceptions about settlers and about what constitutes indigeneity; that both were in flux at this time gives the Western Arctic region a particularly dynamic nature.

As recounted by Robert McGhee in the foundational 1976 *Inuit Land Use and Occupancy Project Report*, "By 1910, the Mackenzie Eskimos were reduced to a few score survivors scattered among the more numerous Alaskan Eskimo immigrants who flooded into the Delta in the company of European whalers and traders."[18] He further recalled,

> After the appearance of the American whaling fleet along the Mackenzie Delta coast in 1889, and with the increasing association between the indigenous population and the whalers wintering at Herschel Island and elsewhere, the effects of disease and the disruption of aboriginal social patterns accelerated rapidly. . . . At the same time as Eskimos were being decimated by disease, local aboriginal culture was being submerged beneath a wave of American and Alaskan Eskimo introductions.[19]

The influx of Inupiat settlers from Alaska in the years that followed ensured continued Inuit demographic predominance and in many ways helped to solidify the cultural and linguistic cohesiveness of the region.[20] The survival and endurance of Indigenous polities moderated imperial ambitions toward the naked land and power grabs experienced elsewhere in the colonial world, most famously with the East India Company, which many scholars describe as the world's first MNC, and in terms of negative consequences to native peoples and polities, perhaps its worst.[21] Indeed, as Nick Robins has argued,

> The East India Company, romantic as it may seem, has more profound and disturbing lessons to teach us. Abuse of market power; corporate greed; judicial impunity; the "irrational exuberance" of the financial markets; and the destruction of traditional economies (in what could not, at one time, be called the poor or developing world): none of these is new. The most common complaints against late 20th- and early 21st-century capitalism were all foreshadowed in the story of the East India Company more than two centuries ago.[22]

Instead, the early northern chartered companies more generally, though not without exception (as was the case in Russian America), collaborated with the Indigenous peoples whose resources they coveted. In time that collaboration gave way to co-management and, later, joint ventures and partnerships between MNCs and the newly formed native corporations that arose from the historic land claims process. That is perhaps why even today, the HBC, an icon of Arctic colonial history, retains its popularity across the North.[23] As Burns reported after the HBC sold off some of its stores,

> To Indians, Eskimos and other Canadians living in communities served by the stores, from Labrador in Newfoundland to Whitehorse in the Yukon Territory, and in dozens of settlements in between, it will mean an end to trips to "the Bay," as the company's stores are universally known. It has been enough to stir protests in some of the affected communities, where the stores have become a symbol of contact with a distant and more comfortable world.[24]

Indeed, that history of contact reaches back centuries, and has become intimately connected with northern life for natives across Arctic North America.

The sale of the HBC's northern stores also riled many northern experts, as Burns further noted:

> Many enthusiasts for the old Hudson's Bay find the situation hard to accept, believing that the sense of abandoning history will be felt well beyond the company. One of them, the Canadian writer Peter Newman, at work on a multi-volume history of the company, has demanded that the parchment carrying the Royal Charter, housed in offices adjacent to the Bay store at Yonge and Bloor streets in Toronto, be transferred to the northern stores' new owners. "They're the ones who are fulfilling the charter."[25]

Peter C. Newman went on to publish his highly regarded three-volume history of the HBC—*Company of Adventurers* (1985), *Caesars of the Wilderness* (1987), and *Merchant Princes* (1991)—in addition to his illustrated history *Empire of the Bay: An Illustrated History of the Hudson Bay Company* published in 1989. *Up Here* magazine similarly turns to Newman to capture the grand historical legacy of the HBC:

> Historians speak in lofty terms of the grand old company's early days—no mere mortal men, the fur traders and explorers were titans, kings, Caesars in birch bark canoes. Through cunning, ambition, skirmish and diplomacy, the HBC gained sovereignty over a third of what is now Canada. It was a trade monopoly, selling furs and other resources back to Mother Britain, but to those living and working under the company's lordship, there was no greater authority, at least on this side of the ocean. "We know only two powers—God and the Company," an old HBC chief factor is quoted as saying by journalist Peter C. Newman in his accounts of the company. At the height of its power, some would flippantly say its iconic initials stood for "Here Before Christ."[26]

But *Up Here* also suggests a darker side to the HBC's intimate embrace of the North:

> While in the south the HBC became the Bay, a high-end department store, it kept its bravado in the North, where those old initials were sometimes explained, with a wink, as the Horny Boys Club. Writes employee George Cleveland in 1924, who is said to have had nine wives and 14 children, "The sounds that float from my kitchen, or out of the fur house on the three nights a week we hold our barn dances, tell the sky and stars that I am the most popular entertainer as well as the leading business man of Repulse Bay." But these days too came to an end, and the company sold off its Northern Stores Division (including its fur trading operations) in 1987. Many of the old Hudson's Bay buildings still scatter the far reaches of the once-great empire.[27]

Even under new ownership, however, the former HBC stores have remained vital to the communities they serve and a welcome and important part of the northern business community, in marked contrast to how the liberated, postcolonial citizens of India felt about the British East India Company and the displaced Aleut/Unangan community felt about the Russian-American Company after their departures—though in the latter case, Alaska's native communities were divided from the very arrival of Russian traders from across the Bering Sea as to whether their arrival was a risk or opportunity for Alaska natives. This division on the merits of economic integration with the fur empires at times led to armed clashes not only between natives and Russian traders but between natives as well.[28] The emergence of a new community of Métis, or Creoles as they are known in the former Russian-American Company territories,[29] further complicated the narrative, with a new, hybrid demographic occupying the middle ground between the foreign MNCs of the colonial era and the Indigenous peoples they encountered.

Compared to the experience of HBC-governed Rupert's Land, Russia's treatment of the Aleuts has been described as "brutal." Sociologist Dorothy M. Jones, in her moving history, *A Century of Servitude: Pribilof Aleuts Under U.S. Rule*, writes,

> In the first fifty years of Russian occupation, the free trade period, Russian fur hunters brutally mistreated the Aleuts and at the same time commanded their labor. The Russians stole the Aleuts' wives, slaves, and possessions, and slaughtered any who resisted their domination. They sent Aleut men on long sea hunting expeditions from which many never returned and during which many women and children, left alone in the villages, suffered severe deprivation. This mistreatment, combined with the diseases the Russians introduced, nearly decimated the Aleut population. In the first thirty years of Russian contact, the Aleut population declined from an estimated 12,000 to about 1,900.[30]

Jones further explained that this brutality by Russian traders catalyzed Russian state interest in its new North American territories, with the foundational charter of the RAC committing the Russian crown to the protection of natives from settlers:

> Eager to prevent ruinous competition between its companies, to regulate traders' treatment of the Natives, and most importantly, to protect and expand its sovereignty in Russian America (its first overseas

> colony), the Russian government in 1799 granted a monopoly to a private firm, the Russian-American Company. The government gave the company not only a monopoly on trade but authority to govern and garrison the new territory. . . . The first charter granted to the Russian-American Company contained no definite regulations about the status and treatment of Natives other than an injunction to treat them amicably and convert them to Christianity. The second and third charters, in 1821 and 1844, however, specified Natives' political status. Aleuts and other Natives under company administration were declared Russian subjects: "Tribes inhabiting the places administered by the company are . . . Islanders, Kurils, Aleuts, and others. . . . As Russian subjects they shall conform to the general laws of the empire and shall enjoy the protection thereof." The protection was hardly forthcoming.[31]

But sovereign protection was nonetheless a principal motivation for Russian imperial interest. HBC, similarly, aspired to provide the natives of what would later become Canada's northern territories with a protective, mutually beneficial, and sustainable relationship based on enduring multigenerational reciprocity with the hunters and trappers of Rupert's Land. In many ways, the HBC (and before its 1821 merger, the North West Company[32]) seeded the primordial north, where subsistence and regional trading networks, which already interconnected with an Indigenous international trade system that stretched across the Bering Sea to Asia,[33] facilitated a lighter and less intrusive form of colonization that did not require the subjugation, displacement, or enslavement of the native population. In fact it called for much more than the mere survival of the Indigenous people of the North: namely, their sustained and supportive participation in the globalized commercial activities that would come to define the northern political economy.

As noted above, John Bockstoce observed that the northern fur trade was already part of a well-established inter-Indigenous/international trading network linking Alaska natives to Chukchi traders across the Bering Sea, who were in turn trading with both China and Russia, when Russia expanded across the Bering and displaced those preexisting Indigenous trading networks with its own. Bockstoce recounts from his journeys through the Arctic,

> I met a number of Natives who many years before had participated in the fur trade, and I began to appreciate what their lives had been like as frontier trappers in an earlier era, when their existence had been simpler and harder, yet to them was remembered fondly. The fur trade of Bering Strait was one aspect of the European expansion into the

> most remote regions of Asia and America. . . . Far beyond the Europeans and Americans who sought to buy furs, ivory or whalebone for the markets of the south, members of fifty native nations provided these commodities to one another—and to foreigners—in return for goods that they required or desired. Manufactured goods, coastal products, inland products, tobacco, tea, alcohol, and hundreds of other things changed hands many times in the immense region between the Kolyma River in the west and the Mackenzie River in the east. No matter which goods were exchanged, these transfers were almost universally regarded as advantageous by both parties.[34]

The experience of natives in Rupert's Land (as the HBC heartland was known) was in marked contrast to the initial Aleut (Unangan) experience in Russian America, whose population was forcibly displaced from their traditional homeland into southwest Alaska (and whose hunting activities on behalf of the RAC would reach as far south as Mexico) and virtually enslaved by the Russian invaders—a tragic experience that still casts a painful shadow across southwest Alaska, where Aleutian islands remain uninhabited and impoverished Aleuts remain in urban exile.

As valiant and well-meaning as such efforts are to postpone development and protect a supposedly pristine North from the insatiable appetites of expanding states and global MNCs, to ignore or understate the long history of Arctic globalization does a disservice not only to history but to northerners who, like others around the world, long for economic independence and the empowerment that comes from economic growth and opportunity. The same is true for those who see in global warming a sudden awakening of global interest in the Arctic and who mark the polar thaw as the start of a sudden "scramble" for the Arctic,[35] when that scramble has been centuries long and ongoing—since the medieval world sent forth the first maritime explorers and the colonists followed. A closer look at Arctic history reveals there have been, in fact, many waves of Arctic globalization dating back centuries, modernizing and integrating its economy and culture with the global economy and culture along the way.

Even after the famed oil strike at Prudhoe Bay a half century ago, with the rapid expansion of multinational Big Oil to the North Slope and the surge in exploration that extended deep into Canadian territory, and despite the construction of the Dalton Highway (also known as the "Haul Road") in 1974 linking the Alaska Highway to the Arctic coast and the subsequent opening of the Dempster Highway soon thereafter in 1978 under Canada's Roads to Resources development policy, the pace of external settlement has remained relatively slow, and Inuit demographic predominance continues to define the coastal region from end to end.[36] Gordon Jones observes how the remote geography and high costs associated with oil and gas exploration in the Arctic precipitated an industry-wide collaboration in the mid-1960s, leading to the formation of Panarctic Oils, Ltd., a consortium of nineteen oil-industry actors and the Canadian government. It is intriguing how collaboration is embraced, over and over again, as an essential value for corporate survival in the Arctic, from the early years of the fur trade to the modern oil age when consortia like Panarctic arose to meet the challenge of northern development. A similar consortium was formed by BP Pipelines (Alaska), Inc., ConocoPhillips Transportation Alaska, Inc., ExxonMobil Pipeline Co., and Unocal Pipeline Co. to harness the opportunity and divide the risk associated with the Trans-Alaska Pipeline System (TAPS)—and across the border in Canada's Western Arctic, Canadian Arctic Gas Pipeline Ltd., with twenty-seven Canadian and American producers including Exxon, Gulf, Shell, and TransCanada involved. Like other consortia of that era, these were broadly representative of petroleum MNCs but lacked a native equity component. But even without an equity stake in these projects, or perhaps because of their exclusion and the sympathy this engendered, natives organized effectively and found passionate defenders among native rights activists and jurists like Berger. Their efforts thus won major dividends: in Alaska, it led to the pioneering Alaska Native Claims Settlement Act (ANCSA) of 1971, which transferred nearly $1 billion in capital and 44 million acres in land title to new Alaska Native Corporations (ANCs), seeding the development of interior Alaska and the empowerment of the native community, after which TAPS rapidly constructed the 800-mile Alyeska Pipeline from Prudhoe Bay to Prince William Sound.

Across the border in Canada's Western Arctic, during the first Mackenzie Valley Pipeline Inquiry in the 1970s, Berger recommended delaying pipeline construction for a full decade to provide sufficient time for natives to pursue land claims and create institutions to ensure their participation in the development of the North. As well-meaning as this was, the cost was high: no pipeline was built, no jobs were provided, no native contractors were engaged. When the Mackenzie Gas Project resurrected the pipeline project after many land claims in the region had been settled—with one-third native equity participation in the pipeline consortium via the Aboriginal Pipeline Group—the consultative process architected by land claims treaties in the Western Arctic, based on the important principle of co-management, ran over seven years from start to finish, a process so exhaustive with a pace so slow that the macroeconomic conditions that had made a pipeline economically viable no longer existed by the time a favorable decision was reached. Meanwhile the Alyeska Pipeline continues to pump North Slope crude south to Valdez and from there to world markets—and despite MNC equity ownership of the Alyeska Pipeline Service Company, it is at heart a northern company—96 percent of its workforce lives in Alaska, one-fifth of which is native, and 70 percent of TAPS contractors are based in Alaska, many native-owned subsidiaries of the ANCs. Indeed, Alyeska's Alaska Native Utilization Agreement and Alaska Native Program illustrate the company's continued efforts to foster native inclusion and opportunity—and its advisory board includes both Alyeska employees and leaders from the pipeline corridor's ANCs (Arctic Slope Regional Corporation, Doyon, Ahtna, Chugach Alaska), contractors, and other Alaska native organizations.[37] Big Oil came to Alaska, and it has fueled not only the statewide economy but the native economy as well, without forever altering the native majority along Alaska's Arctic coast—much the way early MNCs from the colonial era reached northward without an associated tidal wave of settlers.

While an influx of settlers farther south during the Klondike Gold Rush did permanently rebalance the demographics of the Yukon Territory, and to a lesser degree the Mackenzie River valley (where oil has been successfully extracted since the 1920s), the Western Arctic region overall has largely escaped such a fate, as has the Eastern Arctic and Greenland.[38] The oil town at Deadhorse/Prudhoe Bay on Alaska's North Slope is a rare exception to this, as was the whaling community at Pauline Cove on Herschel Island. During the final years of commercial whaling there was interest in Arctic oil by the whaling industry, a precursor to modern Big Oil and a direct catalyst for the British Colonial Office's formal sovereign claim to Arctic territories beyond Rupert's Land in the 1850s, as they accurately perceived the "threat that the American presence posed to British territorial rights."[39] W. Gillies Ross cites a September 5, 1853, Colonial Office memo that "appears to be the first explicit declaration of British sovereignty over any part of the Arctic Islands beyond the limits of Rupert's Land, and it arose directly out of whaling activity."[40] MNC activity, from its very beginnings in the Arctic, has had profound consequences.

In 1968, more than a century after the Russian displacement and enslavement of the Aleuts, when oil was found in bountiful economic quantities on the North Slope, a frenzied exploration boom extended far across the Western Arctic region, with seismic survey crews entering native lands, leaving scars still visible from the air. The threat to native stewardship of the North presented by modern petroleum MNCs catalyzed a movement across Alaska and the Western Arctic for the preservation of native rights, traditions, and lands, culminating in the historic land claims movement. Curiously, it was not just the threat to natives from Big Oil that precipitated the pioneering Alaska Native Claims Settlement Act in 1971; it was also the threat of years of native litigation against Big Oil and its plan to pump oil out of the North Slope to southern markets via the Trans-Alaska Pipeline System that worried the state of Alaska and the US federal government, whose economic and energy security were both tied to the successful development of northern Alaskan oilfields.[41] It was thus the historic convergence of oil, money, land, and power that brought together Big Oil, the Alaska natives, and both the state and federal levels of government to hammer out the unprecedented ANCSA land claim.[42] And despite ANCSA's many initial structural flaws, the Alaska land settlement would nonetheless spark a wave of settling claims that would ripple across the North, flowing across the border to the Western Arctic and on to Nunavut, to Nunatsiavut in northern Labrador, up the Mackenzie River valley into the Dene homeland, and up the Yukon River and its tributaries into the communities governed by the Council of Yukon Indians, now known as the Council of Yukon First Nations.

As the land claims model transformed unsettled lands to the east and south of Alaska's North Slope, it evolved and expanded to include new models of self-governance that radically transformed the land claims model from its original primary intent to economically integrate the North (and thereby assimilate the North into the modern political economy). Many critics felt ANCSA was designed to fail, to precipitate a transfer of lands, wealth, and corporate control from the native community back to the nonnative majority as Alaska Native Corporations inevitably succumbed to bankruptcy. That's because ANCSA only protected native ownership and control for twenty years and did not guarantee "new natives" (Alaska natives younger than eighteen at the time of ANCSA's signing as well as all those born later) shares in the new native corporations and thus in the governance of their lands and resources—a situation described as the "1991 time bomb," named for the year when the Alaska Native Claims Settlement Act was widely expected to collapse.[43]

But in those first twenty years that followed ANCSA, as the clock ticked toward detonation in 1991, two things happened: First, land claims spread across the border via the Inupiat, whose original excitement (and newfound wealth) inspired the Inuvialuit, primarily descendants of the Inupiat from an earlier migration into Canada, persuading them to quickly settle their own land claim—breaking with the Inuit of Nunavut, whose lands were not yet under assault by the seismic crews of Big Oil. The Inuvialuit were able to improve upon the land claims template dramatically, ensuring all "new natives" were automatically enrolled as shareholders and preventing shares from ever being sold to nonnatives, ensuring continuity of native corporate control for perpetuity while also preserving native subsistence and protecting the land and its wildlife. Second, Alaska natives recognized the original ANCSA structure that was agreed to in 1971 was fatally flawed and worked steadfastly to improve it, taking inspiration from their Canadian brethren, who showed an improved model could be agreed upon with government. The 1991 time bomb was thus defused, and in Alaska the native corporations that were reborn from the brink of insolvency caused by their early days of excess and inexperience are now an important stakeholder in Alaska's economy, comparable to the native corporations in Arctic Canada.[44]

All this took place because the corporation came north before the state, in some cases centuries ahead of the state's formal arrival to the North. These crown-chartered companies served as imperial proxies but under the administration of corporate employees rather than political leaders and appointees. Some would govern despotically, while others like the "Company of Adventurers" (as described by Peter C. Newman in his many volumes about the HBC)[45] would govern in a more humane and sustainable manner. Because the Arctic did not have forests to cut or plantations to plant, it did not need an influx of farm laborers to upend the Indigenous demographic predominance. The political economy that emerged was based on the fur trade, which required native participation and continued stewardship over their lands. And so the chartered companies of the North, notably the HBC and the North West Company, would become collaborative partners in its modernization, bringing northern natives along with them into a future that would be defined by their symbiosis.

The corporate model thus adapted and responded to the unique conditions of the North, allowing the corporation to become the foundational unit of governance for northern natives, the gateway toward true self-government rather than a vehicle designed to abolish self-government, as many believe the original ANCSA was intended to do.[46] It was the ANCs born of the land claims process that introduced the northern native leadership to their first experience of governing—and paved the way toward the more balanced form of co-management across the North that has brought native participation in Arctic development activities to a new level. As a model of governance it has crossed over into southern Canada, taking root (in the face of settler opposition) in British Columbia and the Prairie Provinces, even in the absence of settled claims in some cases. Co-management, indeed, is perhaps the most contemporary expression of the economic integration that began with Arctic colonization by those MNCs centuries ago, with its hybrid nature balancing tradition and modernity, subsistence and market forces.[47]

Today's Arctic presents us with a fascinating realm of settled land claims, dynamically evolving systems of Indigenous and regional governance, distinctly Indigenous and collaborative international diplomacy, and flexible balancing of subsistence culture with economic modernization and development, blending two worlds, one traditional, one contemporary—a compelling example of enduring order in the absence of strong state institutions. The experience is far from uniform, with the Russian Arctic and Fennoscandia experiencing a more disruptive wave of demographic transformation resulting from colonial state expansion than experienced in Arctic North America. But even there, pan-Arctic flows of ideas about co-management and native economic participation as well as Indigenous diplomatic activities by the Indigenous Permanent Participants of the Arctic Council have helped nudge governments of those Arctic states to engage with their Indigenous communities, and to commit to more collaborative management.

From Whaling Ships to Trading Posts: Nineteenth-Century Arctic Globalization Hubs

Mary K. Bercaw Edwards, associate professor of English and director of Maritime Studies at the University of Connecticut, is a longtime contributor to the preservation of the *Charles W. Morgan* at Mystic Seaport, the last of an American whaling fleet that once exceeded 2,700 vessels, that was built and launched in 1841. She's also an expert on whaling culture as expressed in American literature, including Melville's classic works, in addition to her contribution to the *Morgan*'s legacy and preservation. Her work on whale ships as platforms of the fast-globalizing culture of the maritime world has helped me to understand the fragile yet persistent role of comparable remote and/or mobile platforms of modernity, world trade, and the increasingly global state in colonial-era

Arctic and subarctic lands and waters. What Melville described aboard the *Pequot*, named for one of the first Indigenous peoples of the new world to be annihilated through war and formally extinguished by treaty into legal and political nonexistence (genocide by treaty), was a frenetic, multicultural, multiethnic, and multinational culture, a middle ground at sea where oceanic peoples and representatives of the global maritime economy and culture came together into one kaleidoscopic community. This was not so different from what was experienced on land at trading posts across the colonial Americas. I will explore these colonial foundations below in my discussion of the fur empires that integrated remote subarctic and Arctic regions of the New World with the Old World. Then, I'll proceed to the HBC and RAC and their interactions with the Indigenous peoples of Alaska and the Canadian North during another fascinating chapter of early Arctic globalization, as new "company-states" emerged that predate the formal integration of these territories into the United States and Canada.

Whether trading posts or whaling ships, for generations prior to the emergence of self-governing Westphalian states in North America, we observe a mix of smaller platforms of sovereignty by proxy that served as the primary (and in some cases sole) interface between the Old World and the New World of the Arctic.

As Bockstoce observed above, the colonial-era fur empires that ruled much of Arctic North America were each powered by a vast network of Indigenous hunters and trappers. Some, like the Unangan/Aleuts, have been described by historians as virtual slaves,[48] displaced from their homelands by the conquering Russians and coerced by them into a lifetime of forced labor, as noted above by Jones. Others, like the trappers of Rupert's Land and the old Northwestern Territory, maintained much of their autonomy even as they became integrated with a globalized economy, and ever more dependent on its trade goods. But as a result of these northern chartered companies' dependency upon native labor, both sides of the Russian/British North America boundary would ultimately only lightly settle their colonial dominions, in part due to their respective efforts to ensure the sustainability of wildlife resources that fueled their economies, with Russian-American Company settlers never totaling more than 5,000 and the Hudson's Bay Company said to have peaked at only 3,000 despite the vast extent of their territories. Both RAC and HBC overharvested and decimated local wildlife populations, forcing their trappers and hunters to seek new, untapped ecosystems farther afield, but like farmlands left fallow, they would return once replenished.

But even the light demographic intrusion of the sort initiated by the early Russian fur traders and the HBC's "company of adventurers" who came to dominate the interior, as well as the coastal whalers who traded seasonally by sea (and occasionally overwintered, though usually not by choice), could bring calamity even without the direct displacement experienced farther south (as exemplified by the Cherokee "Trail of Tears"[49]), particularly as new illnesses carried by fur traders, whalers, or miners devastated native populations, whether smallpox, tuberculosis, influenza, or other diseases. But in spite of these dangers, globalization continued to integrate the northern and southern economies, establishing a pattern that is retained in the post–land claims Arctic, where land claims treaties, at once a tool of state expansion and consolidation and a tool of decolonization and Indigenous empowerment, retained the corporation as a central component of their structure.

The arrival of both mobile maritime and remote terrestrial platforms of trade and proxy sovereign representation had a profound and transformative effect on the peoples and polities of Indigenous Arctic North America. These various platforms, whether on land or sea, established trade links that endured for generations and reshaped empires. Both Russian America and British North America enjoyed quasi-sovereign control for centuries via crown-chartered trading monopolies, representing what Edward Cavanagh calls "company-states."[50] They were not entirely sovereign, but in many important ways they exercised sovereignty—whether in the administration of justice, the management of regional economies and international trade, or in defense and security matters—while serving as proxies for distant sovereigns whose economic, military, and diplomatic support undergirded their efforts. In many ways this model of governance came to define the nature of governance in Arctic North America, with a preponderance of corporate entities, such as natural resource exploration and extraction companies, in quest of resources, reminiscent of the fur empires and whale trade (itself a predecessor of the modern petroleum trade), wielding enormous local and regional influence over relatively weak governing entities.

"Company-States" of the Fur Trade: Colonization by Gentler Means?

The experience was not entirely unlike that experienced elsewhere in the colonial world, such as India under the British East India Company

(EIC), as Nick Robins observed above. And while Bockstoce, also noted above, illustrates the deep and enduring participation by Indigenous peoples in the fur trade and the mutuality of gain enjoyed by both Indigenous and non-Indigenous participants of that trade, critics point out there are fundamental asymmetries and inequalities in the power relationship between the two.

As Melissa Gismondi has noted,

> While the HBC provided urban shoppers with a range of goods, many indigenous peoples maintained a different relationship to the company, particularly those in northern communities where the HBC trading post was the only store around. As related in The Other Side of the Ledger, a 1970 documentary produced by Canada's National Film Board, the HBC operated some 100 stores in indigenous communities into the 20th century. Charging the company with setting low prices for furs and high prices for their goods, a process that kept indigenous consumers in a perpetual state of debt, narrator George Manuel, then-President of the National Indian Brotherhood, notes, "The Hudson's Bay Company has almost complete economic control and through this power governs the lives of our people." Ojibwe artist and scholar Duke Redbird, who appears in the documentary, says this relationship persisted until 1987, when the HBC sold its interests in Canada's fur trade and northern trading posts to an American company.[51]

Gismondi added that the "history of the HBC is messy and complicated. It has moments of compassion, but also competition and contention. It's the history of global capitalism, North American colonialism and the British Empire."[52]

Perhaps the biggest advantage is that it prepared people on both sides of these trading relationships for the world to come. Trading posts set the stage for the Arctic's modernization. While the fur empires brought either mobile maritime or remote terrestrial platforms of sovereignty to Arctic North America, serving as an economic bridge between the old world and new, and even left largely intact ecosystems that would provide much future wealth to the sovereign states that arose on this colonial edifice, as noted they also introduced new risks, particularly pathogens and illicit substances like alcohol or opium, which were opposed by the major chartered companies that shared, relative to some of their less ethical peers, a commitment to the well-being of the Indigenous peoples of the North. While war was a feature of the continental expansion of the young United States as it sought to overpower and dispossess the Indigenous polities to its west, its presence in Arctic North America was the exception and not the rule. But as an exception, it would on occasion disrupt the pacific calm that otherwise characterized the Arctic region and the exceptionalism for which it has become known.

One notorious example took place during the US Civil War, when the Confederate commerce raider, the CSS *Shenandoah*, took aim at the Yankee Arctic whaling fleet. In a desperate gambit to compel the Union to negotiate an end to the war, it attacked the North's lucrative whale oil trade at its source, offshore of what was still Russian America.[53] Thus the colonial world of Arctic trading posts and whaling ships presents us with a fascinating episode, and brief counternarrative, where a rogue state, in this case the Confederate States of America, exploited the absence of effective sea control by the quasi-sovereign crown-chartered companies, particularly the RAC, though unbeknownst to the RAC, the HBC was actively trading in and around Fort Yukon, deep in RAC territory. The RAC possessed nominal sea power assets barely sufficient to sustain the economic bridge between colonial territories and the Russian mainland, leaving no capacity for naval protection of remote coastal regions from a rogue external actor like the *Shenandoah.* And while both the RAC and the HBC exercised reasonable (if not absolute, as HBC's trading presence on Russian-American territory illustrates) territorial control, offshore lay a veritable maritime free-for-all allowing for nearly ubiquitous access by foreign-flagged vessels, making the waters off RAC lands a virtual American lake benefiting the Yankee whalers and, in the case of the *Shenandoah*, drawing the attention of a foreign belligerent. Onshore there were disputes over contested lands as well, as seen during the decades-long battle over the Northwest Territory between the HBC and its rival, the North West Company, which continued by proxy the long Anglo-French rivalry in the New World after the French and Indian War formally concluded. In later years, similar free-for-alls emerged with the various gold rushes in which gold miners, often from the United States, flowed into new areas of the North, regardless of their nationality. The Klondike gold rush of 1898 was thus largely an American economic experience though it took place on Canadian soil,[54] and just off the Yukon North Slope on Herschel Island, overwintering whalers, also predominantly American, established a persistent presence that made Canadian authorities nervous.

The success of the *Shenandoah* mission illustrates the anarchy that can emerge in lightly settled colonial polities, revealing a risk in what is otherwise an advantage to native peoples. Russia, always worried about British expansion toward its flanks, saw in the reunification of the United States in 1865 an opportunity to create a buffer state by selling its colony in Alaska to the newly reunited nation, relying on American power to contain British expansion and leveraging to Moscow's advantage the de facto US presence in RAC waters, which was until then a liability to the RAC. Whale ships were more often challenged by sea ice, the unpredictability of Arctic weather, the vagaries of whale migrations, and the potential for hostile reactions by natives than by war ships, even converted, unarmored, lightly armed raiders like *Shenandoah,* which had more bark than bite. So they remained vulnerable to what was perceived to be an unlikely risk. The undefended nature of the Yankee whaling fleet is what made it such a tempting target to the Confederacy, much like the Saudi oilfields today (as the Iranians realized when launching a cruise missile raid in 2019[55]) and, one might argue, remote oil and gas infrastructure in both Alaska and Siberia. As one sees in the few surviving *Shenandoah* officer diaries that remain today, many a Yankee whaler regretted surrendering to what they thought was a bona fide war ship upon realizing the *Shenandoah* used deception to feign greater destructive capabilities than it really possessed, being at heart a vessel of commerce designed for the Bombay spice trade, and barely armored.[56]

Reflecting upon the colonial-era fur trading companies and their intriguing quasi-sovereign status (as Cavanagh, above, has aptly described with his "company-states" concept)[57] illuminates their historic role in the ongoing process of Arctic globalization, integrating the Arctic interior and coastal regions with the global economy well ahead of formal state presence in, and direct sovereignty over, the region. In many ways, the cruise of the *Shenandoah* smashed this architecture, revealing the weak grasp that European states had over the seas adjacent to their colonies and the maritime free-for-all that resulted. American whaling captains were largely in a position of complete autonomy, as were the officers and crew of the *Shenandoah,* who were later hunted as pirates once their rogue state collapsed as a sovereign entity and was forcibly reintegrated into the United States. For the Russians, the US commercial presence reinforced their perception that American interests in the Arctic and historical rivalry with Britain positioned the United States as an appropriate sovereign for its former Alaska colony, a buffer state whose renewed unity, military power, and economic influence could protect Russia's northeastern flank from British power.

Tribalism Meets Globalism at the Top of the World

If the trading post (or whaling ship) was central to the globalization experience of the Arctic, and their small command, staff and/or crew, and isolation from the traditional tools of power of the state made them uniquely vulnerable as colonizing agents and particularly sensitive to the needs and aspirations of the Indigenous peoples who greatly outnumbered them, it becomes clear why the Arctic's foundational structures were largely commercial in nature, driven by the reciprocity of trade more than by the pursuit of power in the traditional Westphalian tradition. Tribal and global interests had to be reconciled then and there; each post, each ship, was a node in a global network that had to strike a balance between these otherwise contending forces to endure. Consider Benjamin Barber's classic ideas on the clash between "McWorld" and jihad as he described in the 1990s, which help us to resolve this inherent state-tribe dialectic with a synthesis.[58]

Instead of a continuing dialectic between globalization and tribalism, or the triumph of one over the other through military victory as we see in the US expansion, we witness instead the emergence of a synthesis reflected by the proliferation of both "global tribalization" and "tribal globalization." Nationalist backlashes and renewed state rivalry driven by nationalist sentiments may reflect the former (global tribalization), as did the jihadist movements of the global war on terror, including the emergence of the Islamic State as a recent, transnational (if ephemeral) sovereign form. The latter, tribal globalization, is evident in the increasing role of tribal peoples and organizations in contemporary diplomacy, whether at the UN (as illustrated by the UN Declaration on the Rights of Indigenous Peoples) or at the Arctic Council, with its active and influential Permanent Participants. The colonial-era clash of competing (and, for IR theorists, unusual) "levels of analysis" sets up an ongoing collision of dueling polities, some tribal, some global, that evolves into a new synthesis that defines the modern North, as states find there is room for competing power centers, whether corporate, native, or environmental, to become partners, albeit unequal, in Arctic sovereignty.

It is no coincidence that the emergence of the modern Arctic in North America began in 1867 with the Alaska purchase and the formation of the Dominion of Canada that very same year, right when the traditional "first wave" of globalization (as agreed to by a wide range of globalization theorists, even when they disagree on the subsequent waves) is said to have begun. Just as Barber describes the clash between globalist and tribalist forces in contemporary globalization, we see the same duel play out during the first wave, driven by Secretary of State Seward's infamous "folly" (or prescient brilliance, which is often confused with historical folly by contemporaries), as articulated in his 1853 "Destiny of America" speech, which foresaw America's rise as a global power through its dual expansion to the north and south.[59] To ascend as a global power, Seward envisioned, the United States was destined to expand north to the polar regions even as it expanded south to the tropics. So just as forces of modernization arrived in Alaska with the conclusion of Russian colonial rule, forces of tribalism were likewise released from the captive tyranny imposed by Russia on the natives of Alaska. This dualistic rise of modernization (and closer integration to a modernizing world) and renewal of native traditionalism is part and parcel of the underlying clash of values inherent in globalization. Reconciling the traditional and modern worlds became the mission of the United States and Canada in the Arctic from the very start of their domestic experiences in 1867. Globalization theory, through the lens of Benjamin Barber, provides us with a helpful analytical tool pitting McWorld and jihad, which he explained as a clash between economic integration and its homogenized culture of assimilation to modernizing impulses, and a backlash by forces of tribalization, whether literally by tribal peoples, polities, and values or more figuratively by subnational groups and movements as well as nationalist movements opposed to globalization.

If we think about whaling and the fur empires as "platforms of globalization," these platforms can become two-way circuits and be harnessed to serve as platforms of retribalization as well, as we will see when looking at the native rights movements, Arctic land claims, and the introduction of new systems of native governance. As Boraas and Leggett observed with the Dena'ina resistance, the only Athabaskan people inhabiting an oceanic coastal region,[60] a resistance that was mirrored by Aleut, Tlingit, and Haida (whether armed, economic, and/or spiritual in nature), the Russian Orthodox faith, an institution of colonial hegemony and part of a colonial-era culture war, was ultimately harnessed by natives to serve as a vehicle of cultural protection and reappropriation. The Orthodox church became a center of native identity that, curiously, survived the end of Russian rule, in part because, as one can see in the text of the Treaty of Cession, church property was protected as community-owned property and not transferred to US sovereign control, in contrast to most other property asset classes, from RAC offices and warehouses to Russian military bases and forts.

We also see in the literature on the Aleutians, from the Russian invasion to the Japanese occupation to the American liberation,[61] the enduring presence of Russian Orthodox faith as part of modern Unangan (Aleut) identity. Thus churches are, in many of the same ways, platforms of theological globalization as much as platforms of cultural synthesis and native identity restoration—in short, platforms of retribalization, mirroring the same dialectical interactions between globalism and tribalism taking place across the Arctic. Trading posts, too, become more locally controlled, and after native land claims, we see new native corporations rising, gaining in wealth and influence, making the corporation another platform of retribalization, in preview of my discussion of land claims and native cultural restoration, below—but before jumping ahead on this point, let's step back and consider the transition to modernity that was ushered forth, in part, by our Alaska purchase.

In the literature on the HBC, its heritage, and its historic charter articulating its ambitions to discover and utilize new sea lanes to the Pacific and to find, chart, navigate, and commercialize a Northwest Passage in the interest of Britain's strategic interests, it was also a commercial venture to profit from the untapped fur resources in the high North and to extract wealth from the Hudson's Bay watershed. The company established its presence, evolved in fits and starts with the times and in response to more nimble (and in some cases, less ethical) rivals, and in time expanded beyond its original footprint to encompass the Northwest Territory in addition to Rupert's Land, which was transferred to Canada much the way Russian-colonized Alaska was transferred to the United States, allowing for expansion by two North American states into the Arctic and completing the official sovereign map of North America. Intriguingly, these early modern platforms of economic modernization and globalization, the networks of trading posts along the rivers, lakes, and ocean shores of the Arctic region, became platforms of sovereignty as well. In the maritime domain, with the first summer cruise of the Revenue Cutter *Bear*,[62] we see the Revenue Cutter Service, the predecessor of the US Coast Guard, quickly emerging as a principal platform for projecting sovereignty along Alaska's vast coastal and archipelagic regions, from the southeast all the way up to and beyond the Aleutians to the Arctic coast.

Recalling the era of commercial whaling in the age illustrated by Herman Melville's literary imagination,[63] whaling ships served as solitary, mobile platforms of trade and economic globalization all around the world ocean, with each ship containing a diverse demographic mix and providing an interface for trade with isolated peoples worldwide. These platforms of globalization were often disconnected from the world of sovereignty, acting as independent nodes and at times violating laws and regulations placed on commercial enterprises, such as the illicit sales of alcohol and arms to natives, which disrupted much of the economic order that the HBC had sought to foster over its long presence in the interior. Whalers could thus contribute to a state of anarchy in the maritime Arctic, a situation that the CSS *Shenandoah* exploited. Indeed, the *Shenandoah*'s commerce-raiding mission may be viewed as a pre–Alaska purchase emergence of a powerful platform of sovereignty that imposed its own dramatic order on the Arctic via its assault on commercial whaling. The Revenue Cutters, representing the restored and reunified United States, would similarly impose sovereign order on the same region, the same commercial whaling enterprises, and the Indigenous peoples of the coastal and insular regions—but instead of the destructive order of a rogue state at war, or the repressive tyranny of empire, it was a constitutional order of the new sovereign in Alaska, the United States.

When looking at Alaska after its purchase, one can't help but be struck by the vibrant, enduring, and central native presence in this newly acquired US territory, as captured by both the literature and historic photos of that era. And at the same time, one can't help but be equally struck by the dated, and in so many ways imperial views of William H. Seward. He, too, noted the ubiquitous native presence, its diversity and legacy, but he did not seem so concerned with its future or survival. At the same time, the excitement and passion of Seward's words can't be missed, even if his views on diversity are antiquated—he loved the Alaska territory, every square inch of what he saw, a beautiful, resource-rich land for which so many native peoples had served as its stewards for so long, and that the Russians, obsessed as they were (like their HBC counterparts) with furs, had largely overlooked. But Seward's observations do not really consider the mission of protecting native values; instead he foresees, indeed predicts, their eventual displacement. The Seward of 1869 (the year he finally set foot in Alaska), two years after the Treaty of Cession, does not seem much changed from the Seward of 1853, when he presented his aforementioned "Destiny of America" speech with his expansionist vision of America's Arctic manifest destiny.[64] But "Hell-Roaring" Mike Healy, the famed US Revenue Cutter Service captain, presented a more balanced approach, both on the ground and in the seas adjacent to the newly possessed Alaska territory.[65] US Coast Guard Atlantic Area historian William H.

Thiesen described Healy as the "tamer of America's western sea frontier"[66] for his decades of service on Bering Sea Patrols, where he brought to remote Alaska coastal villages all manner of government services, from search and rescue, to food and medical assistance, to the administration of justice. According to the prolific historian of the US Coast Guard Dennis L. Noble, these patrols "became, in effect, the only government known to those who resided in the isolated Bering Sea and coastal Arctic Regions."[67]

What has been made of Alaska since, and what the fur empires left for Alaska and Canada to both develop and protect in more modern times, provide us with further opportunity to consider the historical links between the era of Arctic colonization and the modern world. We will thus examine next the historic re-empowerment of Arctic Indigenous peoples through native land claims, and the resulting renaissance of innovation in governance systems that synthesize Indigenous and state interests, to see how from these early efforts to balance the interests of tribe and state through colonial-era collaborations, co-management by tribe and state would come to define the way the contemporary Arctic would be governed at the local, regional, national, and even international diplomatic levels.

Notes

1. See Zellen, *On Thin Ice*. For a discussion of this structural duality of IPOs, see Zellen, "A 'Fourth Image' at the Interface of Tribe and State: Tribal Homelands, Expanding States, and the Foundation of International Order from the Western Arctic to Borneo," presented at the 14th International Borneo Research Council Conference, August 6, 2018. Additionally, the theoretical contours of the "fourth image" are described in my earlier "Rethinking the Subcomponents," and "Clan, the State, and War," 20–21.

2. Singer, "Level-of-Analysis Problem," 77–92.

3. Ibid., 78, n3.

4. Ibid., n4.

5. Ibid., 89.

6. Beyerchen, "Clausewitz," 59–90.

7. Ibid.

8. Barry Scott Zellen, "State-Tribe Collaboration and the Future of Arctic Cooperation: Moderating Inter-State Competition Through Collaborative Multilevel Governance, from Trading Posts to the Arctic Council," Center for Arctic Study and Policy (CASP), September 2020, https://uscga.edu/wp-content/uploads/2022/05/State-Tribe-Collaboration.pdf; also in the special issue on polar law of *The Polar Journal* 10, no. 1 (September 2020): 113–129. Also see Zellen, "Multinational Corporations," 157–173.

9. See Heininen and Southcott, eds., *Globalization*; Keskitalo and Nuttall, "Globalization of the Arctic," 175–187; Keskitalo and Southcott, "Globalization,"

401–42; Larsen and Huskey, "Arctic Economy," 159–174; O'Rourke and Williamson, "When Did Globalization Begin?" 23–50.

10. Dmytryshyn, "Russian Expansion," 1–25; Hartwig, *Polar and Tropical Worlds*.

11. Cavanagh, "Company with Sovereignty," 25–50; Hartwig, *Polar and Tropical Worlds*; Heininen and Southcott, *Globalization and the Circumpolar North*.

12. Hartwig, *Polar and Tropical Worlds*; Heininen and Southcott, *Globalization and the Circumpolar North*; Jones, *Century of Servitude*; Newman, *Company of Adventurers*; Bockstoce, *Furs and Frontiers*; Boraas and Leggett, "Dena'ina Resistance," 485–504.

13. Bogoyavlenskiy and Siggner, "Arctic Demography," 27–41.

14. Heaton, "Athabascan Village Stores," 133–155.

15. See Berger, *Northern Frontier, Northern Homeland*; Berger, *Village Journey*; Berger, "Conciliator's Final Report"; CBC Radio, "Lawyer Thomas Berger on How Yukon's Peel Watershed 'Was Saved,'" CBC News, February 4, 2018.

16. Carlene Arnold, "The Legacy of Unjust and Illegal Treatment of Unangan During World War II and Its Place in Unangan History" (master's thesis, University of Kansas, Department of Global Indigenous Nations Studies, 2011); Newman, *Company of Adventurers*, and *Caesars of the Wilderness* (Toronto: Penguin Books Canada, 1988); Bockstoce, *Furs and Frontiers*.

17. Bockstoce, *Furs and Frontiers*.

18. McGhee, "Nineteenth Century," 141–152.

19. Ibid.

20. Ibid.

21. Nick Robins, "The World's First Multinational," *New Statesman*, December 13, 2004.

22. Ibid.

23. John F. Burns, "Farewell to the 'Northern Stores,'" *New York Times*, March 29, 1987.

24. Ibid.

25. Ibid.

26. "The Far Flung Corners of the Empire," *Up Here*, July 2015.

27. Ibid.

28. Boraas and Leggett, "Dena'ina Resistance."

29. See Wood, "Russian Creoles of Alaska," 204–208; Smith-Peter, "Creating a Creole Estate," 441–459; and John McGraw, "The Creoles and the Cross: The Role of Russian–Alaska Native Creoles and the Orthodox Church in Colonial Russian Alaska," Undergraduate Thesis, University of Florida, 2011.

30. Jones, *A Century of Servitude*.

31. Ibid.

32. North West Company, "History." northwest.ca/about-us/history.

33. Bockstoce, *Furs and Frontiers*.

34. Ibid.

35. Dodds and Nuttall, *Scramble for the Poles*; Sale and Potapov, *Scramble for the Arctic*.

36. Jones, "Role of Research Associations."

37. Alyeska Pipeline Service Company, www.alyeska-pipe.com.

38. Coates, *Canada's Colonies*; Norman Easton, "History of the Border in White River," presented at the Borders in Globalization Summer Institute, Yukon College, Whitehorse, Yukon, June 21, 2016.

39. Ross, "Whaling, Eskimos, and the Arctic Islands," 33–50. Also see Ross, *Arctic Whalers, Icy Seas*.

40. Ross, "Whaling, Eskimos and the Arctic Islands"; Ross, *Arctic Whalers, Icy Seas.*

41. See Zellen, *Breaking the Ice*, esp. chap. 1.

42. Clifford John Groh II, "Oil, Money, Land and Power: Passage of the Alaska Native Claims Settlement Act of 1971" (honors thesis, Harvard University, 1976).

43. Native American Rights Fund, "1988 Amendments," 1–5; Len Sillanpaa, *Alaska Native Claims Settlement Act: A Living Settlement?* (Ottawa: Indigenous and Northern Affairs Canada [INAC], Circumpolar Affairs Section, December 1988); Thomas, "Alaska Native Claims Settlement Act," 328–329; Worl, "Models of Sovereignty."

44. Freeman, "Persistence and Change," 257–266; McGhee, "Nineteenth Century Mackenzie Delta Inuit"; Native American Rights Fund, "1988 Amendments"; Sillanpaa, *Alaska Native Claims Settlement Act*; Thomas, "Alaska Native Claims Settlement Act"; Zellen, *Breaking the Ice.*

45. Such as his *Company of Adventurers*, as well as *Caesars of the Wilderness* and *Empire of the Bay* (Toronto: Penguin Books Canada, 1998).

46. Berger, *Village Journey*; Native American Rights Fund, "'Fifth Disaster,'" 1–14, 47.

47. Lloyd N. Binder and Bruce Hanbidge, "Aboriginal People and Resource Co-Management: The Inuvialuit of the Western Arctic and Resource Co-management Under a Land Claims Settlement," presented at International Association for the Study of Common Property (IASCP) Common Property Conference, September 26–29, 1991, Winnipeg, Manitoba; Notzke, "New Perspective," 187–209; Nuttall, "Aboriginal Participation," 617–634.

48. Torrey and Krukoff, *Slaves of the Harvest.*

49. See the Cherokee Historical Association website for more information on this tragic episode of forced displacement; www.cherokeehistorical.org/learn-more-about-the-cherokee-indian-removal-and-the-tragic-trail-of-tears.

50. Cavanagh, "Company with Sovereignty."

51. Melissa Gismondi, "The Untold Story of the Hudson's Bay Company: A Look Back at the Early Years of the 350-Year-Old Institution That Once Claimed a Vast Portion of the Globe," *Canadian Geographic*, May 2, 2020.

52. Ibid.

53. Among the many works on the fascinating cruise of the *Shenandoah*, see the following: Hughes, *Confederate Biography*; Whittle, *Voyage of the CSS Shenandoah*, edited by Harris and Harris; Chaffin, *Sea of Gray*; Curry, *Officers of the CSS Shenandoah*; Schooler, *Last Shot*; Waddell, *C.S.S. Shenandoah*, edited by Horan.

54. Coates, *Canada's Colonies.*

55. For more on that, see Tamir Eshel, "Attack on Saudi Oil Refineries Proves the Devastating Potential of Iranian Cruise Missiles," Defense-Update.com, September 14, 2019.

56. For more information on the cruise of the CSS *Shenandoah*, see History Guy, "Last Shots of the American Civil War and the CSS Shenandoah" (Updated), www.youtube.com/watch?v=oJWjTp7DHEk; Karen Percy, "Crew Diary Sheds Light on Shenandoah Links to Melbourne," ABC News (Australia), www.youtube.com/watch?v=yEBX2QTMrQo. The History Guy captures with his trademark intensity and energy this unique naval encounter off Alaska's shores and the fascinating role the Arctic played in the final chapter of the US Civil War, illustrating America's well-established commercial and strategic interests offshore of Russian America near the end of its colonial existence. ABC News (Australia)'s report discusses the most recently discovered officer diary from the *Shenandoah*, which the

Australians purchased at auction in 2016 and which was authored by Lt. Dabney Minor Scales.

57. Cavanagh, "Company with Sovereignty."

58. See Barber, "Jihad vs. McWorld," as well as his book-long elaboration of his classic article, *Jihad vs. McWorld.*

59. William H. Seward, "The Destiny of America," speech at the dedication of Capital University, Columbus, Ohio, September 14, 1853, www.loc.gov/resource/gdclccn.17018137/?st=gallery.

60. Boraas and Leggett, "Dena'ina Resistance," 485–504.

61. See Alaska Extreme, "The Bombing of Dutch Harbor, Alaska, and Aleut Evacuation—75th Anniversary," December 16, 2017, www.youtube.com/watch?v=vnU97AS8iC4.

62. As described in Donahue and Switzer, *Steaming to the North.*

63. Mary K. Bercaw Edwards, "Melville, Whaling and the Arctic," presented at Roundtable 5-C: The Enduring Arctic/New England Nexus, Politics and History Section, New England Political Science Association Annual Conference, April 24, 2021.

64. Seward, "Destiny of America."

65. Noble and Strobridge, *Captain "Hell Roaring" Mike Healy.*

66. William H. Thiesen, "Captain 'Hell Roaring' Mike Healy—Tamer of America's Western Sea Frontier," Background, Search for the U.S. Revenue Cutter Bear, NOAA Ocean Exploration, https://oceanexplorer.noaa.gov/explorations/19bear/background/healy/healy.html.

67. Dennis L. Noble, "Fog, Men, and Cutters: A Short History of the Bering Sea Patrol," U.S. Coast Guard Historian's Office Digital Library, April 20, 2020, https://media.defense.gov/2020/Apr/20/2002284666/-1/-1/0/BERINGSEA.PDF.

3

The Age of Arctic Land Claims

In this chapter, we will examine the evolution of the native land claims movement from its 1971 start in Alaska through the 2005 Nunatsiavut land claim agreement and constitution in northern Labrador, which completed its generation-long trajectory. Each new iteration of an Inuit land claim agreement has generally introduced a new order or structure of Indigenous self-governance (or, in the absence of one, a vision for such an order or structure) in tandem with the land and resource rights negotiated by treaty. And each new land claim accord builds upon the achievements of preceding land claims agreements, a generational process that aligns with the state's increasing recognition of inherent rights of Indigenous self-governance and a deepening faith that increased Inuit self-governance, autonomy, and empowerment strengthen the sovereign bond uniting the remote north with the constitutional center, and that decolonization of northern governance actually increases the sovereign legitimacy of the state in the Arctic rather than undermining it. Efforts to amend and improve prior land claims are often made after the fact, particularly evident in the case of the Alaska Native Claims Settlement Act (ANCSA), which has been legislatively amended numerous times and augmented by the subsistence and land protection provisions of the federal Alaska National Interest Lands Conservation Act (ANILCA) in 1980. Two years earlier, in 1978, the Inuvialuit of Canada negotiated their agreement in principle for a Western Arctic land claim that united the corporate provisions of the ANCSA with resource co-management institutions explicitly integrated into the land claim treaty. It is no coincidence that the US government at that time, during the Carter administration, viewed native subsistence and harmonious interethnic relations in Alaska as part and parcel of the national interest, likewise viewing state-tribe synthesis to be in the interest of the state and its efforts to assert, and augment, its Arctic sovereignty (an effort that continues and is reflected, most recently, in the 2022 US National Arctic Strategy.[1]

And now there is a new effort by Labrador Inuit-Métis of NunatuKavut (southern Labrador) to forge a new agreement with Ottawa. The effort so far lacks the support of the Inuit Circumpolar Council and the Inuit Tapiriit Kanatami, whose constituents in Inuit Nunangat inhabit a region to the north, where there are now resolved Inuit land claim areas and limited non-Inuit population settlement, in marked contrast to southern Labrador, which diverges demographically and shares a history of settlement with the island of Newfoundland and much of the south beyond. This explains to some extent why the more northern settlement areas and their leaders do not recognize the existence of this southern Inuit-Métis group. They view with skepticism the claims presented by southern Inuit-Métis who have not yet earned formal recognition by the other Inuit in Canada, in contrast to the circumpolar Inuit solidarity that has been a hallmark of Inuit politics and diplomacy since the Inuit Circumpolar Council's formation during the 1970s.

To help frame this fascinating treaty history, it can be helpful to recall basic international relations (IR) theory to make sense of the complex, asymmetrical interorganizational relations between the Inuit and the sovereign states of Arctic North America. There is an unsteady and dynamic equilibrium between realist and idealist/liberalist values in regional and world politics and the many sub- and transnational structures that exist between the national and international levels of analysis, which are the focus of regime theory (and more recently, institutionalist IR theory). While globalization theory pits globalism against tribalism, classical IR theory pits realism against idealism, and much of post–Cold War IR theory pits national and international structures against sub- and transnational structures. When considered in parallel, we can glimpse the complexities inherent in the state-tribe dialectic in the Arctic, and how Indigenous land treaties become a cross-level bridge fostering reconciliation of these many disparate tensions all at once. One can think of it as a quantum IR theory for Indigenous regions, where so much is the same and yet so much differs from the Westphalian world.

This will help to illustrate the many intense pressures of the modern, industrial, post–World War II, nuclear-age state on the fragile, largely pristine, and lightly settled Arctic. In the previous chapter, we considered the early platforms of sovereignty and globalization in the Arctic that preceded the arrival of the modern state and paved the way to govern lightly, with minimal reliance upon traditional tools of state power. But during World War II, modern warfare introduced a new level of destructive state capacity, and this new and robust scale of state power did not go away at war's end. With the Cold War increasingly defining the limits and nature of post–World War II international relations, we began to see numerous large-scale state-sponsored projects planned for the Arctic, some that came to fruition such as the Trans-Alaska Pipeline System (TAPS), now known as the Alyeska Pipeline or more commonly Alaska Pipeline,[2] whose construction directly precipitated ANCSA by uniting Big Oil, federal and state governments, and Indigenous interests (at least for the moment necessary to reach agreement).[3] But less well known is the Rampart Dam Project, which was proposed in 1954, but conceived in 1948, and not formally abandoned until 1980 with the passage of ANILCA and the shift in national interests from megaprojects to land conservation, with environmental security offsetting energy security as a pillar of Arctic security well before the end of the Cold War. The project sought to dam the Yukon River and flood the Yukon Flats, which would have created the largest reservoir on Earth with over 3,000 miles of new coastline, comparable in size to Lake Erie. It was so large it was expected to take over thirty years to build and fill with water. Such megaprojects are now perceived as hallmarks of third world development projects, but it was not that long ago that the United States was seriously considering such a project, with all its potential environmental and cultural destruction, for a significant portion of Alaska.

Over on the northwest Alaska coast, where many have long dreamed (and still dream) of a deep water (or as favored today, deep-draft) port, there was Project Chariot, conceived by Project Plowshare and proposed by the Atomic Energy Commission in 1958, to blast a deep water port by atomic demolition at Cape Thompson.[4] Headed by Manhattan Project luminary Edward Teller, the father of the H-bomb, Project Plowshare was dedicated to the art and science of large-scale nuclear construction and demolition projects. The project generated intense native and environmentalist resistance, bringing together two groups that have collided on other Arctic issues (namely, whaling and sealing) and precipitating the project's cancellation in 1962. This successful coalition of native and environmental interests soon turned its attention to the next Alaska megaproject: the Trans-Alaska Pipeline System, a project that could not be stopped but that did successfully court native support through the innovative Alaska Native Claims Settlement Act, using a corporate model to modernize Alaska's political economy and turbocharge native participation in it (at least for the first twenty years after ANCSA was enacted, the period during which native ownership was assured).[5]

Just a few years earlier, with the Cold War intensifying and the nuclear age becoming more dangerous with the horizontal spread of nuclear weapons and the vertical growth in the arsenals of both superpowers, it was determined that an early warning radar system was urgently needed across the Arctic coast. Driven by the perception of military necessity, much like the rapid World War II construction of the Alaska Highway,[6] the Distant Early Warning (DEW) Line quickly went from concept to construction.[7] It was conceived in 1952 by the Summer Study Group at the Massachusetts Institute of Technology, and its 63 manned radar sites stretching 3,000 miles from Alaska to Greenland would be constructed by an army of 25,000 engineers and builders across the entirety of Arctic North America in just three years (1954–1957). It operated continuously until the Cold War's end, when the DEW Line was upgraded to the automated, Canadian-managed North Warning System.

Catalyzed by the coalition formed to ensure the Trans-Alaska Pipeline System could proceed, the Alaska Native Claims Settlement Act of 1971 was the first comprehensive land accord affecting Arctic lands and peoples—and sub-Arctic as well, given Alaska's vast north-to-south geography. It modernized Alaska's political economy with new, native-owned corporations holding title to vast tracts of native lands, and it institutionalized corporate cultural values across village Alaska and soon thereafter, as the land claims model caught on in Canada, across the entirety of Arctic North America from the Bering Sea to Baffin Bay. Like TAPS, the Rampart Dam Project, Project Chariot, and the DEW Line are all Arctic megaprojects emblematic of the modern state's arrival and maturation in the Arctic, with its hitherto unrivaled appetite for resources, energy, and continental security. Shortly after the DEW Line emerged to transform the Arctic coastal landscape, the USS *Nautilus*, the first US nuclear-powered submarine with unprecedented capabilities for extended subsurface operations, reached the North Pole in 1958[8] to great fanfare, helping to redress the perception across the United States of Soviet technical superiority that followed the surprise Sputnik launch the year before. While less disruptive to the people of the Arctic, the *Nautilus* polar mission was an intrusion of the strategic competition between the United States and the Soviet Union to the otherwise pacific region, and an important part of the nuclearization of the Arctic during the Cold War.

What was originally criticized as "Seward's Folly" for acquiring what was perceived as a nonproductive wasteland would in time become a highly valued strategic territory and resource repository for globally expanding US power. Its coastal waters were well known at and before the purchase to offer great riches, and various resource commodity rushes brought a boom/bust realization of its wealth to later generations. But by the twentieth century, the economic and strategic value of the Alaska territory became not only universally recognized but noncyclical, enduring, and increasingly essential to the nation's power, and state interest rose accordingly. In many cases, negotiations for native land and resource rights were driven by governments fearful of their own judicial branches, which could thwart executive-led projects. Increasingly, they found native inclusion could reduce the risk of litigation and judicial intervention.

World War II and the Cold War: From Arctic Megaprojects to Indigenous Gigaclaims

The momentum to settle Inuit land claims just as these megaprojects—both energy-related and nuclear-strategic—came to fruition was no coincidence. Isolated and remote platforms of sovereignty and globalization were now giving way to extensive, permanent infrastructure in places where the state had not hitherto expanded, at least not in Arctic North America, and these indicated that state expansion into the Arctic had achieved a new level. Settling native land claims, to further native participation and inclusion and to solidify sovereign claims to the Arctic by gaining native consent to expanded state governance, can thus be understood through a Clausewitzian perspective tying war to its political context.

Just as the famed Prussian theorist believed war to be a "continuation of politics (or policy) by other means," the settlement of comprehensive Indigenous land claims (utilizing an adaptive model that balanced corporate development and economic equity participation with, increasingly, cultural protection, social development, and resource and land co-management) would likewise be a continuation of state expansion by other means—strengthening the state's military capabilities and sovereign presence in the Arctic by partnering with the region's Indigenous peoples. Thus the age of Arctic land claims emerged directly from the challenges to native sovereignty and survival during the prior (and, to some degree, overlapping) age of Arctic megaprojects, and has been shaped and defined by the native response to the many megaprojects that were perceived as a threat to native lands, health, and culture. Indeed, megaprojects and their unprecedented scale of destruction of natural ecosystems in vast and unspoiled Indigenous homelands helped to catalyze the movement for land claims. The claims sought when possible to prevent such calamitous intrusions that threatened to despoil pristine natural ecosystems upon which subsistence depended, and when prevention was not possible, to at least mitigate the risks and dangers, and to provide remediation and compensation for losses, which started in 1971 with the billion-dollar settlement with Alaska natives, catapulting new native corporations to elite status in their remote rural economies—hence my use here of the term "gigaclaims" to redress the impacts of the many megaprojects of the Cold War Arctic.

With the modernization of the state and its expanded presence in the Arctic during and after World War II, national-level interests and projects also expanded greatly beyond earlier platforms of globalization and sovereignty. More robust and increasingly interconnected networks of sovereignty created a new, more tightly woven national fabric in a region long defined by patchworks of autonomous local and tribal communities. Values, interests, and conceptions of security at the village level by predominantly tribal peoples of the Arctic region were to be frequently challenged by assertions of sovereignty by the post–World War II state, with its capacity to rapidly transform and develop infrastructure and geographically re-engineer large spaces in places that had, prior to the thaw, been largely insulated from such a forceful, industrial-age state presence.

As discussed above, some of these modern megaprojects were dramatic and sensational, such as the Rampart Dam Project that emerged as an idea in 1948 to harness Alaska's vast interior hydropower potential to electrify the state and stimulate its economic modernization. It sought, in the name of economic progress, to flood the Yukon Flats region and would have permanently altered the physical and environmental geography of the Yukon River watershed. Project Chariot proposed to carve deep water ports and cross-peninsular canals out of coastal lands with chains of atomic bombs, and a coalition of environmental and Indigenous actors in Alaska and across the United States thwarted both these projects.[9] But not all megaprojects were doomed to become historical what-ifs. The DEW Line project, conceived in 1952, rapidly constructed (between 1954 and 1957) over sixty state-of-the-art radar sites along the entirety of the Arctic coast of North America and remained in operation nearly half a century with dozens of new, space-age platforms of sovereignty erected across the Inuit homeland. A decade later, the Trans-Alaska Pipeline System connected the North Slope of Alaska with Prince William Sound by an equally impressive engineering marvel partly built on fragile permafrost that wormed its way 800 miles from north to south, which remains in operation today, but which infamously planted the seed for the environmental destruction experienced in the 1989 Exxon Valdez oil spill.

Such large-scale projects remain part of the continuing narrative of Arctic development, and over time they have transformed from state-driven projects imposed on native peoples who had little or no say about those projects to co-managed joint venture projects with the equity and managerial participation of natives, albeit "oil-age" natives who differ culturally from earlier generations. Indeed, many of the more audacious projects were conceived just as Alaska was transforming from a remote territory purchased in 1867 into a bona fide US state in 1959. It was the resulting age of Arctic land claims that fostered the emergence of a new, modern, corporate culture across the Indigenous Arctic, and though this corporate culture is not representative of the entirety of the native community and at times finds itself at odds with traditionalists and practitioners of subsistence, it has provided a growing pool of trained and experienced participants to channel such megaprojects and reshape them to better meet the needs of native communities and values, with native hiring and subcontracting preferences, distributions of royalties and dividends, and commitments to clean up and restore lands to their original state or provide compensation when that proves impossible. Elsewhere in the world, megaprojects on native lands, primarily proposed or constructed dams and pipelines, have led to persistent clashes between native communities and state interests (Borneo, Burma), but also to formal reconciliation processes through land claims treaties (Quebec and the Western Arctic) and/or recognition of Indigenous cultural rights (Hokkaido, where the unprecedented Nibutani Dam case led to judicial recognition of Indigenous cultural rights in Japan, part of a slow but forward process of recognition that continues to the present).

The Alaska experience—harnessing the transformative vision (and risks) of megastate development and channeling these, through political organization, lobbying, and negotiation, into a catalyst for positive change and Indigenous empowerment—directly inspired the rest of Arctic North America. Land claims treaties followed quickly on the heels of the 1971 ANCSA legislation, including the 1975 James Bay and Northern Quebec Agreement (JBNQA); the 1978 Inuvialuit-Canada Agreement-in-Principle, which laid the foundation for the Inuvialuit Final Agreement (IFA) of 1984 that so profoundly redefined the ANCSA land claim model; the 1980 Alaska National Interest Lands Conservation Act (ANILCA) that protected much of Alaska's wilderness and redressed ANCSA's exclusion of subsistence; the 1993 Nunavut land claim agreement and subsequent 1999 formation of the Nunavut territory, after a successful 1992 plebiscite to divide the Northwest Territories in two;[10] and the 2005 Labrador (Nunatsiavut) Inuit land claim—with the potential for one more land claim by the Inuit-Métis of southern Labrador, the NunatuKavut people or NunatuKavummiut, who are now engaged in discussions with Ottawa while facing intense resistance from the other Inuit regions with settled claims and less nonnative settlement.

Land Claims Agreements as State Expansion "by Other Means"

We will next examine the historic sweep of Inuit land claims agreements in Alaska and Canada, starting with the Alaska Native Claims Settlement Act of 1971 (or ANCSA), which was ironically rejected by the Inupiat of the North Slope but still became the catalyst for all subsequent Arctic land claim agreements, including the JBNQA of 1975; the Northeastern Quebec Agreement (NEQA) of 1978; the IFA of 1984, whose terms were agreed to with Ottawa six years earlier in the 1978 Agreement-in-Principle; the Nunavut Land Claim Agreement of 1993, tied to the historic 1999 partitioning of the Northwest Territories, resulting in the birth of a predominantly Inuit eastern territory corresponding to the settlement area of the Nunavut land claim; and the Labrador Inuit (Nunatsiavut) land claim of 2005. Each of these land treaties pioneered new governing structures and expanded the powers of Indigenous claimants beyond the achievements of prior land claims, incrementally strengthening the foundations of Indigenous self-reliance and self-governance across the North.

But once land claims became a fixture in the political geography of the Canadian Arctic, they quickly transformed from a program of rapid economic modernization and integration driven by time-sensitive energy development projects into an innovative program of mutual recognition by Indigenous peoples and the state, strengthening the sovereign fabric of both the US and Canadian states in their northern regions that remained lightly settled and largely undefended. In this section I describe the foundational and transformative role of land claims as a mechanism to help the state expand and consolidate its sovereign claims in an alternative, nonmilitarized fashion while offering the Inuit numerous material benefits in exchange, from cash compensation to co-management and constitutional protection of their gains—with both the state and Indigenous peoples benefiting from this mutual consent and the increasing strength of collaboration that resulted therefrom.

For well over three centuries, there has been an ongoing process of globalization, economic integration, and institutional modernization underway across mainland and maritime Arctic North America that has resulted from the colonial expansion of Westphalian states from Europe into the northernmost reaches of North America (from Alaska's westernmost terminus in the Bering Sea to Greenland in the high North Atlantic), initially via proxies such as crown-chartered companies with monopoly trading rights and offshore fleets of whaling and fishing vessels. Economic dependency and collateral losses from diseases introduced by the representatives of those state proxies were common and notoriously destructive, as were later efforts to forcibly assimilate natives through aggressive social and educational policies—also implemented by proxies, in this case various churches—resulting in generational separation, loss of cultural continuity, and a tragic loss of life only now being recognized as hundreds of unmarked graves are being unearthed across Canada. But without the even more horrific traumas of frontier warfare, known collectively as the Indian Wars in the United States, which continued as the new nation's frontier was pushed by armed force farther and farther West, annihilating all forces of Indigenous armed resistance in its path, it became possible for an alternate history to unfold.[11] That history was shaped primarily by the reciprocity of trade that took place at trading posts, which would in time dot the entirety of Arctic and sub-Arctic North America, whether under the flag of the Russian-American Company in what is now Alaska or the Hudson's Bay Company (and for a time, the North West Company with which it later merged) in northern Canada.[12]

However, the long-term economic interests of these commercial trading entities favored restrictions on the influx of settlers, moderating the colonial experience and mitigating the excesses associated with waves of human settlement as experienced in what became the United States that yielded the kinetic and highly destructive frontier wars between settlers and Indigenous peoples.[13] Colonization by proxy, while not without its challenges and traumas to the Indigenous peoples of the North, offered one notable advantage to the region's first peoples: a near absence of dispossession from traditional lands or loss of political autonomy (though policies of assimilation including residential schooling led to generations of Indigenous students being taken from their home communities, experiencing a separation from their land no less traumatic than dispossession).

Since Arctic land claims began to be settled in the 1970s, far-reaching and transformative structural innovations have been made to the political economies of Arctic North America, from the Bering Sea in the west to Baffin Bay in the east. The multidecade process of settling aboriginal land claims treaties fostered an enduring partnership between the Indigenous peoples and the modern state through a variety of new institutions, including aboriginal regional and community corporations, investment corporations, land administration agencies, a variety of tribe-state co-management boards, plus a complex patchwork of local, regional, and territorial governments created to give a voice to the native interest. As a result of these changes, the Inuit and other aboriginal northerners have become powerful stakeholders in the economic and political systems that govern the Arctic today, and also, importantly, the largest private landowners with direct control over some 10 percent of North America's Arctic territories and indirect influence over a far larger portion of the Arctic land mass.[14]

From the outermost Aleutians to northern Labrador, it has largely been a two-step process: First, land claims negotiations clarify title, helping to identify who owns which lands and thus reconcile the competing interests of tribe and state (representing the collective will of Indigenous and settler stakeholders), and then the region is opened up (or, for sensitive ecosystems and traditional hunting lands, closed off) to economic development with various mechanisms of co-management to keep native and state interests in balance. The second step has been the pursuit of new systems of aboriginal self-governance, taking various forms and employing various structures, with greater powers becoming available over time and earlier policies of assimilation being replaced by more contemporary policies promoting cultural and political renewal. The establishment of municipal or borough governments under existing constitutional law, such as in Alaska in the 1970s, evolved to the creation of newly empowered tribal councils governed by federal Indian law in Alaska and the Northwest Territories in the 1980s and 1990s, or to the negotiation of entirely new systems of governance. The most ambitious of these was by Nunavut, whose comprehensive land claim settlement in 1993 was linked to the subsequent formation of a new territorial government in 1999. They created a complex and potentially powerful system of self-governance by applying a public model to a predominantly Indigenous region for de facto Indigenous self-governance.

After Nunavut, the evolution toward more distinctly Indigenous self-governing structures has continued, as reflected in the Labrador Inuit Land Claims Agreement of 2005 with the very first truly Inuit self-governing structure, whose governing principles were articulated in detail in the 2002 Labrador Inuit Constitution. More recently, in November 2008, the far-flung Danish province of Greenland held a referendum on evolving beyond their "home rule" system of autonomy toward formal state sovereignty and independence, which passed decisively, paving the way forward for the eventual emergence of a formally sovereign Arctic state with a majority Inuit population, with literally revolutionary (or devolutionary) implications for the rest of the Inuit homeland. In the years ahead, we may see even further advances in the process of native empowerment toward increased autonomy and perhaps leading toward the Balkanization of the Arctic into independent (or at least more genuinely autonomous) political units.

Regardless of the jurisdiction, whether in Alaska or Arctic Canada, or beyond the shores of continental North America, Indigenous peoples have shown tremendous ingenuity in their effort to build new systems for self-governance since the land claims movement took root in the 1960s, creatively adapting existing institutions or creating new ones when possible, lobbying for and negotiating to further advance their powers. Ideas and institutions for reconciling the interests of Indigenous northerners and the modern state have evolved, becoming stronger with each new iteration and reversing many of the negative consequences of the colonial experience. The domestic balance of power has been transformed to lean heavily in favor of tribal interests, particularly in social, environmental, and economic matters. This shift in power has increased the capacity of the Indigenous peoples of the North to confront the many social and economic challenges that remain in their communities, and to grapple with the complex challenges and opportunities associated with climate change and a potential Arctic thaw.

Social conditions in Alaska and the Canadian Arctic have been described by many as a fourth world, with third world conditions exacerbated by climate, isolation, and limited infrastructure, including a near absence of roads and rail networks—making seasonal ice roads and summer sea lifts an economic lifeline.[15] Communities are generally small, ranging from just a few dozen people to several hundred, with the larger administrative centers being home to just a few thousand people; their populations are predominantly Indigenous, with subsistence hunting, fishing, and trapping still essential to their nutritional and cultural survival. Fuel costs are high, as are imported foods, making hunting and fishing all the more important. Economic opportunities have been limited, with natural resource development presenting one of the more enduring opportunities, from the nineteenth century's Klondike gold rush to the oil boom of the 1970s, to the diamond rush of the 1990s, to the more recent rush for all manner of Arctic natural resources. Land claims have helped to ensure that when economic development does take place, local concerns and tribal interests are not overlooked, with Indigenous leaders becoming governing partners in assessing environmental risk, mitigating impacts to traditional subsistence, and ensuring economic participation through jobs, training, and resource royalties. This can create deep rifts within the native community, as tradition and modernity collide. But the new governing structures were designed in part to intermediate this collision, converting thesis and antithesis into a truly northern synthesis.

The settlement of land claims and emergence of new structures of self-government have increased the role of Indigenous peoples in the decisions made about the Arctic and its future. One dramatic illustration: In the 1970s, when the Mackenzie Valley Pipeline Inquiry was held by Justice Berger, the struggle was primarily between corporate interests and tribal interests, with the latter excluded from the decisionmaking of the former. During the more recent Mackenzie Gas Project, the Aboriginal Pipeline Group sat with the oil companies as an aboriginally owned equity partner, and the Joint Review Panel examining the environmental and social impacts of the proposed pipeline was empowered by the settled regional land claims, providing an Indigenous perspective on both sides of the table—contributing to a slow pace but a unique review process with Indigenous inputs at all levels.[16]

1971: Alaska Native Claims—Starting the Land Claims Process

When the Alaska Native Claims Settlement Act of 1971 was enacted, it aimed to quickly bring Alaska natives into the modern economy, and at the same time clarify the limits of aboriginal title, making it possible to fully develop the state's natural resources and in particular to build the trans-Alaska pipeline.[17] Because its objectives were largely economic, its corporate model became its defining and most transformative characteristic—but not without controversy, since the corporate model was viewed with some skepticism by Indigenous leaders as a tool of assimilation, and there remains a continuing debate over the appropriateness of the corporate model to the Indigenous north. ANCSA formally extinguished aboriginal rights, title, and claims to traditional lands in the state while formally transferring fee simple title to 44 million acres—or some 12 percent of the state's land base—to Alaska natives, with $962.5 million in compensation for the lands ceded to the state, $500 million of which was to be derived from future oil royalties (as a result of which over half the "compensation" was to be derived from resources extracted from the Inupiat homeland—an irony not missed by Alaska natives).[18] ANCSA also created twelve regional native corporations (and later a thirteenth for nonresident Alaska natives) and over 200 village corporations to manage these lands and financial resources.

These new corporate structures introduced a brand new language and culture, as well as a new system of managing lands and resources that seemed at variance with the traditional cultures of the region and their traditional subsistence economy. The early years of ANCSA were famously described by Justice Thomas Berger as dragging Alaska natives "kicking and screaming"[19] into the twentieth century, and many native corporations approached the brink of bankruptcy, forced to monetize their net operating losses in what many thought would be their last corporate act. A new cottage industry of northern investment, legal, and policy advisors emerged—sometimes to the benefit of their clients, but often not.

In addition to the *corporatization* of village Alaska, ANCSA's original design also had some structural flaws that nearly proved fatal to the land claims experience, including a twenty-year moratorium on transferring shares in native corporations to nonnatives, which many feared would inevitably result in the dilution of native ownership, known as the "1991 time bomb." While critics of the land claims process are correct to point out these original structural flaws and the assimilating pressures introduced by new corporate structures, the land claims model has nonetheless proved resilient and adaptive, as native corporations matured and their boards, managers, and shareholders found ways to better balance traditional and modern values, learning from their crash course in capitalism as they went. Today the native corporations represent a huge economic force in the state of Alaska.

1975/1978: James Bay and Northern/Northeastern Quebec— Strengthening Subsistence and Self-Governance

The James Bay and Northern Quebec Agreement of 1975, and its extension to the Naskapi of Northeastern Quebec in 1978, followed soon after ANCSA and shared some similarities with that pioneering land claim, including its multiethnic nature serving both the Inuit of northern Quebec and the Cree of the James Bay lowlands. While ANCSA's diverse multiethnic constituency served both Inuit and non-Inuit Alaska natives, subsequent Inuit land claims in Canada served only Inuit subregions after initial efforts at a multiethnic land settlement spanning what William Wonders described as the Inuit/Dene interface. Also uniting JBNQA and ANCSA are their swift time frames from negotiation to enactment, under immense pressure from large-scale energy megaprojects (TAPS in Alaska and the James Bay hydroelectric development project in northern Quebec).

But differentiating JBNQA profoundly from ANCSA was its deep philosophical and institutional commitment to extending effective and collaborative governance into Quebec's Indigenous northlands. Tying the land settlement to the establishment of numerous new institutions of local and regional self-governance foreshadowed by a generation the approach taken by Inuit of Nunavut, who chose to delay settling their claim by a decade so that their land settlement would coincide with the formation of Indigenous Nunavut territorial government to fulfill their self-governance aspirations. Another distinguishing feature of JBNQA is its commitment to traditional Indigenous land use, including subsistence harvesting of wildlife on native lands.

JBNQA was signed November 11, 1975, by Cree and Inuit representatives, the governments of Quebec and Canada, the James Bay Development Corporation, the James Bay Energy Corporation, and Hydro-Québec; three years later, on January 31, 1978, the Naskapi signed the Northeastern Quebec Agreement, "establishing similar rights to those acquired by the Cree and the Inuit under the JBNQA" and, combined, introduced "the first comprehensive land claim agreements signed in modern times between the Governments of Quebec and Canada and Aboriginal people."[20] As Crown-Indigenous Relations and Northern Affairs Canada further describes,

> These Agreements include self-government components and lay the foundations for a new relationship between the Cree, the Inuit, the Naskapi and the Government of Canada. The territory covered by the JBNQA and NEQA includes more than one million square kilometres of land in Quebec between the 48th and 62nd parallels. It was once part of a larger federal territory known as Rupert's Land of which two vast stretches were transferred to Quebec in 1898 and in 1912. . . . JBNQA and the NEQA define the land regime applicable to the Cree, the Inuit and the Naskapi, as well as their rights in many areas such as resource management, economic development, policing and administration of justice, health and social services and environmental protection.[21]

In the early 1970s, Quebec—under the premiership of Robert Bourassa—had sought to rapidly develop its hydroelectric potential in the James Bay region and planned a series of dams without prior consultation with the Indigenous communities affected. The Cree and Inuit opposed Quebec's plan and its lack of consultation, and took the province to court in 1972; on November 15, 1973, they won the first round, after which "the parties undertook negotiations aimed at organizing development of the territory and planning for the use of natural resources."[22] As John Ciaccia, special representative to premier Robert Bourassa, recounted, "After two years of intensive, difficult, complicated negotiations, we are on the point of making history. . . . Indeed, it is a major event in the history of North America, for there has never been any other agreement like it."[23] He added, "This Agreement has enabled us to accomplish two great tasks to which the government committed itself. It enables us to fulfill our obligations to the native peoples who inhabit our north, and to affirm finally Quebec's presence throughout its entire territory."[24] In his discussion of the "Integration of Territory into the Administrative Map of Quebec," Ciaccia explained,

> Native communities will have local administrations, substantially in the manner of local communities throughout Quebec, and regional administrations will exercise municipal functions in areas beyond the old established communities. . . . With this agreement 21 new

> municipalities are to be created. . . . Thirteen of the municipalities will answer to the Ministry through an administration to be called the Kativik Regional Government, a novel instrument suited to the conditions of the region. In all, 250,000 square miles of newly organized territory will come under the jurisdiction of the Ministry of Municipal Affairs when this Agreement is implemented.[25]

In addition to JBNQA's structural commitment to governmental administration in its native north, it embraces Indigenous land use, protects the environment, and fosters a balanced and mixed approach to land usage that we now call sustainability, using joint governance mechanisms now known as co-management. The land categories are reminiscent of those described in ANCSA, but with stronger support of traditional Indigenous land use. As Ciaccia describes in "Three Categories of Land," JBNQA introduces "a new regime respecting the most important traditional occupation of the native peoples—hunting, fishing and trapping. There will be a new regime for the protection of the environment. These initiatives will all entail the participation of the native peoples in governmental bodies in the decision-making process respecting these matters."[26] Category I and II lands "are of particular importance in the context of the preservation of the culture and the economy of the Cree and Inuit peoples" and "will be allocated to the native peoples for their exclusive use . . . in and around the communities where the native peoples normally reside,"[27] while Category III lands are those "where exclusive rights or privileges are not granted to the native peoples," but where "they will be able to pursue their harvesting activities—hunting, fishing and trapping—the year round, as in the past."[28]

As Ciaccia further describes, JBNQA's land regime is,

> admittedly, an elaborate regime. It sounds as if we are going to great lengths to deal with special interests and problems. And we are. For while affirming the integrity of its territory, the Government of Quebec has sought to ensure that the native peoples can maintain their traditional way of life, which is the basis of their economy, their culture, and their survival. And what's more, the Government has sought to ensure that this way of life is viable. . . . The native peoples are offered, with this Agreement, reasonable condition for the pursuit of their traditional occupations . . . services that are available to all other citizens of Quebec in their own communities . . . powers of local administration as great as those of other communities in Quebec . . . [and $225 million in compensation] to meet their needs in the economic, social and community domains . . . in the course of the next twenty years.[29]

While ANCSA in 1971 focused primarily on the creation of Alaska native corporations only to later redress the omission of subsistence and environmental preservation with ANILCA in 1980, JBNQA in 1975 (and NEQA in 1978) aimed to foster a balance between governance, culture, environment, and economy. JBNQA/NEQA leapfrogged over subsequent claims in their advances in regional and local governance, owing in part to the efforts of Quebec to extend its sovereignty into its remote north, much the way Ottawa would by wedding the 1993 Nunavut land claim to the formation of the Nunavut territory six years later—a forward-looking revision to the land claim model pioneered in Alaska four years earlier. While ANCSA has been criticized for fostering assimilation into the modern, neoliberal economy at the expense of cultural and environmental protection, and Nunavut (and later regional land claims in the Northwest Territories and Yukon) for restraining the modern, neoliberal economy, JBNQA/NEQA aimed to strike a balance between these two contending economic philosophies. As Ciaccia explained, "Finally, and perhaps most important, the native peoples are offered a choice. They will be free, as individuals, to choose between their traditional occupations and new occupations. In this sense, I think that both the Quebec Government and the native peoples can feel that they are sharing a victory."[30]

1978/1984: The Inuvialuit Final Agreement— Co-Management as a Counterweight to Corporations

Across the border, the Inuvialuit of the Western Canadian Arctic had a front row seat to ANCSA, and they were impressed by all the money that was flowing north, as well as the new corporate structures created, and the sizable land quantum formally transferred to Alaska natives. But they also noted continuing threats to Indigenous culture, and the lack of adequate protections of subsistence rights, traditional culture, and the environment, and they were determined to improve the model. When they negotiated the 1984 Inuvialuit Final Agreement (IFA), largely completed in their 1978 agreement-in-principle, the land claims model was significantly enhanced—in addition to creating new native corporations, the IFA also made an equal institutional commitment to preserve native culture and traditions, to preserve the land and the wildlife, and to empower not just new corporate interests but also traditional cultural interests by creating new institutions of co-management and more powerful hunter and trapper committees. They also made sure all Inuvialuit became shareholders, and that no non-Inuvialuit ever could, learning from the Alaskan experience. The Inuvialuit thus successfully modified the land claims concept so that its structure included internal institutional balancing—not unlike our own balance of powers concept—that has enabled a greater commitment to cultural and environmental protections.

Their land claim entitled the 3,000 Inuvialuit living in six communities to 35,000 square miles of land; co-management of land and water use, wildlife, and environmental assessment; wildlife harvesting rights; financial compensation of $45 million in 1978 dollars, inflation-adjusted to $162 million, for lands ceded to Canada; a share of government royalties for oil, gas, and mineral development on federal land; the formation of new national parks in their settlement area that further protect their land base from development while allowing unhindered subsistence activities; and a commitment to meaningful economic participation in any development in their settlement area. This model has remained largely intact in later comprehensive land claims, showing twenty-five years of endurance as a model for northern development. The Inuvialuit sought to gain increased control over, while fostering increased participation in, the intense oil boom in their homeland; the Inuit of the central and eastern Arctic, where there was less development pressure at the time, sought to gain increasing powers of self-governance, which required another decade to achieve. The Inuvialuit felt they did not have the luxury of time given the frenetic pace of oil and gas exploration in their lands. Nunavut in contrast was far more isolated, providing the Nunavummiut more time to rethink, and renegotiate, the land claims model.

1993/1999: Nunavut— Augmenting Land Claims with Territorial Governance

In the years separating the signing of the Inuvialuit land claim in 1984 and the Nunavut land claim in 1993, much progress was made on the political question, and growing support for aboriginal rights in Ottawa enabled the establishment of a new concept: reshaping political boundaries at the territorial level to correspond to a land claims settlement area and establishing a new government to administer this region, augmenting the land claims with enhanced political power. In 1993, with the signing of their historic accord, the Inuit of Nunavut were awarded $1.1 billion and title to 135,000 square miles of land, including 13,600 with subsurface rights, on top of various co-management boards, clearly defined rights protecting subsistence, and royalty sharing from resource development activities. Nunavut has a population of around 30,000 in 28 communities spread out over 770,000 square miles, or one-fifth of Canada's land mass, including the High Arctic islands and the central Arctic coastal mainland. While its population is tiny, its jurisdiction is vast and its resource base potentially tremendous, and the sea lanes that cross through the territory include the famed Northwest Passage.

The most striking innovation of the Nunavut claim was the way it was formally linked to the division of the Northwest Territories and the formation of a brand new territory, resulting in the 1999 birth of Nunavut. Nunavut has now been up and running for a decade, gaining valuable but often painful experience in self-governance as it confronts some daunting social and economic challenges in one of the most challenging geophysical environments imaginable. There have also been intergovernmental frictions with Ottawa over implementation, and a growing perception of a crisis in Canada's youngest territory. But there is still much reason for hope for the future; the problems facing Nunavut are not likely to be quickly overcome, but the solutions developed can now be northern solutions, rooted in a deep understanding of northern social realities. Since its population is predominantly Inuit, a public government can, at least for now, govern in an Indigenous style—as the principles of the Nunavut land claim and the governing power of the new territorial government mutually reinforce one another. There is a long-term risk the territory could become more like the Yukon, especially if a major mineral strike results in a new mining center. But for now, a public model in an Indigenous context is a creative way to create self-government by other means.

2005: The Labrador Inuit (Nunatsiavut) Land Claim and the Law of Inuit

Half a decade after Nunavut made headlines around the world, the final Inuit land claim along the North American Arctic and sub-Arctic coast—the Labrador Inuit (Nunatsiavut) Land Claims Agreement—was settled. Ratified in December 2004 and put into effect a year later, it represented a new stage in the evolution of Inuit governance, making the two-step process more of a one-step process. It further redefined the limits of self-government within a land settlement area and transcended the public model applied by the Inuit of Nunavut and the Inupiat of the North Slope. The agreement created the 28,000-square-mile Labrador

Inuit Settlement Area with an adjoining 18,800-square-mile ocean zone extending as far as Canada's territorial waters. The settlement area includes 6,100 square miles of Labrador Inuit lands, five predominantly Inuit communities, and 3,700 square miles set aside for the Torngat Mountains National Park Reserve (following a tradition established by prior Inuit land claims to create vast national parks in which subsistence was protected)—with the Inuit retaining special rights in each of these areas. The government of Canada will pay the Labrador Inuit $140 million in 1997 dollars in compensation for lands ceded to the crown.

Just as the formation of the Nunavut territory was the most notable innovation of the Nunavut land claim, the emergence in 2002 of the 159-page Labrador Inuit Constitution was a hallmark of the 2005 Labrador Inuit land claim. As described in section 17.2 of the claim, it "exhaustively sets out the law-making authorities and self-government rights of Inuit," with the newly created government of Nunatsiavut to be based on the "fundamental law of Inuit" as enunciated by the Labrador Inuit Constitution. The constitution, among its many components, included an Inuit charter of human rights, recognized Inuit customary law and its application to "any matter within the jurisdiction and authority of the Nunatsiavut Government," and embraced laws to protect Inuit culture, language, and traditional knowledge. The Labrador Inuit Constitution created a blueprint of Inuit values and a pathway to the rapid formation of a truly Inuit system of government in a region that is adjacent to coastal waters of emergent strategic significance, with active commercial and subsistence fisheries, major mineral deposits such as the Voisey's Bay project, and the prospect of much future economic potential. It also showed a new path toward aboriginal self-government, one that did not require a secession like Nunavut but instead forged a regional subgovernment within an existing province with unique governing principles.

The age of Arctic land claims that got started with ANCSA in 1971 continued to evolve and transform over the next half century. Amendments on the Alaska side in 1976, 1988 (the so-called 1991 amendments that addressed the "time bomb" when native protections would expire[31]), 1995, and 2017, as well as the subsistence provisions of ANILCA in 1980 and the 1993 federal tribal recognition at the culmination of a tribal sovereignty movement in village Alaska, were all part of a complex, piecemeal, multilevel effort to align federal policy and tribal aspirations amid periodic resistance by the state of Alaska and oil, mining, and commercial fishing corporations. Across the border, the process has been less piecemeal. Land claims accords integrated some processes that were addressed outside the land claims process in Alaska, primarily subsistence but also aspirations for tribal sovereignty and autonomy. Numerous implementation challenges along the way revealed that the same fault lines persist between traditional/Indigenous and Western/state values in both Canada and Alaska, as reflected in the 1984 Inuvialuit Final Agreement (agreed to in principle in 1978), the 1993 Nunavut Land Claims Agreement (whose secession to form the Nunavut Territory took place in 1999 but was agreed to by plebiscite in 1992), and the 2005 Labrador (Nunatsiavut) Inuit Land Claim. The latter opted for an ethnic government rather than a public system, rejecting the Nunavut model in the face of intense demographic and economic pressures from the province of Newfoundland and Labrador and the consequent risk of nonnative settlement and large-scale economic development.

What appears in Alaska to be a raw, dissonant clash in the Hegelian tradition of contending processes driven by separate branches of government and competing levels of analysis looks to be a more unified and harmonious synthesis in Canada. But where the Canadian Arctic process achieves such smooth synthesis is primarily in the absence of large settler populations with powerful regional governments embodying their political will—in regions similar to Alaska, with substantial nonnative populations such as the northern portions of its provinces, a very similar and highly dissonant dialectical clash takes place. And even in regions of Indigenous demographic predominance, there has been persistent resistance to federal implementation, insufficiency in training funds and support, and many other skirmishes between native land claim beneficiaries and Ottawa (and in the case of JBNQA/NEQA, Quebec). At times the government has appeared to be a reluctant partner, suggesting perhaps some second thoughts on the extent of devolution in power achieved by the Inuit, and the relevance of Inuit gains to other, more heavily settled Indigenous regions of Canada. But despite these bumps along the road to implementation, the embrace by governments at all levels and by many corporate stakeholders of co-management systems that endeavor to balance these competing interests offers no less important a path toward synthesis of these competing forces. States continue to predominate in terms of material power, but natives still (for now) command demographic predominance in most remote northern regions, becoming essential stakeholders in Arctic sovereignty.

Lazarus and West described ANCSA in 1976,[32] just five years after it was passed, as a "flawed victory" leaving Alaska natives corporatized but lacking sufficient protections to ensure native ownership would endure. The goal at the time was not all that far removed from earlier assimilation policies—in this case the goal was to assimilate natives into a corporate-dominated economy—but in time, corporations not only transformed natives by modernizing their economic systems, but were themselves transformed as the very nature of a corporation became infused with new values. But this "flawed victory" of ANCSA inspired activism for more autonomy at the village level known as the tribal sovereignty movement, which sought greater (and perpetual tribal) protections of native lands, a closer relationship with the federal government modeled on the government-to-government relationships of tribal law in the Lower 48 (to help counterbalance the state of Alaska's interests where they were often pitted against native interests), and the protection of subsistence-oriented cultural traditions. At the same time, the very speed of ANCSA's implementation caused numerous growing pains. A lack of local capacity and experience required disproportionate dependence on external consultants, lawyers, and advisors, whose interests often proved contrary to the native village interests. Many of the new corporations were pushed toward an early risk of failure—a failure the architects of ANCSA may have not only anticipated but sought, since native protections would expire in twenty years, after which lands and resources could be sold to service debts and thus fall into the hands of the nonnative majority. Additional strains included a revenue-sharing requirement between Alaska natives, resulting in a perception of unfair redistribution of native wealth from areas endowed with natural resources to those without—dividing the native community from the get-go.

By the end of the 1970s the new corporations were, more often than not, in distress. In the early 1980s, just a decade into ANCSA's implementation before many of the subsequent amendments, former BC Supreme Court justice Thomas Berger, who had chaired the pivotal Mackenzie Valley Pipeline Inquiry, and his colleagues on the Alaska Native Review Commission traveled village to village across rural Alaska to investigate the seemingly impending collapse of ANCSA. Berger called for a retribalization of native governance to counterbalance the corporate bias built into ANCSA, and to protect native lands from the risk of loss to nonnative interests as a result of economic failures. Wayne Edwards and Tara Natarajan suggest that ANCSA, and ANILCA, which followed in 1980, exhibited a "capabilities failure,"[33] and consider the Alaska land claim from the perspective of "entitlements and capabilities," seeing the land claims experience as a "half-full/half-empty" situation in contrast to Berger's more pessimistic view earlier in ANCSA's experiential history, though perhaps more empty than full. As they conclude, ANCSA and ANILCA

> both provided Alaska Natives with entitlements. Both acts had, on average, positive impacts on Natives. However, the structure of the settlement provided by ANCSA meant that some entitlements were aggregated and subject to the control of corporate directors. Furthermore, the land conveyance process remains incomplete and so the entitlements provided by ANCSA have not been fully transferred, let alone the ultimate capabilities that could have flowed from the entitlements. For ANILCA, the impermanent nature of the traditional lifestyle protection it offers diminishes any capability that might be generated by the entitlement. The positive impacts are short-term rather than permanent. Both acts can therefore be viewed as partial failures from an entitlements and capabilities point of view even if some short-term positive outcomes resulted.[34]

This shift from doom and gloom to just gloom without doom itself illustrates ANCSA's evolution through its many amendments, what some call a "living settlement," whose terms adapt to the times. And the 1980 ANILCA response to ANCSA's exclusion of subsistence arose in part from the pessimism of Berger and much of the Alaska native community, thus inspiring reform, transforming the land claims model to address social and cultural issues and not just economic ones.

The Canadian experience with Arctic land claims, starting with the 1984 Inuvialuit Final Agreement, the first to be settled "north of 60," is a decades-long journey across Arctic North America. Land claims evolved from treaty to governing paradigm, growing far beyond the original template provided by ANCSA,[35] and from land settlement to a broader institutional transformation that expanded organically with each new iteration. Land claims are thus rightfully called "living treaties" or, as Len Sillanpaa described ANCSA in 1988, as "living settlements."[36] Sillanpaa's terminology may fit more precisely, since technically speaking, ANCSA was not a treaty but an act of Congress, while all of the Arctic land claims, whether negotiated treaties or not, are indeed land settlements. Because ANCSA was enacted by congressional legislation, it has had to do its living, adapting, and evolving via subsequent legislative amendment or through a completely new act of legislation (ANILCA), in addition to the interventions by the White House in 1993 granting Alaska native villages tribal status. On the Canadian side, land claims were indeed living treaties, and they would grow and expand with the times through bilateral negotiations between natives and governments. Through the integration of land claims with subsistence protections via institutional representation of hunting and trapping organizations and their formal participation in the co-management process, they achieved greater inclusivity than seen in Alaska, where ANCSA's corporate structures aligned primarily with business interests and continue to do so today.

That's why when ANCSA turned forty in 2011, it was both celebrated and critically scrutinized: Paul Ongtooguk found that "ANCSA is criticized both in terms of product and process, i.e., it yielded too little and the process did not meet a reasonable bar for inclusion. Most condemning for some is that as Native corporations, created in the image of a Western corporate model, have prospered, many Alaska Natives see ANCSA as a vehicle for assimilation."[37] But as Ongtooguk further reflects,

> On the other hand, ANCSA is also celebrated as a David and Goliath story that ended with a victory that far exceeded what anyone could have, would have or should have predicted. . . . Once ANCSA had been legislated into existence, the terms were exceedingly complex, and corporations were formed in an era of high inflation by Native peoples who had little professional training and virtually no talent pool. While there have been painful failures, there have been outstanding fiscal successes.[38]

ANCSA has been criticized in some of the literature, and by many Alaska natives, for its corporate focus. Yet the land claims structure in Canada preserves the corporate structure nearly intact while at the same time correcting the many initial flaws in corporate structure identified by critics of ANCSA, such as its exclusion from shareholdership of natives born after the land claim was settled and its failure to initially ensure native ownership of lands and corporate shares for perpetuity, or to ensure cash compensation was inflation-protected to meet future budgetary needs, or to provide supplemental implementation funds to ensure proper training and capacity development.[39]

As noted above, the age of land claims swept across Arctic North America for a half century after state expansion and modernization ramped up with a surge of new megaprojects following World War II. Some were only planned, like Project Chariot, the Rampart Dam, and the Mackenzie Valley Pipeline, but others were fully implemented (TAPS, the DEW Line, and the James Bay Hydro Project), and they galvanized a wave of Indigenous activism, partnerships with environmental interests, and negotiations for greater empowerment and constitutional integration of Indigenous peoples into the fabric of the modern state. Negotiations for the Inuvialuit land claim, which brought the land claims model into the Canadian Arctic, gained traction during the mid-1970s and came to an agreement with Ottawa in 1978, early in ANCSA's implementation and prior to a wave of corporate governance and fiscal crises in the 1980s. By the time the Inuvialuit Land Claim was signed in 1984, the Alaska Native Review Commission was established in Alaska. Confidence in ANCSA and its corporate model was at a historic low. Most of its worrisome structural problems had not yet been corrected through subsequent amendments, and a grassroots movement for tribal sovereignty, embraced by Berger and his colleagues, called for a "retribalization" of native lands at the village level to counterbalance the new corporate interests that ANCSA favored.

In the Canadian Arctic, the lessons of ANCSA were quickly digested, and Ottawa's conclusion (one shared by most of the native community) was that the corporate model was visionary and innovative, well-suited to the North where trading relationships and networks between the Arctic and non-Arctic worlds go back centuries. But it needed a counterweight to protect traditional native cultural values and interests, reflected in a large degree by the subsistence economy and culture, to avoid the risks and dangers of ANCSA, where the land claims teetered on the edge of failure, and where its original model overemphasized corporations at the expense of native traditions. The final structure of the Inuvialuit Final Agreement thus more equally valued and more equitably balanced its corporate and cultural components, and this revised model would remain part of Canadian Arctic land claims for the next quarter of a century, with little changed in the actual land treaty structure in Nunavut in 1993 or Nunatsiavut in 2005.

But as the Inuvialuit discovered, institutional balancing can only take you so far. While subsistence was better protected by land claims structures in the Canadian Arctic, a new wealth gap still emerged between the native corporations and the more traditional institutions such as the Hunters and Trappers Committees. As this gap grew, corporate influence began to overwhelm traditional interests—and a replay of ANCSA's early growing pains would recur in each of the subsequent land claim settlement areas as beneficiaries of each new land claims agreement struggled to control these newly established institutions, and to keep corporate interests and values in check. Marie Wilson's twenty-eight-minute CBC *Focus North* news documentary ("The Inuvialuit Land Claim"[40]) captures the hope and excitement felt by beneficiaries of the first comprehensive land claim accord in the Canadian Arctic, one achieved by a group (the Inuvialuit) that is, intriguingly, predominantly of Inupiat descent with close ties across the border that helped sustain community commitment to the slow-moving land claims process over the years, and to recognize the protective opportunities a well-structured land claim could provide in the face of intense oil and gas exploration.

The community's celebration of the land claim's finalization suggests why these agreements can be much more than they seem. They are reminiscent of the old "numbered treaties" between native tribes and the British crown that were perceived by the native negotiators to be, at heart, nation-to-nation treaties of friendship and alliance, even if the crown itself viewed the treaties as (or transcribed them to become) treaties of surrender and constitutional subordination of the natives to the state. In Wilson's report, we're introduced to Nellie Cournoyea, who rose from an Inuit Land Use and Occupancy Project (ILUOP) field worker and negotiator for the Committee for Original Peoples' Entitlement to serve as a member of the Legislative Assembly representing Nunakput and then as premier of the Northwest Territories. She returned to the Inuvialuit Settlement Region after a financial scandal at the Inuvialuit Regional Corporation led to a grassroots movement to rescue and renew the IFA under her leadership. In her remarkable journey from grassroots activist to national icon and powerful political and corporate leader, we can gain a sense of the importance of the individual level of analysis, where just one charismatic leader can play a supersized role in shaping Arctic history and policy, and ascend from the local to the national (and in some cases international) level and back to the local level again when needed. Reporter Wilson went on to serve as head of CBC Western Arctic in Yellowknife and later as a commissioner of the Truth and Reconciliation Commission of Canada, helping to shape the historic process of reconciliation between Ottawa and Indigenous peoples—another example of influential individuals and Indigenous communities transforming national politics and expanding the role of Indigenous peoples in the governance of the Arctic.[41]

After the IFA, a new lesson was learned by the Inuit: a land claim treaty, even a well-balanced one, greatly benefited from the formation of a parallel institution of self-government that would ensure balance between the corporatists and traditionalists was maintained, and that parallel government had to represent the collective native interest in order to ensure a synthesis of these dialectical internal forces. So, when the Nunavut Land Claim Agreement was signed in 1993, it included a political accord that promised to create a new public but predominantly Inuit territorial government, the Nunavut Territory, by 1999, building on the gains achieved in 1975 with JBNQA and 1978 with NEQA, while greatly extending the scale of political power well beyond the municipality or region. In 2005, when the Labrador Inuit of Nunatsiavut signed their land claim, it also created a mechanism for Inuit self-government, but instead of a public territorial government, it was accompanied by a Labrador Inuit Constitution that created ethnic, Inuit self-governance of the five villages in the Nunatsiavut settlement area, one that could never be undermined by a change in demographics (which Nunavut, as a public and not an ethnic system of government, is not as well-structured to withstand).

Another important consideration is in the process utilized to negotiate a land claims treaty. With ANCSA, the land claim was achieved through a lobbying process in the halls of Congress as stakeholders sought to influence the outcome; it was not the direct result of a community-based bilateral negotiation process with the government. In Canada, however, land claims were indeed treaties that resulted from such bilateral negotiations and in the end were approved by referendum. ANCSA was approved by vote of the Alaska Federation of Natives; there the Inupiat famously voted against the land claim but were ultimately outvoted by the majority of Alaska natives who stood to gain more from ANCSA's revenue sharing provisions than the Inupiat, whose oil fields would power not only the state economy but the native economy as well. In Canada, the Inuvialuit communities voted overwhelmingly in support of the land claim, which had evolved greatly beyond the ANCSA model and had, with the breakup of a broader, more diverse and expansive land claim from the western to the eastern Arctic, become a regional land claim serving the more ethnically cohesive Beaufort-Delta region (with its own divisions and variation but less complexity and diversity than seen in ANCSA, which was state-wide in its scope, or JBNQA/NEQA, which included Cree and Inuit [in 1975] and extended to include Naskapi [in 1978]).

Nunavut came into existence through a more complex process, since it was associated with the subsequent secession of the eastern Northwest Territories (NWT) from the western NWT, requiring multiple referenda. A 1992 referendum process preceded the formal 1993 signing of the Nunavut Land Claim, whose first step was a May 4, 1992, plebiscite on whether to divide the Northwest Territories, which then included the Eastern and High Arctic, allowing Nunavut to peacefully secede from the NWT with 54 percent of the NWT-wide vote. The next step was a referendum held November 3–5, 1992, in what was set to become the new Nunavut territory, with 69 percent of this regional vote in the Eastern and High Arctic communities. By the time of the formal birth of Nunavut on April 1, 1999, it was a huge historical moment for the Inuit, for Canada, and for the history of land claims in the Arctic. The birth of Nunavut was comparable, in geographic scale, to the 1867 Alaska purchase: as the latter brought Alaska under US sovereignty after its Russian colonization, thereby initiating a slow multigenerational process of political reform that would, over time, better serve Indigenous interests, the former more strongly aligned Inuit and national interests by augmenting the land claim with the secession of a majority Inuit territory, enabling Indigenous governance through a public model. So, while the latter integrated the former RAC colony into the US constitutional polity and brought an end to Russian sovereign claims on mainland North America, the former redrew internal territorial boundaries so that the Inuit could effectively govern themselves as the overwhelmingly predominant ethnic group of the new territory, instead of as a minority in the more demographically complex predivision Northwest Territories. Each has contributed to the political stability of Arctic North America and enhanced the sovereign claims of their respective states (the United States and Canada). In many ways, each has enabled those two states to more ably assert and project sovereignty in the Arctic, though the purchase of Alaska was only the first step. It required ANCSA a century later to provide Alaska natives with additional avenues for meaningful participation in the governance of the state, followed by ANILCA and numerous subsequent legislative amendments tweaking and improving ANCSA, right up to and beyond the 1991 amendments (technically enacted in 1988, in anticipation of the looming 1991 time bomb,[42]) just around when the Nunavut land claim was finalized and the path to the 1999 Nunavut secession mapped out.[43] Both cases were thus pivotal but unto themselves incomplete forward steps toward Indigenous inclusion in Arctic governance.

As we see with JBNQA/NEQA in 1975/1978, the Inuvialuit in 1978/1984, Nunavut in 1993/1999, and Nunatsiavut in 2005, settling an Indigenous land claim is just the starting point in the broader process of native empowerment and inclusion. It has become universally accepted as an essential first step, and has retained the native corporation (and its division into regional and village or community corporations) as its economic engine, but it has grown into a more complex and ambitious project, starting with the institutional inclusion and protection of subsistence as part of the land claim and the restructuring of government on a massive scale to ensure native democratic participation and control at the local, regional, and territorial levels.

The Inuvialuit have long aspired to create a structure of regional self-governance to augment their land claim, but they are nonetheless proud that they solved the subsistence challenge that ANCSA had left unresolved across the border. Indeed, it appears to be more than just a coincidence that the blueprint of the 1984 Inuvialuit Final Agreement, the 1978 Agreement-in-Principle, precedes the Alaska National Interest Lands Conservation Act (ANILCA) of 1980 by two years, which, in addition to offering federal protection to much of Alaska's wilderness lands also addressed subsistence, and while separate from ANCSA, it is often perceived to be the completion of the journey that ANCSA started. Moreover, it appears to be more than just a coincidence that in 1993 Ottawa agreed not only to the Nunavut Land Claim settlement but also to the subsequent formation of the Nunavut territory, while at the same time in Alaska the federal government recognized Alaska native villages as tribal entities, counterbalancing the state of Alaska's long opposition to tribal government and further strengthening the native position with regard to the feared losses of native equity control and ownership associated with the 1991 time bomb.[44] These historic alignments in native empowerment on both sides of the border suggest that aspirations and realities would reset in both Alaska and Canada, with the former happening much more piecemeal and the latter, over time, happening in a more unified manner but also with the need for additional steps to ensure implementation lived up to expectations. So, there was progress on both sides of the Alaska-Canada border, and setbacks too, along with decades-long efforts at institutional reform and improvement. Institutionally, the land claims experience in the Canadian Arctic has benefited from the passage of time and the expansion of opportunities, from subsistence to self-government to territorial governance. Also beneficial have been the general absence of large concentrations of nonnative residents, in contrast to many parts of Alaska, and a more favorable constitutional environment, particularly with Canada's constitutional recognition of the inherency of aboriginal rights.

In the famed 1961 documentary on the Barrow Duck-In[45] that preceded ANCSA, Indigenous rights leader Charlie Etok Edwardsen refers to the rise of "Eskimo power" in response to Alaska's modernization and expansion of governance as a state, and the state's early insensitivities to Indigenous rights issues. The pioneering transformation of Arctic governance fostered by ANCSA, its many (and seemingly never-ending) amendments, in combination with ANILCA that reintroduced federal protection of subsistence, created new institutions for Indigenous governance and economic participation, primarily the

Alaska Native Corporations, and for federal management of subsistence on its vast Alaskan wilderness lands, protecting those lands from development pressures by corporate and state-level governmental interests. The result was a complex system of contending levels of governance with a more robust role for Indigenous stakeholders than ever before. That role would continue to grow as Alaska natives looked beyond the corporation as a model and revisited tribal government as an option for achieving their aspirations at the local level. As the late Thomas Berger chronicled in his writings on his "village journey"[46] into native Alaska in the mid-1980s, that movement arose in widespread community opposition to ANCSA's corporate structure but gained federal support during the Clinton administration with the tribal recognition of Alaska native communities in 1993, when Ada Deer served as assistant secretary of the interior and head of the Bureau of Indian Affairs. Deer was a member of the Native American Rights Foundation steering committee, and during 1984–1985 she participated with Berger in the Alaska Native Review Commission (ANRC), which called for a "retribalization" of Alaska native lands. The ANRC was sponsored by the Inuit Circumpolar Conference (now Council), one of the pioneering institutions behind the founding of, and a Permanent Participant on, the Arctic Council and part of a broader movement to augment and internationalize the voice of the Inuit. The fact that it helped to foster the movement for greater tribal sovereignty in Alaska as it worked to create new autonomous forms of governance in Canada through its evolving land claims model is one more indicator of the transnationality of the Inuit and their approach to strengthening autonomy and self-governance within the constitutional parameters of their various sovereign states. This fascinating journey of Inuit transnationalism and diplomacy and associated institutional innovations in co-management and collaborative governance that followed, introducing a new paradigm for Indigenous self-governance in Arctic North America and beyond, is fundamentally rooted in the land claims experience.

Co-management as a Model for Collaborative Arctic Governance

The balance struck between modernization of the Arctic and its retribalization, as called for by the ANRC, manifests itself primarily in the mutual commitment to co-management, which was enshrined as a governing paradigm and key structural component in all subsequent Arctic land claims and appended to the corporate systems in Alaska by ANILCA, which reintroduced federal oversight of the co-management process on federal lands. Co-management has come to reflect the philosophy of Arctic governance in the age of land claims, as it is not only built into the structural fabric of Canadian Arctic land claims but has taken form outside the framework of land claims (as it did in Alaska) in regions where land claims remain unsettled, and where non-Indigenous interests complicate negotiations due to the multiplicity of third-party interests absent from most of Arctic North America. And with the formation of the Arctic Council in 1996, co-management has emerged as a paradigm for joint tribe-state governance in not only domestic politics but international politics as well. Indeed, in the structure of the Arctic Council we see a very similar balance of Indigenous and state interests, with state interests continuing to hold most of the cards when it comes to decisionmaking powers, but Indigenous peoples having a seat at the table and enjoying a prominent position from which to influence those decisions.

In the literature on land claims and co-management in Arctic Canada, we see in general a more balanced approach between the processes of modernization and retribalization than we see in the literature on the Alaska land claim. The name of the federal intervention to protect Alaska's wilderness lands and subsistence upon them, the Alaska *National Interest* Lands Conservation Act, is telling and illustrates how the conservation of land was perceived by the White House in 1980 as a pillar of the national interest, and as such, a component of national security. In Canada, there was a similar recognition of the importance of land conservation as not just a tribal interest but a national interest as well, and this mutual recognition made it possible for Ottawa to agree with Inuvialuit aspirations to protect subsistence and the lands upon which it depended.

Milton Freeman's retrospective discussion, "Looking Back—and Looking Ahead—35 Years After the Inuit Land Use and Occupancy Project," reflects on this watershed moment in land claims history, which commenced in 1973 at the request of the Inuit Tapirisat of Canada (now known as the Inuit Tapiriit of Canada).[47] The ILUOP introduced the use of individual "map biographies" as a methodology for countermapping (using maps to restore Indigenous rights and to counter claims by settler states), and in so doing illustrates a remarkable victory by the Inuit, who persuaded Ottawa to accept their own narrative on land use and occupancy to establish the historical foundation for the land claims process (based on unsubstantiated claims of Indigenous land use and occupancy "since time immemorial"). In thinking about the levels of analysis, this is an interesting illustration of how the individual level of analysis became central to the groundwork for the land claims process, which ultimately was accepted by the national government to remap the northern third of Canada—evidence of effective individual agency in national politics. In the ILUOP, Inuit-hired anthropologists (led by Freeman) and a team of Inuit field workers—including Nellie Cournoyea, the future NWT premier and long-serving chair of the Inuvialuit Regional Corporation—mapped Inuit land use by relying on the oral testimony of Inuit hunters and trappers. These oral testimonies were accepted by Ottawa to verify Inuit claims to most of the Canadian Arctic, even lands that were only recently settled by Inuit in the twentieth century and that had not been used historically by Inuit. How was this possible, one might wonder? Inuit recognized Ottawa's ultimate weakness: in not effectively occupying its own Arctic lands, Ottawa needed a proxy to do so on its behalf. Indeed, in the 1950s, Ottawa had relocated several dozen Inuit to the High Arctic, in a resettlement program that has gained much controversial notoriety in the years since. The Inuit conclusion was that Ottawa needed Inuit consent to its sovereign claims, which put them in a position of negotiating strength despite their relatively small numbers and lack of material wealth or power. Their gambit succeeded when Ottawa accepted the ILUOP in full without revision. When looking at the High Arctic exiles, the relocation effectively strengthened Ottawa's claims to the region, which would soon be further strengthened with the 1956 extension of Canadian citizenship to Inuit. That Ottawa insisted on Inuit bringing subsistence to the High Arctic islands while offering little economic assistance supports the notion that Ottawa wanted Inuit to use as much of their new insular lands as possible, a necessity to survive without government support. This was a very different kind of colonization, one that strengthened subsistence, a foundation of Inuit culture, but at very high personal cost and risk to the families and very low cost to Ottawa. The later human rights blemish and demands for compensation offset this somewhat, but not so much considering how much land Ottawa gained from it.[48]

What did Ottawa get in return for granting Inuit so much autonomy, and such a large land base (with significant economic compensation too)? Just as a land claim treaty transfers land title of what has been, on average, around one-tenth of the land inside a settlement region to the native beneficiary group, the native group in turn agrees to surrender any aboriginal title to that land to the national government. So, in settling the IFA, Ottawa in essence gained recognition for its claim of crown title to the other 90 percent of their oil-rich region that the Inuvialuit did not gain title to; and in Nunavut, over a much larger and more strategic area, the ratio was much the same. So, Ottawa gained title (or at least something akin to title insurance, in that the Inuit agree not to contest Ottawa's title in exchange for the clarity of fee simple title to the 10 percent they did receive) to the remaining 90 percent of a region that it has historically not occupied, settled, or developed, in exchange for its grant of fee simple title and cash compensation (and other rights and powers) to an Indigenous group willing to consent to Ottawa's sovereignty—the essence here is diplomatic and strategic.

In Alaska, it was much the same: ANCSA provided Alaska natives with closer to one-ninth of the land, but at the same time the state gained immensely from the surrendering of native title to those other lands—ensuring federal and state ownership of the rest of Alaska was, at last, undisputed. The state, in short, receives undisputed title to the vast majority of the Arctic land mass, with the consent of the Indigenous people of the Arctic to assert sovereignty over those lands. And on the Canadian side, this is ultimately rooted in those individual map biographies claiming individual use and occupancy of the land that laid the foundation for Canadian Inuit land claims. For areas where Inuit did indeed use and occupy these lands historically, the mutual exchange between the state and the Indigenous community does seem to legitimize sovereign claims over those lands. But what of those areas where Inuit land use and occupancy are relatively new, and thus lack such a historic foundation? And what if claims of individual land use and occupancy in the map biographies are exaggerated as users of productive lands around the world will often do? Fortunately for both Inuit and Ottawa, neither has suggested there is anything shaky at all to the foundation of either Canadian Arctic land claims or Canadian Arctic sovereignty.

Moreover, the structures created by land claims established a solid institutional foundation that, over time, brought effective and meaningful governance to these regions—so even if the historical foundation was shaky, the postsettlement institutional development has strengthened Arctic sovereignty through the emergence of a diverse patchwork of governing institutions. The IFA established not only six community corporations and the Inuvialuit Regional Corporation but also an Inuvialuit Development Corporation, Inuvialuit Investment Corporation,

Inuvialuit Land Corporation (to hold title to Inuvialuit-owned lands), and Inuvialuit Land Administration (to administer those lands), as well as six Hunters and Trappers Committees, an Inuvialuit Game Council to which all six belonged, and five subsistence-oriented and environmental co-management organizations upon which they served: an Environmental Impact Screening Committee, Environmental Impact Review Board, Fisheries Joint Management Committee, and two Wildlife Management Advisory Councils (one for the NWT and one for the Yukon North Slope, despite strenuous objections by the Yukon territorial government that neither Ottawa nor the Inuvialuit had a right to govern anything inside the Yukon—a view Ottawa and the Inuvialuit rejected). The Inuvialuit made seasonal use of the North Slope for whaling but had no permanent communities along the slope past the Alaska border.

The inclusion of such a diverse array of structures of co-management with specific functions established what we can think of as a separation-of-powers system between the economic and the cultural—giving equal institutional consideration to both corporate and subsistence practices while also embedding in the land claims structure an unprecedented degree of granularity matching each function to its own institution. While most of the money, and thus economic influence, stayed on the corporate side, creating an imbalance in influence within the Inuvialuit Settlement Region during its first decade, the co-management structures would become, in essence, governing institutions that ensured a native voice was heard on matters of supreme importance to the Inuvialuit traditionalists, and in the rare case this voice was ignored by the federal government, the government felt enormous pressure to disagree respectfully—knowing the courts could, and would, intervene if Ottawa did not abide by the promises enshrined in the land claim treaty, which had constitutional protection. Even though Ottawa retains final decisionmaking authority, the landscape had forever changed, ensuring native participation in all decisions affecting the land claim settlement area.

With co-management embedded into the land claim's structure, it gained the same constitutional protections that the native corporations receive, becoming a mechanism to protect Indigenous rights as guaranteed by the Canadian constitution. Elsewhere, as noted above, co-management is separate from such constitutional enshrinement, but as we see in ANILCA, subsistence gained federal protection even if not additional constitutional protection, which has proven to be powerful and enduring, if a source of ongoing friction with the state government in Juneau. In southern Canada, co-management has emerged separate from the land claims process, but because of Ottawa's constitutional protection of Indigenous rights, these co-management systems enjoy a special quasi-governmental status even in jurisdictions where natives remain a minority. The paradigm of collaborative governance articulated through the co-management experience has been a process of continuous dialectical interaction and refinement between thesis and multiple antitheses on a long, multigenerational road toward reconceptualizing Arctic sovereignty that dates back to the fur empires.

Much has been gained, but these gains should not be conflated with a presumption of ease, as the process has been with numerous challenges, requiring lifetime commitments of dedicated individual actors who have through their determined efforts literally kept the state in check and constrained its worst excesses as seen in the Westphalian world. These challenges are hauntingly conveyed in *Kajutaijuq: The Spirit That Comes*,[49] set and filmed in Nunavut and featuring Inuk actor Johnny Issaluk. This 2014 film captures the intensity of horror and the depth of anger unleashed by the spirit world, directed (paradoxically, it seems) against an Inuk hunter who is trying to maintain a balance between the worlds of tradition and modernity. What provoked the wrath of the spirit? The spirit's anger is not aimed at those who embrace modernization fully, or seemingly those who reject modern ways in favor of tribal traditions, but instead those like the protagonist who try to reconcile and balance the dueling thesis and antithesis of globalization as described by theorist Benjamin Barber around the time the decades-long negotiations for Nunavut were successfully concluded—as an epic and ongoing clash of globalism and tribalism. And this leaves a mystery for us to grapple with: Why is it angry at this synthesis?

Considering the Inuit journey from land claims toward restoration of Indigenous sovereignty, one wonders if the spirit in this film was enraged by Inuit efforts to live simultaneously in both the modern and the traditional world, or perhaps by the hubris of humanity to build modern political structures in a remote and distant land where tradition and nature have always held the upper hand. Going forward, perhaps we can expect further reconciliation, but this will require continued dedication and sacrifice—hopefully not as great as that of the protagonist, but of both time and energy.

Inuit have regained so much—land, wealth, political power, autonomy. True, they have yet to regain formal sovereignty over their land as an independent nation, but this does not seem to me to be the reason for the spirit's anger. The victim of the spirit's wrath listens dutifully to an audio recording in his language, about the restoration of traditions.

He is practicing Inuit traditions of hunting and living off the land, central to Inuit identity for millennia. What was his transgression? Was it his reliance on outside knowledge and technology? Was it his and his people's aspiration to live in two worlds at once, when the spirit commands obedience to just one? Or is it a reminder that in the end, nature is and always will be unforgiving—and the ultimate arbiter of whether we live or die? Or is it a metaphor for Nunavut's failure to fully live up to the hopes and dreams of the Inuit, and the continuation of so many of its challenges (in part due to continuing tugs-of-war with Ottawa over sufficient resources and political will)? The film is not explicitly political, and as a pure horror film can bring a class of content-oversaturated students, hardened by the ubiquity of Netflix, to a primal and unstoppable scream. Nowhere in it is Nunavut or its challenges discussed specifically, but it captures this paradox perfectly, and hints that the reconciliation of tribalism and globalism by land claims is itself an imperfect solution to the challenge of balancing tribal and state interests in the Arctic. But considering reconciliation's historic endurance, and notable gains, and continued momentum, it is perhaps the best we can achieve in this world, contextualized by our shared history. As settled Arctic land claims continue to mature, and time brings greater capacity with a new generation taking over the helm of governance in the Arctic, it is a fruitful intellectual exercise to continue to wonder what it was that provoked the wrath of the spirit in *Kajutaijuq* and endeavor to proactively address this through further structural innovation and collaboration in Arctic governance.

Land claims and co-management have transformed Arctic governance, setting it on a course of collaboration between tribe and state and allowing the state to expand with the consent of the Indigenous people, a new experience for North America and a welcome alternative to the history of war and conflict seen elsewhere, including in the Lower 48, with their history of Indian warfare. But like Clausewitz reminds students and practitioners of war, war is at heart a continuation of politics (or policy) by other means—the original word in German is *politik.* Land claims are thus an alternative to war, a kinder, gentler, and more respectful path to state expansion, but in the end, they facilitate the state's expansion and the consolidation of its sovereignty over the Arctic. This is the ultimate compromise that drives it forward, bringing tribal and state interests to the negotiating table over and again across the ages, much as we saw at the parley table at Old Fort Niagara, where native commanders and their colonial counterparts met to discuss the ebbs and flows of the French and Indian War, known in Europe as the Seven Years' War—a war for control of the interior of North America. Colonial warfare did not extend into the Arctic except on occasion, allowing instead for the future of the Arctic to be determined by contemporary parleys, which evolved into the negotiations between tribe and state that brought forth the age of Arctic land claims. In this way, war has been averted, but the clash of interests between tribal polities and peoples on the one hand and Westphalian states on the other continues, channeled into negotiated settlements, peace treaties without war. It is my hope, in the years ahead, that the spirit's wrath we come to know in *Kajutaijuq* is in the end tamed, and that the compromise achieved through this novel and innovative alternative to armed conflict becomes a model for the world.

The Arctic land claims model, with its subsequent modifications, has become an inspiration to many, proof positive of what can be gained through a determined, forward-looking effort to rebalance and modernize the relationship between the Indigenous people of the North and the modern state. As with any land reform effort, changes in land tenure can have a profound impact on the domestic balance of power, shifting not just title to land but the wealth created from that land, resulting in concentrations of economic power in the hands of a small Indigenous population numbering in the thousands or tens of thousands. In Alaska and the Canadian Arctic, the Inuit have become owners of vast tracts of land, making them a landed elite with control over numerous economic and, increasingly, political levers. While not formally sovereign, they are poised to become increasingly influential stakeholders, partners in the consolidation of state sovereignty, and in the economic development of the northern frontier. A comparable situation exists in the post-Ottoman Middle East, with extended tribal families and clans sitting at a powerful and lucrative nexus of land ownership, natural resource wealth, and political power. While natives in Arctic North America are not in command of the ultimate levers of sovereign state power, such as military forces or national treasuries, they do have in their possession or within reach many tools of regional power, making them dominant regional elites. As the climate warms and the Arctic basin yields more natural resource wealth, the economic resources in their possession will also increase, and with that political influence.

In November 2008, Greenland held a historic but nonbinding referendum on increasing the island's autonomy and eventually restoring its sovereign independence, which was approved decisively, showing how the desire to be self-governing is universal across the Arctic.[50] Denmark has shown a unique openness to the possibility of Greenland becoming formally independent (in contrast to the other Arctic states, which attach great economic, strategic, and emotional/ideological significance to their Arctic territories), and if Greenlandic independence is achieved, it would mark perhaps the final stage in the process of Indigenous re-empowerment that began with ANCSA nearly half a century ago, with the full restoration of sovereignty to an Arctic nation. Some question whether Greenland, with a population just north of 55,000, can be a viable sovereign state, but other microstates have achieved sovereignty (even when unable to defend that sovereignty unilaterally)—from Singapore in maritime Southeast Asia to Iceland in the high North Atlantic, each seeking strong strategic partnerships with larger states and alliances to ensure their sovereignty can be protected. So why not Greenland?

What a sovereign Arctic state will look like, how it affirms traditional Indigenous values and balances modernization with tradition, will be fascinating to observe. The risks are real: Iceland's economic collapse in 2009, Nunavut's persistent social challenges since its formation, the near collapse of the Alaska Native Corporations (ANCs) before the 1991 time bomb was defused are all cautionary tales to consider. The move toward Greenlandic independence, combined with the economic challenges faced all across Arctic North America, has encouraged policymakers north and south to rekindle their efforts to better integrate the Arctic into the political, economic, and strategic architecture. This became evident in the final days of the George W. Bush administration, when US Arctic policy was updated in a sweeping manner, and while the policy was largely overlooked outside the Arctic as issues more salient and immediate from the global economic crisis to the rise of ISIS dominated the news cycle and many people's attention, it marked an important reaffirmation (by one of the most unilateralist of administrations) of a lasting reconciliation between tribe and state at the top of the world, ensuring Arctic exceptionalism was sustained.

An Alternative World System: Arctic Exceptionalism as a Model for International Relations

In the Arctic, we witness an alternate historical narrative defined by a historic reconciliation of tribe and state, a restoration of Indigenous land and cultural rights, and a rise in native participation in both economic and international relations at the regional level. Here, ideas and insights from the Alaska land claims process of the 1970s flowed across the international boundary and into the Canadian Arctic and sub-Arctic, where they were rethought, refined, revised, and reapplied. The result was a stronger, more resilient, and ultimately more scalable model for northern development that reflected a deep and enduring commitment to collaborative cross-border management, intergroup (and international) partnerships, and constructive transboundary relationships that present a compelling model for how the world can and should be governed. All this has its roots in that very first wave of northern development fostered by colonial-era chartered companies, and in particular the blueprint for northern development crafted by the HBC (as discussed above). Because of the distinct challenges of the northern landscape, and its concentrated dependence upon natural resources (including wildlife resources that would be vulnerable to demographic influxes of settlers), multinational corporate survival required native survival, and corporate success ultimately depended upon native success. As recalled by Inuit historian Mini Aodla Freeman,

> I think I can say that Hudson's Bay Company made easier lives for Inuit since 1670. They were in my home area of James Bay long before I was born, fur trading with my ancestors. Inuit have always traded amongst themselves either for short period or long period of time. Inuit understood the trading systems of the Hudson's Bay Company. I think one of the reasons why Inuit welcomed the Hudson's Bay Company was the fact that the company never tried to change Inuit ways of behaving or thinking. Yes, they changed our equipment, to better steel knives, steel saws, steel nails, steel axes and manufactured cloth. Inuit understood it was furs that the Hudson's Bay Company were after. Inuit hunters had employment through the Hudson's Bay Company. It was the familiar job Inuit enjoyed. We still hear older Inuit today saying that the Hudson's Bay Company is most useful in Inuit lands. They did not interfere with lifestyles of Inuit.[51]

HBC-administered lands were perhaps no Shangri-la, but it was a far cry from the excesses witnessed in other parts of the colonial world. Its legacy was the enduring collaboration between the Indigenous peoples of the North and the governing entities that asserted sovereignty over it. It was not always frictionless collaboration, since there were times and issues where interests can and did clash.

But despite these very real and recurring collisions of values between native, environmental, settler, and resource-extractive interests—as we've seen ever since the oil strike in Prudhoe Bay catalyzed the rapid formalization of the Alaska Native Claims Settlement Act in 1971—collaborative efforts between natives and northern governments, and between neighboring native communities that stretch across the border, have remained ongoing, helping to mitigate those inevitable conflicts when they do arise. On the Canadian side of the international boundary that divides Arctic North America, lessons from the Alaska land claims experience and its initial structural flaws were closely studied, and this cross-border flow of ideas and insight influenced a new model for land claims settlements that ensured native lands and corporations would always remain in native hands, that young natives would be automatically enrolled as shareholders upon adulthood, and that subsistence would forever be protected on both native-owned lands and adjacent government lands. The Alaska experience thus proved critical in guiding Canadian natives forward in their quest to assert, and protect, their aboriginal rights.

By the time the states turned to the Arctic region for sovereign expansion, they did so more gently and less muscularly than they did in other parts of the world—guided not only by their still maturing recognition of Indigenous rights, which has enabled them to expand while integrating northern Indigenous peoples largely intact into their constitutional structures, but also by the enduring, resource-based northern political economy long in place, having been nurtured by the chartered companies that enjoyed their quasi-sovereign dominion over their northern territories for so many generations. The northern tranquility observed by so many to define the Arctic region as a whole, known widely as Arctic exceptionalism, owes much then to the mutual reciprocity of commerce embraced by the HBC, with its deep historic roots dating back centuries.

Land claims and their evolving integration of native corporations and tribe-state co-management organizations, augmented by increasingly powerful institutions of native self-governance, have taken this reciprocity even further. The Inuvialuit land claim presents a substantial evolutionary leap beyond the Alaska land claim that inspired it, with many prescient and enduring advances in collaborative management and stronger protections of native lands and traditions missing from the original, unamended Alaska land claim. Had the Inuvialuit not so enthusiastically embraced and constructively improved the land claims model, with native corporations still at its core (but not to the exclusion of traditional cultural values, as had been experienced with ANCSA), the many original structural weaknesses of the Alaska land claim[52] might well have doomed the model altogether. Blatchford has noted how the very land claim model that has transformed the political economy of Alaska, Yukon, NWT, Nunavut, and northern Labrador would ultimately be rejected by Indian Country in the Lower 48 as a flawed model, where corporations are viewed with much more skepticism and as contrary to native values. But in the Arctic, it has become a central and evolving blueprint for strengthening the bond between First Nations and the state, and a defining feature not only of the Western Arctic but the entire Arctic littoral of North America, a vast and increasingly strategic region of not only the North, but of the world. This embrace of, effort to improve, and continuing process of reforming the land claim model as it flowed from the Inupiat to the Inuvialuit and on to Nunavut and Nunatsiavut are a reflection of the collaborative mechanism that defines the Western Arctic, reminiscent of the integrative and symbiotic dynamic of HBC and North West Company trading posts, and reflective of the very dynamic and integrative "Dene/Inuit interface" of the Mackenzie Delta as described by William C. Wonders in his pioneering work on the region a generation ago.[53]

When the land claims movement swept across the Arctic coast, starting in Alaska in 1971 and culminating with the birth of the Nunavut territory a generation later in 1999, it illustrated the Arctic region's propensity for accommodation and reconciliation, for welcoming diversity and inclusion, and for collaborative problem-solving and co-management with both governments and MNCs. This reflected the enduring legacy of the early pioneers in northern economic development, and some of the world's very first MNCs—the chartered companies of the North. This collaborative spirit is evident today in the close relationship between the Inupiat and the Inuvialuit, who have partnered on numerous cross-border issues—including the Inuvialuit-Inupiat Polar Bear Management Agreement in the Southern Beaufort Sea, and the Inuvialuit-Inupiat Beaufort Sea Beluga Whale Agreement—and whose collaboration extended to the resumption of bowhead whale harvesting by the Inuvialuit during the 1990s, when community-to-community exchanges ensured the transfer of traditional knowledge required for a successful and safe restoration of bowhead hunting.

It is equally evident in the complex (and occasionally prickly) relationship between the Inuit and the Dene along Wonders's aforementioned Dene/Inuit interface, a history that includes intertribal warfare as evident at the tragedy of Bloody Falls during Samuel Hearne's fateful expedition but that also includes a long-standing (but ultimately unsuccessful) effort, forged across decades of collaboration between traditional adversaries, to form a Western Arctic Regional Municipality (WARM) to jointly govern Dene and Inuit in the Western Arctic. While WARM never came to fruition, the many regional land claims settlements in the Dene territories to the south of the Inuvialuit Settlement Region closely resemble the Inuvialuit Final Agreement, and in the wake of these settlements we've witnessed a resurgence in MNC activities across the Dene homeland, in both the mining and petroleum sectors, as evident in the ongoing Dene collaboration and participation in Canadian Zinc's—now known as NorZinc's—efforts to bring the Prairie Creek Mine back into production in the Deh Cho region, as well as in the diamond mining sector north of Great Slave Lake, and the Mackenzie Gas Project, discussed below.

After Land Claims: Tribes and States Share the Stage and Increasingly Unite Their Voices

Indeed, one generation after Berger's Mackenzie Valley Pipeline Inquiry effectively blocked resource development in the Western Arctic for a decade, the Mackenzie Gas Project held a new round of consultations in the very same communities of the region starting in 2004, now with native groups sitting at both sides of the table, representing both the corporate stakeholders hoping to extract and transport natural gas up the Mackenzie River to southern markets (as one-third equity owners via the Aboriginal Pipeline Group) and the local communities grappling with the impacts to traditional subsistence activities and the natural environment. The effort proved successful procedurally, and in 2011 the Mackenzie Valley Pipeline was granted federal cabinet approval, and the National Energy Board granted a Certificate of Public Convenience and Necessity. But by then gas prices had collapsed, and the economics of the pipeline were no longer attractive, so the victory was pyrrhic—the consultation process was now beyond doubt inclusive of native interests, as diverse as they were, but it had consumed so much time that the underlying market fundamentals had transformed before its favorable conclusion, ultimately dooming the project. But it remains no less illustrative of the tectonic shift in the Arctic's political economy that had taken place since the first Mackenzie Valley Pipeline Inquiry in the 1970s, and of the maturation of native participation in Arctic development (now as a major equity participant). Like WARM, the Mackenzie Gas Project would ultimately fall short on its ambition to construct a native-owned pipeline, built by native-owned contractors, transporting natural gas from native-owned lands to the global marketplace, but it nonetheless illustrated that a fundamental transformation of the northern political economy had occurred, with the emergence of natives as dominant economic actors with interests that now aligned, at least in part, with MNCs and not just as opponents to the development efforts of the MNCs.

Just as the early chartered companies of the colonial-era Arctic found their success tied to the survival of native peoples and their cultures, contemporary MNCs, whether mining or oil extractive industries, recognize that the post–land claims Arctic has endowed native communities with powerful levers requiring continued collaboration and engagement in order for MNCs to operate, and this has resulted in a slew of environmental reviews and assessments, and countless economic participation agreements bringing much needed training and employment to native communities. Native-owned subsidiaries enjoy preferential contracting obligations that ensure meaningful native economic participation, providing further economic stimulus to native communities across the Arctic and sub-Arctic. It's not just in mainland North America, but across Baffin Bay in Greenland where we see a continued effort at integrating natives in the many economic opportunities of the globalizing Arctic, including commercial fisheries as well as mineral and petroleum extractions. While there is room for further progress in Russia and Fennoscandia to achieve parity in native participation with Arctic North America, the ideals of co-management and the realities of Arctic globalization converge there as well, with MNCs and the state increasingly cognizant of, and respectful toward, native participation in their ventures.

Indeed, such a collaborative model based on co-management of lands and resources and meaningful participation in economic activities (with equity ownership of native corporations) has now been emulated all across the Arctic, as is evident in many of the more recent land claim settlement areas, from Nunavut and Nunatsiavut to the east to the many Dene and Yukon First Nations settlement areas to the south. While there is much room for improvement farther east in the partitioned Sami homeland of Fennoscandia and the homelands of the many Indigenous peoples of the Russian Arctic, many of the collaborative structures pioneered by the Alaska land claim, refined by the Inuvialuit land claim, and augmented by the Nunavut land claim with the formation of the Nunavut territory, have now been emulated even farther east, as evident in the evolving consensus between Greenlanders and Denmark on expanding autonomous home rule in the direction of more independent self-rule. All this is rooted in the symbiotic relationships not only between neighboring natives (such as the Inuit and Dene) but between natives and settlers, whose interrelationships have evolved over time to become increasingly mutual, fostered by the reciprocal, transactional, and collaborative foundation of the colonial-era chartered companies like HBC and Royal Greenland Trading Department, and to a lesser degree the Russian-American Company. Their networks of trading posts dotted the Arctic landscape and seeded the emergence of today's governing structures and native corporations, embracing the same mutual commitment to collaboratively governing and co-managing the North that dominates the globalizing political economy of the Arctic today.

It is this foundation that sustains and ensures the endurance of Arctic exceptionalism, and positions co-management between tribe and state as a paradigm not only for domestic governance at the local and regional levels but for Arctic diplomacy (as seen at the Arctic Council), and over time, for a restoration of Indigenous sovereignty itself. With such long, deep, and lasting roots, one can safely predict that Arctic exceptionalism will endure future tests and challenges, as it did when competitive dynamics and tensions began to rise in 2014, culminating in 2022 after Russia invaded Ukraine, when universal circumpolar Arctic cooperation with Russia's full and active participation came crashing to a halt.

Notes

1. For a detailed discussion of the historical context for ANCSA and ANILCA, see Barry Scott Zellen, *Breaking the Ice: From Land Claims to Tribal Sovereignty in the Arctic* (Lanham, MD: Lexington Books, 2008).

2. For an excellent history of the Alaska Pipeline, see Mary Clay Berry, *The Alaska Pipeline: The Politics of Oil and Native Land Claims* (Bloomington: Indiana University Press, 1975).

3. Clifford John Groh II, "Oil, Money, Land and Power: Passage of the Alaska Native Claims Settlement Act of 1971" (honors thesis, Harvard University, 1976).

4. For a detailed history of Project Chariot, see Dan O'Neill, *The Firecracker Boys* (New York: St. Martin's Press, 1994). For a multimedia introduction to the project, see Jennifer Slingerland, "Project Chariot: A Brief Overview," August 4, 2017, www.youtube.com/watch?v=BvxB0JCXHLM.

5. See Berry, *Alaska Pipeline*; Groh, "Oil, Money, Land and Power."

6. See Stan Cohen, *The Great Alaska Pipeline* (Missoula: Pictorial Histories Publishing Co., 1988), and for a multimedia introduction, see The Milepost, "The Building of the Alaska Highway," September 25, 2013, www.youtube.com/watch?v=BkQKQN26JMI.

7. See Center for Environmental Management of Military Lands (CEMML), Colorado State University, *Annotated Bibliography: Distant Early Warning (DEW) System, Alaska* (United States Air Force, 611th Civil Engineer Squadron, Elmendorf Air Force Base, September 2008); for a multimedia introduction to the DEW Line,

see The History Guy, "The DEW Line and Forgotten History," www.youtube.com/watch?v=924n_cAx5zs.

8. British Movietone, "Nautilus Under the Pole," July 21, 2015, www.youtube.com/watch?v=0JZlWyS8Sm0.

9. For a detailed history of Project Chariot, see O'Neill, *Firecracker Boys*. A newer 2007 Basic Books edition of O'Neill's impressive account sports a new, and apt, subtitle: *H-Bombs, Inupiat Eskimos, and the Roots of the Environmental Movement*. For a multimedia introduction to the project, see Slingerland, "Project Chariot: A Brief Overview."

10. See CBC News, "Nunavut: New Territory Born at Midnight," *Focus North*, April 1, 1999.

11. Zellen, "Clan, the State, and War: Lessons from the Far North," 20–21.

12. For a colonial history of the Canadian side of the fur empires, see the following: Newman, *Company of Adventurers*; Newman, *Empire of the Bay*; and Coates, *Canada's Colonies*. For the Alaska/Russian-American side, see Bockstoce, *Furs and Frontiers in the Far North*; and Jones, *Century of Servitude*. For a multimedia introduction on Russian America and Russia's northward and eastward expansion in pursuit of furs and later sovereignty, see Eurasia, "Russia's American Colonies—Russia's Alaskan and Californian Colonies in America," June 9, 2019, www.youtube.com/watch?v=SJ4sc_QDlus; and All About Russia, "The Aleuts," June 2, 2020, www.youtube.com/watch?v=h5e0hHY0xsE. For more background on HBC, see Explore, "Hudson's Bay Company History," January 12, 2017, www.youtube.com/watch?v=DEZtO0xlBmk; Canadian Film Day, "Excerpts from The Romance of the Far Fur Country (1919)," https://canadianfilmday.ca/film/the-romance-of-the-far-fur-country; and HBC Heritage, "Romance of the Far Fur Country Trailer," www.hbcheritage.ca/classroom/videos/archival-footage/romance-of-the-far-fur-country.

13. For more details on this history, see Zellen, "Multinational Corporations in the Arctic," 157–173.

14. Zellen, *Breaking the Ice* and *On Thin Ice*. The historical and structural details of Inuit governance and land claims accords in the following pages present an overview of the case studies discussed in this work. Also see Barry S. Zellen, "The Arctic Land Claims Journey," *Tundra Telegraph*, April 6, 2010; and Richard Condon, "Canadian Inuit Land Claims and Economic Development," *Alaska Native News* 1, no. 10–12 37, and *Alaska Native News* 1, no. 12: 16–18, 40.

15. Hall, *The Fourth World*.

16. See the website of the Mackenzie Gas Project and the Aboriginal Pipeline Group at https://www.sciencebase.gov/catalog/item/5771b8b5e4b07657d1a6cf30.

17. I first synopsized the Arctic land claims journey in *Arctic Doom, Arctic Boom: The Geopolitics of Climate Change in the Arctic* (Westport, CT: ABC-Clio/Praeger Security and the Environment, 2009), boiling down the longer, more detailed history that I presented in *Breaking the Ice*, and I updated this condensed sweep of Arctic land claims history in "Stability and Security in a Post-Arctic World: Toward a Convergence of Indigenous, State, and Global Interests at the Top of the World," in *The Fast-Changing Arctic: Rethinking Arctic Security for a Warmer World*, edited by Barry Scott Zellen (Calgary: University of Calgary Press, 2014). Here, I strengthen my discussion of the 1975 James Bay and Northern Quebec Agreement (JBNQA) and 1978 Northeastern Quebec Agreement (NEQA), which hadn't featured as prominently in my past discussions of land claims history, and which deviates from my original description of a "west-to-east" arc of iterative and cumulative land claims history, challenging

the simplicity of my original conceptualization on the geographical flow of ideas across the North.

18. Tundra Times, *A Scrapbook History: Alaska Native Claims Settlement Act*, 1991; Zellen, *Breaking the Ice*.

19. See Thomas R. Berger, *Village Journey: The Report of the Alaska Native Review Commission* (New York: Hill and Wang, 1985).

20. Government of Canada, Crown-Indigenous Relations and Northern Affairs Canada, "The James Bay and Northern Quebec Agreement and the Northeastern Quebec Agreement—Annual Reports 2008–2009/2009–2010," www.rcaanc-cirnac.gc.ca/eng/1407867973532/1542984538197.

21. Ibid.

22. "JBNQA and NEQA," Hunting, Fishing, and Trapping Coordinating Committee website, http://cccpp-hftcc.com/jbnqa-and-neqa.

23. John Ciaccia, special representative of Premier Robert Bourassa, "Philosophy of the Agreement," James Bay and Northern Quebec Agreement (JBNQA), signed November 11, 1975, 2.

24. Ibid.

25. Ibid., 3.

26. Ibid., 5.

27. Ibid.

28. Ibid.

29. Ibid., 6.

30. Ibid., 7.

31. Worl, "The 1991 Time Bomb Defused," 1.

32. Lazarus and West, "Alaska Native Claims Settlement Act," 132–165.

33. Edwards and Natarajan, "ANCSA and ANILCA," 69–97.

34. Ibid., 93.

35. ANCSA: Caught in the Act, "Program 4: Plain and Simple," www.youtube.com/watch?v=-rseAvUu2-o.

36. Len Sillanpaa, *Alaska Native Claims Settlement Act: A "Living Settlement?"* (Ottawa: Circumpolar Affairs Division, Circumpolar and Scientific Affairs Directorate, December 1988).

37. Paul Ongtooguk, "ANCSA at 40: Where Are We and Where Are We Going?," *Anchorage Daily News*, March 16, 2012.

38. Ibid.

39. For a presentation on the corporate culture that emerged inside the Inuvialuit Settlement Region after the IFA was enacted, see the following Inuvialuit Development Corporation video: www.youtube.com/watch?v=sB2CLAKabmk. This flashy, corporate promotional video perfectly captures the enduring corporatization of the land claims model. It pivots seamlessly from a celebration of traditional culture to an embrace of modernization, globalization, and innovation.

40. For a detailed discussion of the historical context for the IFA, see Zellen, *Breaking the Ice*. For a multimedia introduction to the IFA, see Marie Wilson, "The Inuvialuit Land Claim," CBC *Focus North*, www.youtube.com/watch?v=IjGZDOoJOBw.

41. For a historical discussion of the ILUOP, see Freeman, "Looking Back—and Looking Ahead—35 Years After the Inuit Land Use and Occupancy Project," 20–31.

42. Worl, "The 1991 Time Bomb Defused," 1.

43. See CBC News, "Inuit Vote for New Territory," *CBC National*, 1992.

44. Worl, "The 1991 Time Bomb Defused," 1.

45. Uncivilized Films, *The Duck In*, 1961, www.isuma.tv/uncivilized-films/the-duck-in.

46. Berger, *Village Journey*.

47. Freeman, "Looking Back."

48 See McGrath, *Long Exile*, for more details of this controversial chapter in Canadian Arctic history.

49. Scott Brachmayer, dir., *Kajutaijuq: The Spirit That Comes*, 2014, https://vimeo.com/103292836.

50. Martin Fletcher, "Greenland Referendum Offers Break from Links to Denmark," *Irish Independent*, November 26, 2008.

51. Freeman, "Persistence and Change," 257–266.

52. Berger, *Village Journey*; Edgar Blatchford, "Alaska Native Claims Settlement Act and the Unresolved Issues of Profit Sharing, Corporate Democracy, and the New Generation of Alaska Natives" (PhD diss., University of Alaska, Fairbanks, 2013); Edgar Blatchford, "The Unintended Consequence of the U.S. Congress and the Alaska Native Claims Settlement Act: The Demise of Corporate Democracy, and the Threat of Native Ownership of the Land Base" (master's thesis, Dartmouth College, 2009).

53. Wonders, "Dene/Inuit Interface," 245–260.

4

Inuit Sovereignty in a Westphalian World

Successful collaboration between the Indigenous peoples and the sovereign states of Arctic North America has helped to stabilize the Arctic region, fostering meaningful Indigenous participation in Arctic governance through the introduction of new institutions of self-governance at the municipal, tribal, and territorial levels, and successful diplomatic collaborations at the international level through the Arctic Council and other international fora. Undergirding each of these pillars of Indigenous participation in Arctic governance is a mutuality of commitment to the principle of collaborative governance that has united Indigenous peoples and the state across Arctic North America through their shared experience of post–land claims settlement co-management of lands and resources, from the Bering Strait to Baffin Bay. Indeed, co-management has become so widely and reciprocally embraced by tribal peoples and states alike that it now presents us with a viable and scalable foundation for bridging the Indigenous, transnational world with the Westphalian world of states and statecraft across the entirety of the circumpolar Arctic. The bonds forged between the two through their joint governing experience at the domestic level inform and inspire the relationship between the eight founding member states of the Arctic Council and the six Permanent Participant organizations representing the Indigenous peoples of the Arctic. This became particularly evident with the October 2022 update of the US National Strategy for the Arctic Region, discussed in further detail in Chapter 7, which formally acknowledged co-management as a cornerstone to a secure Arctic and a key foundation of US Arctic security, the first time co-management has been so explicitly recognized in a US Arctic strategy document (though earlier US Arctic strategy documents recognized the importance of tribal inclusion, engagement, and consultation, which are essential components of the co-management process).

The stability, and over time increasingly reciprocal institutional and intergovernmental relationships between sovereign states and Indigenous stakeholders at the domestic level, has contributed increasingly to a shared spirit of international collaboration often referred to as Arctic exceptionalism, embedding it in multiple levels of governance, from the local and tribal to the regional and territorial all the way to the national and international levels. These institutions are increasingly reflected in a growing and important literature on the influence of nonstate actors, primarily Indigenous organizations and self-governing institutions, but other actors are included such as nongovernmental organizations (NGOs) and subnational governments, illuminating the underlying complexity of the Arctic domestic and international political system, and informing comparative studies of Indigenous-state relations across the region.[1] There is also a subset of the subnational literature that looks specifically at institutions and mechanisms of co-management, which brings together the competing stakeholders and their divisions along the cooperation/conflict, homeland/frontier, and sovereignty/security continua that traditionally divide the field of Arctic international relations.[2] There has also emerged in recent years a substantial and growing body of work on the Arctic Council, which provides much detailed insight and understanding of the cooperative mechanisms and outcomes of Arctic diplomacy, that is largely separate from the literature on co-management, the latter of which is traditionally viewed through the lens of domestic governance and the historic reconciliation between natives and states in the Arctic.[3]

In this section I will weave together ideas drawn from these two distinct traditions in the literature, one conceptualizing Indigenous Arctic governance as primarily a domestic process, and the other that conceptualizes international Arctic cooperation and Indigenous-state relations as an international and diplomatic process, in an effort at uniting them through their mutual recognition of Indigenous Peoples' Organizations (IPOs) as key actors that have strengthened the constitutional bonds linking local, regional, and national governments while at the same time strengthening the diplomatic bonds between Arctic states through their transnational dynamics. This dual strengthening of the collaborative underpinnings of both domestic and international orders explains the recurrence and endurance of Arctic exceptionalism as a defining attribute with deep historic roots that predate the formation of the Arctic Council in 1996. This is because Arctic Indigenous peoples, as cross-border and transnational communities, exist simultaneously in both domestic and international political contexts—experiencing multiple levels of analysis at once—and what may look to be a structural ambiguity is actually a powerful duality that harmonizes insights from and experiences at the domestic and international levels.[4]

My ongoing research in Arctic international relations theory has sought to contextualize and integrate this entire range of analytical levels, from the local, Indigenous, and regional to the national and global. But rather than surrender to the temptation of complexity theory, which accepts but does not resolve many of these important ambiguities inherent in the triangular relationship between domestic, international, and transnational entities, I look at commonalities in their interconnections and interactions to determine how domestic structures, when examined granularly down to the village level, assert (in their aggregate) an ordering tendency on international relations in the Arctic. Complexity thus does not yield to chaos, but instead unearths a deeper understanding of the underlying Arctic order—its history, its institutional development, and its many debates, fault lines, and axes of conflict—revealing not just the flow of ideas across the Arctic in both temporal and spatial dimensions,[5] but salient and influential ordering patterns that influence higher levels of governance, starting with the local and tribal, on up to the regional, territorial, state, and international levels.

The normative condition of Arctic exceptionalism is thus not just wishful thinking by one school of thought on one side of an idealist/realist or cooperation/conflict continuum, but a pattern of recurrent behavior that is rooted, structurally, in the progressive reforms of native empowerment that start at the village level and that are amplified over time as those reforms increasingly reshape policies and strategies at higher levels. So the resolution of Arctic land claims, which started in Alaska in 1971 and largely percolated eastward along the Arctic coast (with the one exception being northern Quebec, which jumped the queue under the immense pressure of hydroelectric development during the early to mid-1970s), in turn inspired greater representation of native interests in territorial and state governance while at the same time aligning with transnational paradiplomatic efforts on the global level. These in turn found partners in international diplomacy at the highest levels of state, which in turn created a culture of cooperation and collaboration in Arctic international relations that survived numerous international disputes that in other regions resulted in a preponderance of conflict.

Arctic exceptionalism has come under relatively new threats from a recent renewal of great power competition in the Arctic. Updated and revised strategic policies by numerous states with Arctic interests or aspirations assert a more nationalist perspective in contrast to prior decades of mutual commitment to Arctic collaboration. As Arctic competition between states rises, I believe the multitude of co-management systems and the multilevel, intergovernmental, and interorganizational relationships they have nurtured across the region that began as mechanisms for joint resource management at the local and regional levels between Indigenous peoples and the state can help to neutralize the new threats to Arctic exceptionalism by injecting into international political discourses a steady collaborative counternarrative largely insulated from new interstate pressures. This is because localized co-management has itself evolved into both a mechanism of, and powerful paradigm for, trans-Arctic diplomacy that fosters greater domestic unity not only between tribe and state but between states as well, catapulting mechanisms designed for domestic resource management to the international stage, as illustrated in the research on Inuit transnationalism by Jessica Shadian, among others.[6]

Amid this evolution from the domestic to the international stage, Inuit leaders brought to global politics their insights and experiences from co-managing their homelands with their respective sovereigns. One might argue that the structure established by the formation of the Arctic Council in 1996 borrows from the domestic co-management experience the unique and successful roles and responsibilities of the Permanent Participants of the council, which are the regional Indigenous organizations newly empowered at the domestic level in Arctic North America. Like their domestic co-management counterparts, they lack decisionmaking power but gain access, influence, and, importantly, a prominent seat at the table with their sovereign counterparts (the eight full members of the council representing its eight founding sovereign states), a compromise that both satisfies the sovereign members of the council and meets the needs and aspirations of the Permanent Participants despite the asymmetry in formal power. This asymmetry in power and formal influence has been acceptable to Arctic IPOs in large measure because it has worked so well at the domestic level through decades of co-management implementation and allowed for significant informal influence, moral suasion, and effectiveness.

Domestic Foundations of Arctic Exceptionalism

Largely as a result of Indigenous political foresight and perseverance—from the municipal level all the way up to the international level through diplomatic bodies such as the United Nations and more regionally focused ones like the Arctic Council—a solid foundation for Arctic exceptionalism has been established, buffering the region from contending influences such as renewed nationalism and state rivalry, preserving regional order even as states face off against one another in an increasingly contested Arctic.

It was the historic 1971 passage of the Alaska Native Claims Settlement Act (ANCSA)—the pioneering land treaty transferring 44 million acres of land title and nearly $1 billion in compensation to Alaska Indigenous peoples—that established the foundation stone for the emergent system of co-management and collaboration across Arctic North America. The model was embraced as Inuit land claims built upon ANCSA's foundation in the Canadian Arctic, strengthening it by enhancing the breadth and diversity of collaborative structures while retaining the system of Indigenous village and regional corporations that defined ANCSA, with Inuit land corporations gaining title to nearly one-tenth of their traditional land base much the way the regional corporations did in Alaska, and with new co-management structures increasingly enabling Indigenous participation in resource management and in environmental screening and review processes.

What started with ANCSA continued to evolve and transform over the next half century, both in Alaska and across the international boundary in the Canadian Arctic, a process watched closely by Arctic Indigenous peoples in other regions as well as Indigenous peoples from elsewhere in the world. This dynamic, living, and evolutionary process is illustrated well by Alaska's many post-ANCSA amendments designed to correct perceived flaws in the original land claims structure (starting in 1976, with more to come in 1980, 1987, 1995, and as recently as 2017) and by the subsistence provisions included in the 1980 Alaska National Interest Lands Conservation Act (ANILCA), designed in part to redress ANCSA's omission of subsistence protection. In addition, the 1993 federal recognition of Alaska's Indigenous communities as tribes was the culmination of the 1980s' tribal sovereignty movement that spread across village Alaska in reaction to ANCSA's corporate focus; a piecemeal, multilevel effort to align federal policy and tribal aspirations, amid occasional resistance by the state of Alaska, has played out in the years since.

From Co-management to Collaborative Arctic Sovereignty: Transnationalism as a Bridge

Despite these bumps along the road to implementation, co-management offers no less important a path toward synthesis of these competing forces, with regional, largely internal "regimes" balancing Indigenous and non-Indigenous interests. States continue to predominate in terms of material power, but Indigenous peoples command demographic predominance in remote regions, becoming essential stakeholders in Arctic sovereignty. It is this fundamental demographic truth, and the minimal settler populations in the Canadian Arctic, that has ensured the progress continued and that Inuit empowerment was strengthened, even as Ottawa began to display a more conflicted view through its foot-dragging on implementation and occasional assertions of a desire to streamline the highly localized patchwork of co-management regimes into something more industry (and federal government) friendly. Recent history presents numerous case studies that illustrate the increasing success of co-management as a paradigm for Arctic governance and, perhaps paradoxically, increasing conflicts between the very unequal parties that co-management has structurally united.

While co-management has experienced its share of challenges, with federal foot-dragging on making appointments to co-management boards and funding programs to which Ottawa committed during land claims negotiations, and resistance to decisions driven by traditional over Western knowledge systems or cultural over economic/commercial values, these challenges and associated tensions are, in their own way, proof of co-management's success—establishing a hybrid governing structure to mediate the many differences in values, aspirations, and policies between tribes and states. Co-management structures, in addressing and attempting to find synthesis in the state-tribe dialectic, reflect a transformation in the concept of Arctic sovereignty as states increasingly root their sovereignty in the consent of Indigenous peoples, and as much as possible, align their interests with them (not unlike how HBC governed earlier in history, with its overlapping sovereign layers, devolving much autonomy to its Indigenous partners and proxies while maintaining an iron fist over its monopoly trading rights). Modern states have been no different, with a steadfast desire to maintain their monopoly control over foreign and defense policy, and their economic supremacy, even while showing accommodation on other issues.

It was the co-management table that provided the structural model for the Arctic Council, where an observer state like China sits in the back of the room while the six Indigenous Permanent Participant groups sit at the main table next to the eight member states; though decisions reside with the sovereigns, the informal influence of the Indigenous groups is near parity with the member states, and with close connections to their state counterparts, enduring policy alignments between the Indigenous groups and their host states augment the voice of the Permanent Participants even further. But at the working group level, funding disparities limit areas of Indigenous engagement, providing more space for observer states and NGOs to influence Arctic policy through their technical expertise. On the whole, however, at the co-management table, Arctic Indigenous peoples and states not only share the stage but increasingly unite their voice. Thierry Rodon and Aude Therrien describe the land claims process and the consequent institutional development of co-management boards for jointly governing Indigenous homelands as a complex governance arrangement necessitated by the demands for multilevel governance in the Arctic,[7] and despite complaints from numerous political and business leaders on the resulting "Balkanization" of decisionmaking in the post–land claims Arctic and ongoing efforts to streamline this process through regional integration, Indigenous leaders have to a large measure defended their gains and protected existing co-management structures, with their intensely localized geography, particularly in the ethnically diverse Northwest Territories and Yukon. In Nunavut, which is ethnically unified by its overwhelming Inuit majority population, concern is less with a Balkanized decisionmaking landscape of the sort that has emerged in the more western territories of the Canadian North and in Alaska, and more with the government of Canada's lackluster commitment to land claims implementation, and what Inuit perceived to be a half-hearted commitment to co-management in Ottawa. As a result, Arctic North America's lands, waters, and resources remain governed by a continent-wide patchwork of co-management boards that is, by its very nature, institutionally complex.[8]

This complexity is highlighted in Mark Nuttall's insightful research on the co-management of Arctic oil and gas resources in Alaska and the

Western Arctic,[9] and Claudia Notzke's pioneering research on resource co-management across Canada,[10] which chronicles both the proliferation and increasing saliency of co-management as a governing principle for natural resource management, even in regions without settled land claims. Indeed, so well-suited is co-management to resource-rich frontier regions, it has been decoupled from the land claims process, finding its way into resource-rich parts of Canada's settled southern provinces where competing claims to the land have precluded successful negotiation of land claims, becoming a salient microlevel of analysis independent of land claims settlement. Derek Armitage et al.'s 2011 study of co-management in the Canadian Arctic demonstrates how co-management provides a mechanism for both Indigenous peoples and governments to learn how to be adaptive amid "uncertainty and environmental change," and thus to benefit from social learning for developing adaptive capacity to resolve complex challenges collaboratively, resulting in new levels of knowledge coproduction that successfully synthesize Indigenous and Western approaches—introducing a "collaborative process of bringing together a plurality of knowledge sources" that both reconciles asymmetries in power and overcomes the many complexities of multilevel governance.[11]

The rich literature on co-management, particularly as it has evolved in Arctic Canada from resource management to governing paradigm (and, in my view, the foundation stone of Arctic exceptionalism), provides us with additional theorizing as well as a wide selection of co-management case studies on various issues across the Arctic (and beyond), helping to fill in some of the details on how co-management has evolved over time and why. Armitage suggests the "why" has to do with the integrated knowledge coproduction that has resulted increasingly from co-management, fostering a synthesis of traditional knowledge and Western science. In the years since Armitage et al. published their article, this synthesis has grown even stronger, as echoed by participants from both the native community and the science community at Arctic Science Summit Week conferences. Nowadays, the native side is no longer represented by just the traditional knowledge side of the dialectic but also by the Western science side, with a growing cohort of PhD-wielding native scientists, researchers, and project directors. With the native community now straddling both sides of the knowledge coproduction process, synthesis is more readily achievable and reduces the original clash between two distinct and at times competing pillars of knowledge. This mirrors the findings in Mark Nuttall's research on the two Mackenzie Valley oil and gas pipeline inquiries, the first in the mid-1970s and the next in the 2000s, where the tribe-state dialectic grew more complex post–land claims settlement as native equity participation was introduced to the industry side, which had hitherto represented nonnative economic interests opposed to traditionalist native cultural interests.[12]

Armitage et al. illustrate the dialectical dynamic of co-management's knowledge coproduction with case studies drawn from three northern land claims—Inuvialuit (beluga whale management), Gwich'in (Dolly Varden char management), and Nunavut (narwhal management), showing how co-management systems create opportunities and incentives for collaboration, creating conditions for social learning to help mediate a changing environment, boosting the adaptive capacity of both sides, and fostering the collaborative resolution of complex challenges.[13] Armitage et al. describe the presence of double- and triple-loop learning cycles, the emergence of shared understanding and "sense-making" processes, and the emergence of new networks better optimized for adaptation to change. Making it all work, they argue, is the growing capacity of both sides to participate in adaptive governance and social learning, introducing an ability to learn how to learn "through uncertainty and environmental change—or learning to be adaptive."

Co-management institutions, over time, become more flexible and collaborative as trust barriers from the past—when native and government positions were more adversarial and Western science more often displaced traditional knowledge, often leading to decisions that harmed native interests—begin to fade, allowing co-management boards to become "bridging organizations" linking separate levels of analysis much the way regimes do in world politics. Several interesting case studies of co-management are presented, involving the Fisheries Joint Management Committee established by the Inuvialuit land claim, the Nunavut Wildlife Management Board, and the Gwich'in Renewable Resource Board and their respective experiences at coproduction of knowledge, including knowledge gathering, sharing, integration, interpretation, and application. Co-management, the authors argue, provides a mechanism for learning and adaptation, with traditional knowledge increasingly integrated into the understanding achieved, enabling both multilevel and adaptive governance that has proven itself to be well suited to the challenges of the Arctic environment. At the same time, the authors recognize that co-management is not a panacea, and caution that it remains to be seen if co-management can adapt to climate change quickly enough, particularly in light of the sluggish pace of co-management processes, the Mackenzie Gas

Project being a cautionary tale worthy of much consideration. On the other hand, time does strengthen the process and move the participants closer to synthesis, since more co-management experience tends to yield increasing inclusivity and more meaningful consultation, which in turn fosters increased knowledge coproduction over time. Co-management establishes a mechanism for learning about how to co-manage, transforming governance into a hybrid process that transcends separate levels of analysis.

Claudia Notzke presents an equally compelling analysis of the co-management model, arguing that it has catalyzed a "redefinition" of the relationship between natives and nonnatives in Canada (and thus between tribal peoples and the state) and a "restructuring" of power and responsibility in the management of natural resources, and that it has become so successful that it has been embraced far beyond the land claims settlement areas of the Far North and applied to issues well beyond wildlife management. Like Armitage et al., Notzke sees the fundamental strength of co-management as its "recognition of the strength and potential contributions of two systems of knowledge," one Western and one Indigenous. She distills the co-management process into seven distinct levels that are akin to "rungs of a ladder": information, consultation, cooperation, communication, advisory committees, management boards, and community control and partnership.[14] As with Armitage et al., she introduces several insightful case studies. These are drawn from the Inuvialuit experience co-managing grizzly bears, polar bears, beluga whales, and oil exploration; the Nunavut experience co-managing beluga whales and polar bears; cross-border co-management of barren-ground caribou along the provincial/territorial frontier in Canada's subarctic; moose co-management in Manitoba, where, in the absence of a settled land claim, co-management as a paradigm of joint governance has emerged prior to and separate from the land claims process; and national park co-management, which has become a feature of all the Canadian Arctic's land claims. Establishing a national park, even one so remote few actually ever visit, helps to assert formal land use over a vast expanse of territory without any real requirement for effective or meaningful occupancy, providing states with a very efficient and inexpensive way to assert sovereignty with minimal investment or effort while at the same time preserving natural ecosystems and wildlife populations, making tribal peoples equally enthusiastic once trust barriers are lowered and park officials demonstrate support for native subsistence, as evident in Canada's northern national parks.

Notzke identifies "strategic" co-management as a new governing paradigm, which she describes as an emergent constitutional right that is evolving into hybrid governance defined by adaptive management, joint stewardship, and cooperative management. All these are detachable from the land claims process itself, offering a new model for joint native/nonnative resource management in places where concluding a land claims treaty is particularly difficult if not impossible, due to large settler populations such as those seen in the northern and western provinces of Canada where "regional environmental management councils" for cooperative environmental or industrial management can be formed in the absence of settled land claims treaties, spawning the birth of what we might think of as a brand new level of analysis! This new level of analysis bridging the native and nonnative worlds has much in common with international regimes, albeit in a domestic context where Indigenous populations and settler populations would otherwise face numerous land use conflicts that inhibit the settlement of comprehensive land claims, offering hope for hybrid state-tribe governing institutions even in the absence of the demographic predominance enjoyed by natives farther north in the Arctic, and perhaps offering us a model for hybrid governance of tribal regions all around the world.

Co-management may thus be understood as something much more interesting theoretically, and perhaps more important to world politics, than the wildlife or environmental governance mechanisms co-management started out as. It challenges our understanding of sovereignty and transforms it from the Westphalian model of a unitary state to a more nuanced and collaborative hybrid state that incorporates not only all the levels of governance we've come to expect from the municipality on up to the federal government but also emergent systems of Indigenous governance, whether self-governing institutions that mirror traditional levels of governance for Indigenous regions or joint governing structures that juggle tribal and state interests as co-management structures do. Looking at the evolving role of co-management in governing Arctic North America, and the persistence of co-management as a paradigm embraced by both tribe and state interests, we see an alternative and inherently more collaborative model of sovereignty itself, with Indigenous as well as state components.

Just as co-management strives to institutionalize the dynamic balancing of competing traditional and modern values, and of tribal and state powers, it can also help to establish a foundation for the diplomatic balancing of Arctic interests and non-Arctic interests alike. Indeed, it can serve not only as a metaphor but as an institutional foundation to manage the emergent state competition gathering steam in the Arctic. Armitage, Berkes, and Doubleday's 2007 volume, *Adaptive Co-Management: Collaboration, Learning, and Multi-Level Governance*,[15] illustrates the globalization and universalization of co-management far beyond the Arctic region with case studies that include both northern examples of co-management (from Alaska, Yukon, Northwest Territories, and Nunavut) as might be expected, as well as case studies of co-management in places as diverse as Vancouver Island, Southern Ontario, and Atlantic Canada in Canada's southern provinces in addition to Wisconsin in the United States; Belize, Barbados, and Grenada in the Central America/Caribbean region; and Sweden in Europe. Indeed, co-management has truly gone global. Former Yukon premier Tony Penikett has observed that Indigenous treaties in Canada are themselves "inherently international because natives are transnational," even if governments in North America still view Indigenous peoples as "domesticated" components of their constitutional polities. To Penikett, land claims treaties and the many co-management structures they yielded are "remarkable nation-building achievements," helping to offset earlier colonial-era treaties and the "historic failures of colonizing states to respect their treaty promises."[16]

Looking to the future, as the polar thaw drives continuing Arctic globalization, and non-Arctic states show increasing interest in Arctic lands, waters, and resources, the strategic center of mass for engaging in Arctic enterprises of all sorts will be the myriad co-management structures that define the post–land claims political topography of Arctic North America. Indeed, the Arctic Council emerged from the many structures of multilevel and multilateral co-management developed within the Arctic at the domestic level—between tribes, local governments, state and territorial governments, national governments, economic entities, and environmental actors, both within and across borders.

The Arctic Council, to be discussed further below, with its unique and evolving structure bringing Indigenous Permanent Participant groups and the founding member states together, joined by observers (both state and nonstate that increase in number and diversity over time), is well designed for the inclusion of new voices, particularly competing voices. After all, it was the moderation of competing voices between tribal and settler interests that, through the land claims process, transformed Arctic governance into a meaningfully collaborative process, finding consensus despite the many asymmetries of the actors involved. The Arctic Council is thus firmly rooted in collaboration, bringing co-management to the global stage and elevating its Indigenous actors to a near sovereign level, active diplomatically at the international and transnational levels while essential to Arctic governance from the local to the national levels. This positioned the Inuit to reclaim sovereignty in many ways and at many levels, transforming sovereignty itself from unitary state sovereignty in the Westphalian tradition to what Jessica Shadian has described as post-Westphalian sovereignty. This transformative journey suggests that Arctic exceptionalism is as much about the willingness of Arctic states to encourage and support Indigenous aspirations for sovereignty as it is about cooperative relations between the Arctic states themselves.

While inherently multilateral and collaborative, conceptualizations of Inuit sovereignty take many forms: a transnational form as the Inuit Circumpolar Conference (ICC) has embraced, which Jessica Shadian has thoughtfully examined;[17] a more traditional Westphalian form that aligns, as Hannes Gerhardt argues,[18] with the movement for Greenlandic independence; or a hybrid form of sovereign duality and mutual interdependence, with Inuit consent essential to Canadian Arctic sovereignty and a sine qua non for Ottawa's successful assertion, without external challenge, of sovereignty over its largely unsettled north, as Terry Fenge has elegantly argued.[19] Njord Wegge, in his panoramic tour of international relations (IR) theory as applied to the Arctic, illustrates that sovereignty in the Arctic is a dynamic, complex, shape-shifting phenomenon with diverse and distinctive levels of analysis, including both transnational and subnational polities distinct from the traditional Westphalian state.[20] Shadian further clarifies how the Inuit have pioneered a new, "post-Westphalian" form of sovereignty that is sui generis and provides a new, powerful vision for a world governed collaboratively and multilaterally based on the Inuit model for the Arctic. Shadian's doctoral thesis examined the ICC as an example of a transnational polity that effectively utilized diplomacy to assert its values and interests, and that adopted this post-Westphalian model of sovereignty freed of the confines and limitations of the nation-state, allowing Inuit to transcend national borders while maintaining their unity as part of a distinct political entity or polity. Shadian uses the term *polity* to capture the ambiguity of the ICC's dual trans-state/substate status, and while the organization lacks formal sovereignty, it offsets this with a shared supranational identity.

In contrast to many other theorists who view the ICC, and later the Arctic Council, as examples of regime theory in action, Shadian sees in the ICC not just as a stepping stone to the Arctic Council but also as a deeper example of an alternative sovereign model, one well suited for peoples whose subdivided homelands leave them at once substate and trans-state. Instead of viewing this transnationality as a hindrance, Shadian views it as an opportunity to transcend the nation-state as a model for sovereign expression. She traces the history of the ICC back to the instrumental and foundational role played by Inupiat leader Eben Hopson, who among his many positions of leadership served as executive director of the Arctic Slope Native Association, and who played an outsized role in much recent Inupiat history, including the founding of the ICC in 1976.[21] Hopson brought to his efforts a practicality that balanced tradition and modernity, protecting cultural values while at the same time leveraging commercial opportunities. Hopson's vision for "Eskimo sovereignty within the democratic traditions of our national government"[22] was mirrored by Inuit in other jurisdictions, especially Canada. The ICC never sought formal Inuit independence, nor was it tempted by secession from the state, and this enabled Inuit to remain partners in confederation with their national governments while operating as a transnational polity on the world stage—a collaborative dynamic that set the stage twenty years later for the formation of the Arctic Council in 1996. In the years between the ICC's founding in 1976 and the Arctic Council's formation, the ICC's "amassed legitimacy built up over time," providing Inuit with the "authority for determining the shape and direction in which Arctic development is defined and proceeds," and the ICC thus reflected a "culmination of a more complex and multidimensional narrative of Arctic international relations."[23]

Rodon also examines the evolution of Inuit diplomacy since the 1970s, and like Shadian salutes Hopson's "visionary" leadership, and also his pragmatism. Under his guidance, "Inuit were able to move to the international scene, first by seeking the recognition as an NGO by the United Nations and later by lobbying very effectively for the construction of an Arctic space through the creation of the Arctic Council," part of the ongoing transnational effort by the ICC at "reframing spaces and redefining narratives," evolving into a "clear vehicle toward fostering Inuit autonomy," and later redefining sovereignty in the Arctic, as illustrated by the 2009 Circumpolar Inuit Declaration on Sovereignty in the Arctic.[24] The tug-of-war between regional and transnational forces is evident not only in the above-noted tensions but also in the very model of sovereignty reflected by transnational bodies like the ICC, whether post-Westphalian as Shadian describes, or more traditionally Westphalian, of the sort that Hannes Gerhardt describes as emerging in Greenland as its autonomy process paves the way forward toward eventual independence.[25]

Instead of focusing on the transnational nature of the ICC as an Inuit polity asserting post-Westphalian sovereignty, longtime Inuit advisor and prolific scholar Terry Fenge describes a collaborative sovereignty linking Inuit consent to the Canadian state's ability to assert sovereignty over its vast Arctic.[26] Fenge spent much of his career as a policy advisor to the Inuit leadership and helped to build and sustain the alignment of interests between the Inuit and Ottawa that has enabled Ottawa to successfully assert sovereignty over its largely unsettled and underdeveloped Arctic territories through the consent, support, and partnership of the Inuit. Their "shared space" is not just the institutions of co-management but the physical territory of the Canadian Arctic where two claims to the land are mutually strengthened by the reconciliation and reciprocal recognition achieved through the land claims process, and where joint governance is made possible despite the fundamental asymmetry of the governing partners in population and material power.[27] Because Inuit maintain local and regional demographic predominance in Nunavut, they serve as effective proxies for Canadian sovereignty assertion where Ottawa itself lacks a robust industrial, infrastructural, or settler presence. During the colonial era, states often gained title to Indigenous lands through treaties of cession and surrender, whether negotiated or after military victories, prior to formal settlement (though in reality, settlers often were already there and treaties ratified the new demographic realities); in the age of Arctic land claims, states grant title to Indigenous lands (and compensation for lands ceded to the state) without a demographic influx—winning instead Indigenous recognition of the state's sovereign claims, with the Indigenous peoples serving as the representatives of the sovereign. It is this dynamic that has enabled land claims treaties to empower, rather than disempower, Indigenous beneficiaries.

Inuit Diplomacy: Indigenous Values in International Relations

Next we will look beyond land claims and co-management as domestic processes and phenomena and view them as part of the dialectical clash of sovereign ambitions by tribal peoples with the states that expanded into tribal homelands as the Westphalian world merged with the Indigenous Arctic. Diplomacy has traditionally been a function of sovereign statehood, though public diplomacy and interest group participation through NGOs and regimes in world politics have provided an outlet for formal diplomatic interactions between state, substate, and nonstate entities of the sort we have seen since 1996 in the Arctic Council and that date back at least to the 1970s with the formation of the Inuit Circumpolar Conference (later renamed Inuit Circumpolar Council). Inuit diplomacy has been an enduring and effective endeavor since then, reflecting both the transnational nature of the Inuit as a people whose homeland has been subdivided by four sovereign states—not entirely different from the Indigenous Dayak peoples on Borneo, whose homelands were partitioned by colonial states and remain divided today,[28] or even the Kurds, and other stateless peoples, with homelands extending across sovereign borders due to various whims and fortunes of history—and the enduring aspiration of the Inuit to reassert control over their homelands and shape regional and international politics to align with their principles and values, much the way any state does.

Some literature looks at Inuit sovereignty in its different forms—whether transnational (as the ICC has embraced), Westphalian (which better aligns with the movement for Greenlandic independence), or the mutuality of "sovereign duality" (a phrase I coined about a decade ago in my first book, *Breaking the Ice: From Land Claims to Tribal Sovereignty in the Arctic*) reflected in the way Inuit consent has enabled Ottawa to assert its undisputed sovereignty over its largely unsettled north by proxy. There's also a theoretical literature on Inuit sovereignty, notably Jessica Shadian's post-Westphalian sovereignty that aligns well with my conception of Inuit sovereign duality.

In her 2006 article, "Remaking Arctic Governance: Construction of an Arctic Inuit Polity,"[29] Shadian describes the history of the ICC, and the instrumental role played by Inupiat leader Eben Hopson. He brought to his efforts a practicality that balanced tradition and modernity, protecting cultural values while at the same time leveraging commercial opportunities. Shadian recounts the history of Arctic land claims, and the catalyzing effects of externally driven resource development in the Arctic by multinational corporations, which led Hopson to aspire not only to regulate such activities but to participate in them as a principal stakeholder, and to bring Inuit voices and values to the world stage to counter the voices of the predominant actors in world politics. Thierry Rodon, whose work on co-management is discussed above, examines the evolution of Inuit diplomacy in his 2014 article "Inuit Diplomacy: Reframing the Arctic Space and Narrative."[30] He notes Hopson "used money from oil development to assemble Inuit living throughout the circumpolar region" in 1977, the first meeting of what would be incorporated formally as the ICC in 1980.

The Inuit thus became, through the ICC, an international actor with its own approach to diplomacy, while at the same time the Inupiat under Hopson's leadership asserted loyalty to the US constitutional system—recognizing that national governments could be allies and supporters of the Inuit if the Inuit did not challenge their sovereignty. But after they were excluded from the inaugural Arctic Ocean Conference, held in Ilulissat, Greenland, in May 2008, they began to reconsider the viability of this approach, and in response issued their first formal declaration on Arctic sovereignty. (The ICC was not the only important Arctic stakeholder to be so excluded from the Ilulissat summit; also excluded were the Arctic states Iceland, Sweden, and Finland—all founding members of the Arctic Council, and all troubled by this new division of the Arctic into Arctic rim states, on the one hand, and all others [states and non-state], on the other.) In response to being snubbed, the ICC's 2009 Circumpolar Inuit Declaration on Sovereignty in the Arctic asserted in section 4.2,

> The conduct of international relations in the Arctic and the resolution of international disputes in the Arctic are not the sole preserve of the Arctic states or other states; they are also within the purview of the Arctic's indigenous peoples. The development of international institutions in the Arctic, such as multilevel governance systems and indigenous peoples' organizations, must transcend Arctic states' agendas on sovereignty and sovereign rights and the traditional monopoly claimed by states in the area of foreign affairs.

Note that multilevel governance systems are mentioned here, reinforcing my belief in their importance to the stability of a collaborative international Arctic order. As Rodon describes, this is "clearly a challenge to the classical views on sovereignty," a challenge that was initially welcomed by Secretary of State Hillary Clinton in 2008 (as noted below), who publicly supported the Inuit aspiration for greater inclusion. Rodon describes how Inuit "have actively reframed the Inuit and the Arctic spaces through the concept of Inuit Nunangat, a space based on a culture and a way of life," and have "also actively participated in the construction of the Arctic region, first around environmental issues, but now also around the Arctic people and Inuit rights," as now "institutionalized through the Arctic Council." Inuit have also "actively redefined the narrative on climate change from an environmental issue to one that includes Inuit rights and shared sovereignty between states and the Inuit," while at the same time waging an ongoing struggle against "the global animal rights movement that has threatened their way of life."

Gary Wilson also considers the evolution of Inuit diplomacy in his 2007 article "Inuit Diplomacy in the Circumpolar North,"[31] and in particular the role of the ICC in "representing and defending Inuit interests in international fora" as part of a "multifaceted" effort drawing upon the "considerable strengths" of individual champions of Inuit rights. These include Greenland's Aqqaluk Lynge, Canada's Mary Simon and Sheila Watt-Cloutier, and, I would add, Alaska's own Dalee Sambo Dorough. The ICC emerged as what ICC Greenland leader Aqqaluk Lynge calls "the watchtower of what's going on, what governments do, what the businesses do," achieving what Mark Nuttall describes as "international visibility and credibility through their participation in policy dialogue and decision-making processes at the regional, national and international levels," with the ICC being both "an active participant and agenda setter in a number of important circumpolar initiatives" over the years. Wilson chronicles numerous Inuit policy initiatives as well as comprehensive Arctic policy and its successful efforts at (citing Peter Jull) "changing the attitudes of the public and the Canadian representatives with regards to the Arctic," including on recognizing "environmental protection and sustainable development as part of the same process." Wilson views the ICC as a "successful transnational organization" that has both "united the Inuit" and played "an increasingly important role in Arctic politics" and as the "cornerstone of Inuit diplomacy." But its successes have not been entirely free of setbacks and challenges, including increasing bilateralism and regionalism that counter transnationalism. As Wilson describes, there was (at the time of his writing) "a sense that the ICC is less relevant to the Alaskan Inuit than it was in the past," owing in part to the growing self-sufficiency of the Alaska Native Corporations in addition to periodic tensions between Canada and the Danes that have adversely affected relations between Nunavut and Greenland. Wilson explains, "As the autonomy of Inuit regions grows stronger and bilateralism becomes an increasingly important feature of their circumpolar relations, the question that arises is whether this makes the Inuit regions less reliant on and interested in collaboration through multilateral organizations such as the ICC." This could indicate that the ICC may be less of a post-Westphalian reflection of Inuit sovereignty and more of a transitional structure on the road to increased Inuit autonomy and perhaps even independence at the regional level. This does not diminish all that the ICC has achieved, nor does Wilson expect it will challenge the ICC's prominent role "at the forefront of Inuit efforts to respond to political, economic, and social challenges facing the circumpolar north."

As Gerhardt puts it at the start of his 2011 article, "The Inuit and Sovereignty: The Case of the Inuit Circumpolar Conference and Greenland,"[32] "Inuit strategies for greater self-determination are taking various approaches to the issue of sovereignty. Whereas the Inuit Circumpolar Council (ICC) seeks to partly transcend the Westphalian imaginary, the Greenlandic self-rule government is largely embracing it." Gerhardt takes an informed look at sovereignty through the lens of IR theory, in an effort to both recognize and transcend the "theoretical limitations, and the inherent slipperiness, of the very notion of sovereignty itself." After noting several luminaries like John Mearsheimer, Ken Waltz, Hans Morgenthau, and Hedley Bull, he notes Shadian's "non-state form of sovereignty based on the case of the Inuit Circumpolar Council," and also kindly even cites my 2008 and 2010 work on similar issues, though without mentioning my original "sovereign duality" concept that seems to fit well. Gerhardt differs fundamentally from Shadian, arguing that "the ICC is not declaring an all-out attack on the still very present Westphalian imaginary held by key state actors," an imaginary also held by more regionally bound Inuit movements and polities, most notably in Greenland where its self-rule system, "unlike the ICC, is much more focused on a bounded Greenlandic nation as opposed to a suprastate Inuit one" (of the sort described by Shadian). This is "distinct, although not necessarily exclusionary of, a broader suprastate Inuit nationalism" and "is ultimately grounded on an acceptance, even an unquestioned assumption, of the validity of a Westphalian political ontology."

Gerhardt suggests the Inuit are pulled in both directions at once. Rather than aspiring to replace state sovereignty with their transnational vision, they more pragmatically hope to partner with states to ensure that sovereignty is achieved more by consensus than imposition—as illustrated by their 2009 Circumpolar Declaration on Arctic Sovereignty, in which "the ICC is making clear that any state based sovereignty claims that may be made ultimately depend on the presence of functioning habitation in these regions, and this habitation is provided first and foremost by the Inuit. Furthermore, to make this habitation successful, and to enable it as a foundational pillar of sovereignty claims, it needs to be supported through effective governance," and this "can only occur with the incorporation and respect of the Inuit's customs and their region-specific knowledge." Finding common ground with Shadian, Gerhardt explains that "even if the more radical departures from the traditional Westphalian state-centered system are circumvented by the Arctic rim states in their scramble to lay claim on the Arctic, the ICC still manages to place the Inuit people as pivotal and integral players in the process." Gerhardt expects we will have much to learn from Inuit efforts to reconcile this dialectical clash between Westphalian and post-Westphalian models of sovereignty, and suggests that as the "process unfolds, it will be very interesting to see how Greenland's and the ICC's ambitions evolve and influence each other" in what will "undoubtedly be intense discussions and debates as the Inuit people continue to define and defend themselves in the lands and waters they call home."

Just as Gerhardt and others see in the 2009 ICC Circumpolar Declaration on Arctic Sovereignty an effort to restrain state assertions of sovereignty through an insistence upon Inuit consent, Terry Fenge likewise observes an alignment between Inuit and the state on the affirmation through Inuit consent of state assertions of Arctic sovereignty in Canada. The aligning mechanism that binds the Inuit and state together under a shared conception of sovereignty is somewhat akin to what I call sovereign duality but is actually much more causally intertwined and interdependent than mere duality conveys, linking us back to our prior discussion on the age of Arctic land claims and co-management.

Much like I argued above, co-management becomes the meeting place where Inuit transnationality and state sovereignty are reconciled, bringing us to a grand historical synthesis. I borrow the term "shared space" of co-management from a 2018 article by Jamie Snook, Ashlee Cunsolo, and Aaron Dale to convey this meeting place where tribe and state hammer out, in large measure together, the details of their shared sovereignty, a process with much local variation albeit with many common threads.[33] This "shared space" is not so different from the "third space" of third space theory or the "middle ground" of middle ground theory, which postulated a hybrid native/settler society along the expanding frontier in colonial America. There a similar, and evolving, synthesis took place as the two worlds, Indigenous and colonial, forged something altogether new and distinct, sometimes quite literally as with the birth of the Métis people as well as their Creole counterparts in Russian America, but more often in the institutional evolution of governing systems and adaptations by military organizations to frontier warfare.

Instead of focusing on the transnational nature of the ICC as an Inuit polity asserting post-Westphalian sovereignty as Shadian has done, Fenge looks more bilaterally at the relationship of the Inuit to the Canadian state, more akin to Gerhardt. Many conflicts around the world are defined by two claims to one land, which can lead to protracted, multigenerational conflict, but in the Canadian Arctic, and to limited degrees other parts of the circumpolar North where reconciliation has achieved institutional form, two claims to one land has instead fostered political stability and reduced conflict. Perhaps that is because the state in this particular case is the more vulnerable party with a weaker claim to sovereignty and limited capacity to defend against external challenges to its claims—instilling in the state an appreciation for partnership and collaboration that is otherwise quite rare and conflicts with so many of the Westphalian presumptions of state monopoly on power, force, diplomacy, and foreign policy, all the activities that comprise statecraft. In the Canadian Arctic, statecraft is by default a more collaborative affair.

Fenge's thesis, in essence, is that through their land claims accords (in particular, Nunavut), the Inuit support Ottawa's claim to Arctic sovereignty, and that "Canadian Inuit and Canada's Arctic sovereignty have invariably been linked." Inuit leaders have long recognized this, and they have steadfastly lobbied their counterparts in the Canadian state and society to appreciate it with much success. Former ICC chair and Canada's current governor general Mary Simon did a nationwide tour of Canada in 2007 "to engage and inform Canadians on Arctic sovereignty and to gain public support for Inuit involvement on this issue." And a year before, the president of Nunavut Tunngavik Inc. (NTI), established by the Nunavut Land Claim Agreement (NLCA), "wrote a six-page letter to the Prime Minister . . . suggesting how the agreement could be used to bolster Canada's Arctic sovereignty and inviting the Government of Canada to use it for this purpose," noting in particular the Nunavut land claim's "environmental monitoring provisions (article 12) and establishing the Nunavut Marine Council (article 15)."

However, Fenge notes the "Prime Minister did not reply, and the obvious sovereignty-supporting provisions of the agreement remain unimplemented," a problem Fenge finds to be persistent across Canada as Ottawa continuously "fails to carry out various treaty obligations," to the detriment of Canada's assertion of sovereignty over its remote northland. To the dismay of the Inuit, and Fenge, "We seem now to be in a bizarre situation" where "NTI is in court attempting to force the Government of Canada to implement the NLCA, including articles that support Canada's Arctic sovereignty," and thereby Ottawa "effectively ignores or perhaps even forfeits the opportunity to use the agreement for sovereignty support purposes." Fenge counsels strongly that Ottawa must change its approach and recognize that its sovereign assertions over much of the Arctic are now deeply intertwined with its stewardship and governance of the Arctic in conjunction with the Inuit, owing to its obligations under the NLCA, and that it is now time "to bring Arctic considerations into the heart of our domestic and foreign policy."

This is something that the Inuit have strongly argued for decades, ever since the formation of the ICC. It echoes Eben Hopson's pragmatic approach to sovereignty, recognizing the constitutional authority and supremacy of the state while at the same time calling upon the state to collaboratively and effectively govern the North. It is evident in the efforts of Canadian Inuit leaders at Inuit Tapiriit Kanatami and of the many post–land claim settlement institutions such as Nunavut's NTI. As noted earlier, Arctic exceptionalism does not exist as a default condition born of Arctic geography; though enhanced by the insularity and remoteness of the Arctic, it's the by-product of a collaborative relationship between the state and tribal peoples whose sovereign aspirations align in a complementary and not contested manner. The key word here is *relationship* and, like all relationships, it requires hard work and mutual respect by all stakeholders at all salient levels of analysis. As Fenge noted with dismay, Ottawa's respect was found wanting during Nunavut implementation, just as Thomas Berger found when called to the Arctic once again to serve as conciliator more than a decade ago, even if it was abundantly evident during the negotiations toward, and agreement of, the Nunavut Land Claim Accord. To sustain Arctic exceptionalism thus may require nurturing within the Arctic states every bit as much as between them. For Americans, this means not losing sight of the needs of Alaska natives, and for Canadians, it means putting as much heart into Nunavut's implementation as in its imagination.

Above, I sought to weave together the regional and largely domestic land claims journey with the emergence of co-management as a paradigm for joint governance of the Arctic (and in some cases non-Arctic regions) by the state in collaboration with Indigenous peoples. Despite the many asymmetries (in land, population, and material power) between state and tribal polities across the Arctic, we observe a strengthening of Inuit empowerment and enrichment as land claims became augmented by parallel processes (and in particular new and increasingly powerful institutions of self-governance). This parallelism is similar to the separate legislative processes in Alaska, aligning land claims (ANCSA) with subsistence (ANILCA) in tandem, further augmented, in the case of the North Slope Borough, by utilization of a municipal governing structure rooted in Article X of the Alaska state constitution, and, in the case of the various Indian Reorganization Act tribal councils, using federal recognition of tribal status to enhance native governance with self-governmental authority.

So while Canada integrated land claims and subsistence within the land claims structure, institutions of self-governance have remained largely external to but in parallel with the land claims settlement process all the way from Nunavut to Nunatsiavut, offering continuity with the parallelism we see in Alaska. The Canadian revision of the land claims model thus provides further progress on synthesis of culture and economy, but neither perfection nor full synthesis with governing powers. Greenland, just offshore from Nunavut, with its potential to decolonize its way to formal independence in the years ahead much the way Iceland achieved formal sovereign independence, may present us with the first full integration of all of these once separated processes. That we see stirrings of Inuit sovereignty and articulations of sweeping policies and pronouncements by Inuit leaders and organizations from the 1970s onward suggests to me that the land claims process has been harnessed by a broader movement for Inuit restoration, even if it was conceived initially as a modernization of Alaska's native economy independent of additional processes of cultural or political empowerment, and structured in a manner that courted an eventual loss of land and corporate native control.

Whether this restoration is of a truly Westphalian state sovereignty or an alternative conception of Indigenous sovereignty remains to be seen. Just as the land claims process came to increasingly reflect Indigenous values as time passed, and fostered a greater balancing and reconciliation of state and tribal interests, we can expect to see sovereignty similarly hybridize with time, with elements of formal, Westphalian state sovereignty as well as distinctly Indigenous elements of what Shadian described as a post-Westphalian conception of sovereignty as discussed above. Shadian sees Inuit sovereignty marking a break from the Westphalian tradition, with a greater emphasis on transnationalism than on state sovereignty as a unifying principle of their polity that reaches into the Arctic peripheries of Russia, Alaska, Canada, and Greenland. Others, such as Gerhardt, as illustrated by his 2011 article, "The Inuit and Sovereignty: The Case of the Inuit Circumpolar Conference and Greenland," see a dialectic within the Inuit community between the two contending visions of sovereignty—an Inuit transnational vision as articulated by the Inuit Circumpolar Council and a Westphalian vision articulated by the architects of Greenland's future independence. And yet others, such as Fenge, one of the intellectual architects of Nunavut, see Inuit sovereignty as an affirmation and reinforcement of Canadian

Arctic sovereignty, providing Ottawa with a form of proxy sovereignty that strengthens Ottawa's sovereign claims to the Arctic through the collaborative governance and mutuality of recognition and, importantly, Inuit consent.

We'll next consider several conceptions of Inuit sovereignty. One is articulated by Charles Etok Edwardsen, the famed Inupiat native rights activist and scholar who in the classic documentary *The Duck-In* proudly discusses Barrow's unified display of "Eskimo power" with enthusiasm, and who in his many writings declared the Inuit a republic in all but recognition (such as he did in his 1993 "Science and the Indigenous Arctic," discussed below). Etok's declarations were universally ignored by the Inupiat leadership, but at times, such as in 1992 during a particularly intense battle over subsistence with Juneau, they emulated his dramatic style of declarations by publicly threatening to secede from the state of Alaska and proposing such a secession be seriously studied—causing much heartburn in Juneau over the potential Balkanization of Alaska at the very same time in history that Yugoslavia broke into subnational units with much bloodshed. Juneau was less worried about the potential for armed conflict, which was at most a slim possibility, than the potential loss of oil revenue upon which the state budget has long depended. However, only a year or two earlier, Quebec Mohawks rose up in armed violence against the municipality of Oka and laid siege to the Mercier Bridge across the Saint Lawrence River connecting Montreal to its suburban southern townships for the entire summer of 1990, only ending with a military operation to liberate the bridge from armed militants—so such a possibility was nonetheless real, even if unlikely.

While Etok's declarations were solitary statements and visions of an Inuit republic without a movement of political support to achieve such a vision, if we jump ahead to the 2000s, amid the rapidity of the polar thaw, we see the Inuit Circumpolar Council begin to enunciate various policies on sovereignty that, as Jessica Shadian has suggested, embrace Inuit transnationalism more than Westphalianism. At the same time they serve as a reminder to the Arctic states that their Arctic sovereignty is achieved (as Terry Fenge explains in his writings) with Inuit consent, without which proxy support all of the Arctic North American states (United States, Canada, and Denmark) would be exposed to vulnerability resulting from their respective failures to fully assert meaningful and undisputed sovereignty in their Arctic regions. If the Inuit were to withdraw their consent, and assert their own sovereignty, the situation could change.

And so a close reading of the ICC's two policy declarations, their Circumpolar Inuit Declaration on Arctic Sovereignty (2009) and their Circumpolar Inuit Declaration on Resource Development (2011), reveals an implicit threat (delivered most respectfully) that the assertions of Arctic sovereignty, and sovereign rights to extract resources from the Arctic region, are intimately tied to, and forever intertwined with, Inuit consent to be so governed. As the movement in Greenland for a more sovereign form of autonomy and the public conversation with the support of Copenhagen on a mutually accepted path toward future independence from Denmark indicate, there is an alternative model, and such an alternative model would not necessarily favor the interests of the Arctic states. The caution being urged by the ICC in their declarations is this: don't take us for granted, since it is our consent to be governed that makes your claims of Arctic sovereignty legitimate. This puts the Inuit in a position of some strength considering their small demography (40,000 in Nunavut and 150,000 globally, compared to hundreds of millions of nonnative citizens of the combined Arctic states), and helps to explain their success in utilizing transnationalism in lieu of Westphalian nationalism as a unifying principle for their most interesting trans-state (and according to Shadian, post-Westphalian) polity.

The 2007 UN Declaration on the Rights of Indigenous Peoples (UNDRIP) in so many ways internationalizes the principles enshrined by Arctic land claims, since many Indigenous activists the world over look to the UN for support in their struggles to assert and defend Indigenous identity from state encroachment, including Dr. Dalee Dorough, the international chair of the ICC and a longtime Inuit rights activist and protégé of ICC founder Eben Hopson. A brief summary of UNDRIP has been formulated by Erin Hanson on the University of British Columbia's Indigenous Foundation's web page:[34]

> The United Nations Declaration on the Rights of Indigenous Peoples (UNDRIP) is an international instrument adopted by the United Nations on September 13, 2007, to enshrine (according to Article 43) the rights that "constitute the minimum standards for the survival, dignity and well-being of the indigenous peoples of the world." The UNDRIP protects collective rights that may not be addressed in other human rights charters that emphasize individual rights, and it also safeguards the individual rights of Indigenous people. . . . The Declaration goes on to guarantee the rights of Indigenous peoples to enjoy and practice their cultures and customs, their religions, and their languages, and to develop and strengthen their economies and their social

> and political institutions. . . . Article 26 states that "Indigenous peoples have the right to the lands, territories and resources which they have traditionally owned, occupied or otherwise used or acquired," and it directs states to give legal recognition to these territories. The Declaration does not override the rights of Indigenous peoples contained in their treaties and agreements with individual states, and it commands these states to observe and enforce the agreements. . . . UNDRIP was adopted by 144 countries, with 11 abstentions and 4 countries voting against it. These four countries were Canada, the USA, New Zealand, and Australia. Since 2009 Australia and New Zealand have reversed their positions and now support the Declaration, while the United States and Canada have announced that they will revise their positions.

It seems unlikely a coincidence that all four of these states opposed to UNDRIP in 2007 began as British colonies built on Indigenous homelands that each expanded to become sovereign states, in some cases by conquest and in others by treaty. With their shared origins as colonies carved from Indigenous homelands, largely without full and informed native consent, fully reconciling the aspirations of UNDRIP with the historic and constitutional realities of these colonial origins and aggressive continental expansions remains a challenging process. The Inuit continue to work to complete their restoration of lands and rights, and their impressive gains over the past half century are a testament to their determination and persistence. To continue their advance in Indigenous restoration, they've long recognized the necessity of organizing on the international stage and acting diplomatically as other sovereigns do, entering into treaties and helping to foster the evolution of international law and diplomacy in the continued direction of Indigenous rights recognition. That is why the Inuit Circumpolar Conference (now Council) began as an international forum in the 1970s, reuniting the Inuit regions that had been absorbed into separate Arctic states through colonial history and state expansion into the Inuit homeland (mirroring the Nordic expansion into the Sami homeland and the Japanese and later Russian expansion into the Ainu homeland). In time, with the contributions and leadership of the ICC and partnerships with their respective Arctic states, the Arctic Council emerged as an international/transnational forum to forge new Arctic policies through its various working groups, where national, transnational, subnational, and nonstate entities and polities can work together, united by their commitment to the preservation of Arctic exceptionalism.

How the Inuit can triangulate their way toward increased autonomy and incrementally reclaim most pillars of Indigenous sovereignty from their unique vantage point as a subnational, transnational, and supranational actor in partnership with the national states that assert formal sovereignty over their respective sectors of the Arctic, remains a work in progress, something the Inuit have worked diligently on for decades. In recent years, we have seen a renewed Inuit effort to assert their own voice in foreign and not just domestic policies. I'll next turn to two aforementioned ICC declarations—the Circumpolar Inuit Declaration on Arctic Sovereignty (2009) and the Circumpolar Inuit Declaration on Resource Development (2011)—in addition to an earlier decolonial articulation of Inuit sovereignty by Charlie (Etok) Edwardsen in his 1993 "Science and the Indigenous Arctic," to help understand this range of sovereign aspiration, both within the tribe-state synthesis of a co-managed Arctic and external to it in a decolonized Arctic (a view that aligns better with the current Greenlandic approach of a negotiated decolonization than the co-managed approach that remains predominant in mainland Arctic North America).

The Circumpolar Inuit Declaration on Arctic Sovereignty

On April 28, 2009, a delegation of Inuit leaders from Greenland, Canada, Alaska, and Russia presented the Circumpolar Inuit Declaration on Arctic Sovereignty in Tromsø, Norway, where the Arctic Council was meeting. This historic declaration represented the Inuit response to their exclusion eleven months earlier at the May 2008 Ilulissat Summit of top foreign policymakers among the Arctic rim states, and reflects a formal, if not aggressively forceful, rejection of the modern state's latest effort to shape the destiny of the Arctic without the participation of the Inuit. Fast-forward another eleven months, and the Inuit received the strong endorsement of US Secretary of State Hillary Clinton, who on March 29 attended a meeting in Ottawa of the same Arctic rim states that came together in May 2008 at Ilulissat—Canada, Denmark/Greenland, Norway, and Russia—to continue their discussion of environmental and economic issues related to the thawing Arctic.

But instead of reaffirming the sovereign primacy of the Arctic rim states on issues relating to the Arctic basin as was done in Ilulissat in 2008, Secretary Clinton surprised her host by scolding Ottawa for its continued exclusion of not only Sweden, Finland, and Iceland, but also the nonstate but nonetheless diplomatically savvy Inuit. This was as huge a win for the Inuit as it was an embarrassment for the United States' closest neighbor, largest trading partner, leading supplier of oil, dedicated coalition partner in the bloody war on terror, and loyal ally to the north.[35]

The 2009 Circumpolar Inuit Declaration on Arctic Sovereignty updated Inuit policy on sovereignty for contemporary times, responding not only to the pressures of the changing climate but to their continued diplomatic exclusion from Arctic security and diplomatic affairs as experienced at Ilulissat the year before. While members of the Arctic Council, an international advisory body, the Inuit have been denied a seat at the table when it comes to military, strategic, and diplomatic affairs, which have long been viewed as affairs of state. The Inuit nonetheless aspire to shape policies in the Far North relating to military, security, and diplomatic issues, and during the Cold War endeavored to denuclearize the Arctic basin and to help unify East and West through northern displays of collaboration and cooperation.

The April 2009 Inuit declaration asserted that "central to our rights as a people is the right to self-determination," which "is our right to freely determine our political status, freely pursue our economic, social, cultural and linguistic development, and freely dispose of our natural wealth and resources. States are obligated to respect and promote the realization of our right to self-determination." Section 2 addressed the "Evolving Nature of Sovereignty in the Arctic," and noted sovereignty "has often been used to refer to the absolute and independent authority of a community or nation both internally and externally" but that it remains a "contested concept, however, and does not have a fixed meaning." Further, the declaration noted, "Old ideas of sovereignty are breaking down as different governance models, such as the European Union, evolve," where "sovereignties overlap and are frequently divided within federations in creative ways to recognize the right of peoples."

The Inuit observed that in "exercising our right to self-determination in the circumpolar Arctic, we continue to develop innovative and creative jurisdictional arrangements that will appropriately balance our rights and responsibilities as an indigenous people, the rights and responsibilities we share with other peoples who live among us, and the rights and responsibilities of states," and that in "seeking to exercise our rights in the Arctic, we continue to promote compromise and harmony with and among our neighbors." The Inuit noted the May 2008 Ilulissat declaration had pledged the Arctic rim states to "use international mechanisms and international law to resolve sovereignty disputes," but pointed out that it "neglected to include Inuit in Arctic sovereignty discussions in a manner comparable to Arctic Council deliberations."

The Circumpolar Inuit Declaration thus serves as a reminder that the "inclusion of Inuit as active partners in all future deliberations on Arctic sovereignty will benefit both the Inuit community and the international community," and that "extensive involvement of Inuit in global, trans-national and indigenous politics requires the building of new partnerships with states for the protection and promotion of indigenous economies, cultures and traditions." These partnerships, the declaration contends, "must acknowledge that industrial development of the natural resource wealth of the Arctic can proceed only insofar as it enhances the economic and social well-being of Inuit and safeguards our environmental security." Anything less will be rejected by the Inuit, and with their many settled land claims accords, regional and territorial governments, and numerous mechanisms of co-management and environmental regulation, proceeding without the full support of the Inuit would be unwise. That's why the Inuit have drawn a line in the tundra and why they are so vocally insisting that their exclusion from the table at Ilulissat must now be redressed—so that the future development of the Arctic is a truly joint effort, not just between the Arctic states but between the states and the Inuit.

The Circumpolar Inuit Declaration emerged from the first Inuit Leaders' Summit held on November 6–7, 2008, in Kuujjuaq, Nunavik, in Northern Quebec, where they "gathered to address Arctic sovereignty" and "expressed unity in our concerns over Arctic sovereignty deliberations, examined the options for addressing these concerns, and strongly committed to developing a formal declaration on Arctic sovereignty." In Kuujjuaq, the Inuit leadership had noted with disappointment that the "2008 Ilulissat declaration on Arctic sovereignty by ministers representing the five coastal Arctic states did not go far enough in affirming the rights Inuit have gained through international law, land claims and self-government processes." In many ways, the declaration was a direct response to the foreign ministers of the Arctic rim states for the exclusion of the Inuit at Ilulissat, and it counters this exclusion with a strong argument for a central Inuit role in determining the fate of the Arctic. As the ICC observed at the start of their effort in November 2008,

> Sovereignty is a complex issue. It has a variety of overlapping elements, anchored in international law. But fundamentally it begins with the history and reality of Inuit use and occupation of Arctic lands and waters; that use and occupation is at the heart of any informed discussion of sovereignty in the Arctic. Arctic nation states must respect the rights and roles of Inuit in all international discussions and commitments dealing with the Arctic.

When Secretary Clinton surprised her hosts in Ottawa, headlines quickly hit the newswires and virally circled the globe as news of her public rebuke of Canada spread. Reuters correspondent David Ljunggren, in "Clinton Rebuke Overshadows Canada's Arctic Meeting," noted that Secretary Clinton

> delivered a rare public rebuke to close ally Canada on Monday, criticizing it for excluding key nations from a meeting to discuss the resource-rich Arctic. Canadian Foreign Minister Lawrence Cannon gathered his counterparts from Russia, Norway, the United States and Denmark for three hours of talks on Monday on the grounds that they were the only nations with Arctic coastlines. The decision prompted unhappiness in Sweden, Finland and Iceland, who are also members of the eight-nation Arctic Council—traditionally the body where most important decisions on the region are taken.

In "Clinton Decries Exclusions from Arctic Meeting," Associated Press reporter Rob Gillies reported that

> Clinton said she had been contacted by representatives of indigenous groups who were disappointed they were not invited, according to prepared remarks for Monday's Arctic Coastal meeting. She also said that Sweden, Finland and Iceland—the three Arctic states not represented—had similar concerns. "Significant international discussions on Arctic issues should include those who have legitimate interests in the region. And I hope the Arctic will always showcase our ability to work together, not create new divisions," Clinton said.

And in "Clinton Rebukes Canada at Arctic Meeting," *Washington Post* reporter Mary Beth Sheridan observed, "It was supposed to be a meeting of polar pals. But a high-level session on the dramatic changes in the Arctic turned chilly Monday, as Secretary of State Hillary Rodham Clinton rapped Canada for leaving out several players." As BBC News reported, "Some concern has been expressed that the five coastal states at the Canadian meeting were forming an 'inner core' at the expense of other parties." According to BBC, Secretary Clinton told her fellow diplomats, "We need all hands on deck because there is a huge amount to do, and not much time to do it. What happens in the Arctic will have broad consequences for the Earth and its climate. The melting of sea ice, glaciers and permafrost will affect people and ecosystems around the world."

Canadian Press reported "Clinton's Arctic Comments Cheer Inuit," noting,

> Inuit groups are declaring victory and Arctic experts are warning that Canada's approach to the North will have to change after remarks by U.S. Secretary of State Hillary Clinton. "I can only express my support for her comments and her views," said Duane Smith, head of the Canadian branch of the Inuit Circumpolar Council. Whitney Lackenbauer, a University of Waterloo historian and Arctic expert, agreed. "The clear message from Hillary Clinton is, 'You need to recognize that this Arctic is not just the private sea of the five coastal states.'"

Canadian Press also cited UBC law professor Michael Byers (author of *Who Owns the Arctic?*): "Clinton is going to bat for indigenous people. This will be celebrated by indigenous people across the North." Even Al Jazeera noted, in "Controversy over Arctic Conference," that the five Arctic coastal states "have agreed to work more closely to safeguard the region's environment and resolve disputes over territory. However, the meeting between Canada, Russia, Denmark, Norway and the US on Monday was overshadowed by the exclusion of three other countries with Arctic territories and representatives of indigenous nations."

The significance of Secretary Clinton's controversial embrace of the Inuit interest, and her public rebuke of the United States' closest neighbor and loyal coalition partner—whose losses in Afghanistan, on a proportional basis, greatly exceeded US losses at the time—was surprising in its prioritization of a substate Indigenous actor over an allied nation-state. But the dramatic, if unexpected, nature of her defense of Inuit rights was at heart a very public affirmation by America's top diplomat of the importance of the multilateralism embraced by the Obama administration, as articulated in the latest US Arctic policy. The policy was first promulgated by President Bush in the final days of his administration but had an echo of the later Obama Doctrine woven into its text, one that respected all sovereign components, not just the highest level of the state but also the local village and tribe. For the Inuit, it was a welcome assertion of US respect for the still predominantly Indigenous nature of the far northern frontier.

Seal of Disapproval

Ironically, America's European allies, themselves a diverse mosaic of cultures and languages that are truly kaleidoscopic in nature, have been engaged in a festering cold war with the Inuit over the Inuit right to hunt, trap, and trade in marine mammal products as they have done for millennia. This has resulted in a very odd diplomatic tension between a substate tribe and the very European states whose fur empires led to the colonization of the Far North in centuries past and introduced the Inuit to the globalizing economics of the fur trade. Once proudly independent and with a subsistence economy perfectly aligned with the unique Arctic landscape, the Inuit became part of a new, European-centered, global economy that actively solicited their fur wealth, whether the muskrat, sea otter, or even the seal.

But today's Europeans no longer have the same zeal for soft, marine mammal fur, or the taste for Arctic country foods. During the February 2010 meeting of G7 finance ministers in the Canadian Arctic, Nunavut leaders generously hosted their international visitors with a community feast that included traditional northern cuisine, including a staple of their subsistence diet: seal meat. But as Andrew Clark reported in *The Guardian*, "None of the visiting ministers chose to attend a feast on Saturday night, laid on by the local Inuit community, at which raw seal was on the menu. Canada's Jim Flaherty was left to chow down on some seal meat alone."[36] This was ironic, since it was their own forefathers who expanded the fur empire into the Far North centuries earlier, but they appeared to no longer feel any responsibility for the economic or cultural well-being of the descendants of the Inuit who had welcomed their forefathers in peace and whose hard work contributed to the success of those early globalization efforts. The refusal of the European G7 finance ministers to dine with the Inuit at a feast they hosted in their honor in Nunavut's capital city of Iqaluit was certainly not Europe's finest moment.

The Inuit may be few in number, but they control many local economic and political levers, and their interests are now fully backed by Ottawa—their partner in land claims, self-government, and northern development. Resolving lingering tensions between Europe and the Inuit is a necessary step to ensure the tranquility of the Far North, and the Inuit are demanding a seat at any table where their future and the future of their homeland are being negotiated. This was recognized by Secretary Clinton, and though she rebuked Ottawa, the European states seemed more deserving of her diplomatic tongue-lashing, since Canada, more than any state, has welcomed its Indigenous peoples into a participatory political life with not only open hearts and minds but a rare commitment to devolving governmental power to the people of the North. In Arctic Canada, the Inuit have largely been content with increasing their domestic power, under the sovereign umbrella of the Canadian Constitution. Secretary Clinton's rebuke thus seems to have been aimed at the wrong target. After all, it was Ottawa that elected to hold the G7 finance meeting in Iqaluit this past winter, hoping to foster reconciliation between the Inuit and the Europeans. Ottawa's efforts should be lauded, not subjected to diplomatic pot-shots from a nation whose own history is one where Indigenous peoples were ruthlessly exterminated by war, assimilated by cultural conquest, and concentrated into destitute reserves, and that only recently has sought to atone for its past misdeeds. Canada never so brutally assaulted its own Indigenous peoples and has gone much further than the United States in its efforts to empower them—recognizing the strategic importance of Indigenous harmony thirty years ago when the Oka standoff was settled without further bloodshed, and a host of land claims settlements, governmental innovations, and royal commissions sought to redress historic injustices in a holistic manner.

But even if misdirected at the United States' friendly ally to the north, Secretary Clinton's intention, to show her support for the Inuit and for their central place in the emergent Arctic order, was surely intended to unite, and not further divide, the Arctic. She more than most senior diplomats understands that it really does, as her 1996 book was titled, "take a village" to heal a planet. With Secretary Clinton warmly embracing the Inuit point of view and calling upon the other nations along the Arctic coast to do the same, one could expect to (and ultimately did) see increasingly visible and meaningful Inuit participation at subsequent meetings on Arctic issues, and decreasing exclusion of the Indigenous perspective going forward more than a decade, as the US embrace of Arctic cooperation (and in particular, Indigenous-state reconciliation and collaboration) was so publicly and visibly reasserted. While that would not necessarily restore formal sovereignty to the Inuit, it did at least provide reassurance that their interests and values had an increasingly secure place at the table.

A Bold Response to the Inuit Exclusion from Ilulissat

The ICC's Circumpolar Declaration on Sovereignty in the Arctic is a fascinating document for what it asserts, and what it does not. It's not a formal declaration of independence, nor does it assert any formal sovereign claims on behalf of the Inuit in the Westphalian tradition. But at the same time, it was issued in response to the exclusion of the Inuit by the Arctic 5 (the five member states with Arctic Ocean coastline) at Ilulissat the year before, and may be viewed as an Inuit rebuke of the Arctic rim states for their exclusion, which was viewed rightly as a betrayal of the collaborative dynamic cultivated during decades of land claims negotiations and co-management implementation. While the exclusion of the

Inuit was a setback, it also provided an opportunity to reassert the Inuit view, and a teachable moment for the Inuit to remind their state counterparts that Arctic exceptionalism was not exclusive to the Arctic states but inclusive to the multiplicity of actors who govern the Arctic region.

The declaration was a dramatic reminder to the Arctic states that the Inuit remain essential to the Arctic's political order and stability, and their consent is a sine qua non for effective and meaningful sovereignty in Arctic North America. And so while it's not a declaration of Inuit sovereignty, the declaration does assert that the Inuit and their consent are essential to uphold state sovereignty in the Arctic, and if that consent is lost, state sovereignty itself could well lose its legitimacy—putting the stability of the Arctic into question.

An Enduring Sovereignty, "From Time Immemorial"

Curiously, one problem of historical veracity that we see in the historical claims made in several of the Inuit land claims agreements, of continuous use and occupancy of the Arctic by the Inuit since "time immemorial"—even parts of the Arctic that the Inuit did not use or occupy until quite recently, and mostly only after state interventions such as relocations or migrations that coincided with state expansion—appears in the declaration, right near its start.

In Section 1.2, the declaration asserts, like many other Inuit documents, that "Inuit have been living in the Arctic from time immemorial," a stock phrase common to all of the Inuit land claims that are based in part on historic indigeneity of the claimant group to their respective settlement region. The ICC declaration further adds that Inuit "use and occupation of Arctic lands and waters pre-dates recorded history."[37] However, much of Inuit land use and occupancy in the Arctic are entirely modern, not ancient. For instance, Inuit appear to have migrated to Greenland in the sixteenth century, no earlier than the first Viking settlers, suggesting neither group is any more Indigenous than the other. And, only in the nineteenth century did Inuit migration into the north of Greenland conclude, quite a modern settlement. Additionally, the migration of Inupiat into Canada's Western Arctic took place in the early twentieth century after a series of epidemics decimated the population of the resident Inuit population at Kittigazuit. And the high Arctic communities of Resolute and Grise Fiord were not settled until the 1950s, and only as a result of federal relocation policies that were later criticized by the Inuit on human rights grounds. None of these come close to "time immemorial," and when scrutinized appear to undermine the cornerstone to Inuit claims of Arctic sovereignty, collaborative or not.

On the other hand, while the history of Inuit migration and settlement may be more contemporary than the Inuit would like us to believe, they don't use their historical claim to assert a right of sovereign control over the Arctic. Instead they make a claim of a more nuanced, multilevel, and collaborative form of Arctic sovereignty that aligns well with the states of Arctic North America, for whom collaboration and multilevel governance are mutually embraced across an impressive stretch of time (1994–2016 for the United States, which is an eternity when it comes to foreign policy continuity of any kind—across multiple presidential administrations from both political parties and various wings of those parties, from the start to the end of the post–Cold War era when Russia was viewed as a partner if not an outright ally). But let's refrain from a critical assessment of the Inuit narrative on use and occupancy of the Arctic since time immemorial and the modernity of many of their migrations to areas now defined as part of the Inuit homeland, even though one could fairly assess those as new Inuit colonial territories achieved through expansion just the way states expand their sovereign dominions through aggrandizing new territories—though to the best of my knowledge, it was only later waves of Inuit expansion into Greenland that brought them into contact with another people that had arrived contemporaneously, the Viking settlers, so their other expansions were largely onto unoccupied and unsettled lands that were effectively terra nullius even when claimed by other sovereigns. (There are many Inuit legends of extraordinarily strong, superhumanly capable "super little men" who already inhabited their lands, who were rarely seen, living in the shadows, perhaps a trace memory of an earlier people whose use and occupancy of the Arctic predates the arrival of modern Inuit—so there does remain the potential that the Inuit did displace an earlier group of authentically Indigenous peoples, though most experts suggest these populations had died out centuries before the arrival of the modern Inuit.)

Multilevel Governance as a Foundation for Arctic Exceptionalism

The heart of the Circumpolar Inuit Declaration on Arctic Sovereignty is its embrace of multilevel governance and collaborative sovereignty, providing a perfect alignment with the Arctic states and their various approaches to Arctic collaboration, whether shaped as I argue in the co-management and land claims experience of Arctic North America or in the tradition of Nordic cooperation that took root on the other side of the Arctic, predisposing all Arctic stakeholders, Indigenous and state, to work together.

The declaration describes Inuit as being "united as a single people" whose "sense of unity is fostered and celebrated by the Inuit Circumpolar Council,"[38] what one could consider a seed for later claims of nationhood, if not, as Shadian might expect, of a transnational, post-Westphalian form of sovereignty. Further, they are an "indigenous people"[39] retaining, in language that is similar to Etok (discussed below), "the right to self-determination"[40] as affirmed by the UN Declaration on the Rights of Indigenous Peoples,

> all of which are relevant to sovereignty and sovereign rights in the Arctic: the right to self-determination, to freely determine our political status and to freely pursue our economic, social and cultural, including linguistic, development (Art. 3); the right to internal autonomy or self-government (Art. 4); the right to recognition, observance and enforcement of treaties, agreements and other constructive arrangements concluded with states (Art. 37); the right to maintain and strengthen our distinct political, legal, economic, social and cultural institutions, while retaining the right to participate fully in the political, economic, social and cultural life of states (Art. 5); the right to participate in decision-making in matters which would affect our rights and to maintain and develop our own indigenous decision-making institutions (Art. 18); the right to own, use, develop and control our lands, territories and resources and the right to ensure that no project affecting our lands, territories or resources will proceed without our free and informed consent (Art. 25–32); the right to peace and security (Art. 7); and the right to conservation and protection of our environment (Art. 29).[41]

Most of these rights parallel those achieved through Arctic land claims accords and lay a foundation that UNDRIP could effectively internationalize, in part from the efforts of Arctic Indigenous leaders.

Further contextualizing Inuit political identity is their geographically defined identity stemming from their being "an indigenous people of the Arctic"[42] whose identity is shaped "within the unique geographic, environmental, cultural and political context of the Arctic," which "has been acknowledged in the eight-nation Arctic Council, which provides a direct, participatory role for Inuit through the permanent participant status accorded the Inuit Circumpolar Council (Art.

2)."[43] Their rights that derive from their identity as Indigenous people of the Arctic are further enhanced by their citizenship rights at the national level, as citizens of the United States, Canada, Russia, or Denmark, with "rights and responsibilities afforded all citizens under the constitutions, laws, policies and public sector programs of these states"[44] that "do not diminish the rights and responsibilities of Inuit as a people under international law." In addition to their citizenship in these national sovereign polities, the Inuit are also more granularly "indigenous citizens of [those] Arctic states"[45] with "the rights and responsibilities afforded all indigenous peoples under the constitutions, laws, policies and public sector programs of these states,"[46] in addition to also being "indigenous citizens of each of the major political subunits of Arctic states (states, provinces, territories and regions)."[47] (While not mentioned, and never really a priority for the ICC with its transnational post-Westphalianism, there is also the tribal status recognized in the United States for Alaska natives, which since 1993 has added another dimension to the already granular diversity of Indigenous multilevel governance, in addition to the quasi-governmental dimensions of their equity participation in the native corporations created through land claims negotiations in both Alaska and Canada.) Part and parcel of the multilevel governance systems that manage the Indigenous Arctic is a multilevel political identity that mirrors the institutional complexity that surrounds them. It is amid this multilevelism that we see Inuit conceptions of sovereignty take shape.

Sovereignty as a "Contested Concept"

Indeed, in section 2 of the declaration, the Inuit assert that sovereignty itself is a "contested concept"[48] and is not, by default, inherently Westphalian as the community of nation-states might assume. In so doing, they go far to decolonize the very conception of sovereignty for the Arctic. Indeed, section 2 is titled "The Evolving Nature of Sovereignty in the Arctic."[49] Coming out swinging, the ICC quickly demolishes the Westphalian conception of sovereignty and its presumptions of sovereign absoluteness: "'Sovereignty' is a term that has often been used to refer to the absolute and independent authority of a community or nation both internally and externally."[50] Further discoursing on sovereignty in what sounds like it could be an academic seminar:

> Sovereignty is a contested concept, however, and does not have a fixed meaning. Old ideas of sovereignty are breaking down as different

> governance models, such as the European Union, evolve. Sovereignties overlap and are frequently divided within federations in creative ways to recognize the right of peoples. For Inuit living within the states of Russia, Canada, the USA and Denmark/Greenland, issues of sovereignty and sovereign rights must be examined and assessed in the context of our long history of struggle to gain recognition and respect as an Arctic indigenous people having the right to exercise self-determination over our lives, territories, cultures and languages.[51]

Sovereignty is thus perceived in an inherently transnational manner, as collaborative phenomena that "overlap and frequently divided . . . in creative ways," and in a manner that is informed by the Inuit experience across the Arctic, where collaborative sovereignty, through co-management and land claims, has been part of the constitutional landscape for some time, to varying degrees—as detailed in section 2.2, which states, "Recognition and respect for our right to self-determination is developing at varying paces and in various forms in the Arctic states in which we live."[52] Section 2.3 further elucidates the multilevel governing mechanisms developed in the Arctic, noting, "In exercising our right to self-determination in the circumpolar Arctic, we continue to develop innovative and creative jurisdictional arrangements that will appropriately balance our rights and responsibilities as an indigenous people, the rights and responsibilities we share with other peoples who live among us, and the rights and responsibilities of states."[53] Moreover, the ICC notes that "in seeking to exercise our rights in the Arctic, we continue to promote compromise and harmony with and among our neighbours."[54]

Looking to the evolution of Inuit governance across the Arctic, the ICC observes a shift from local to national and increasingly to international participation in decisionmaking, with the collaboration of co-management at the local and regional levels flowing upward to the territorial level (as seen in Nunavut) on to the international, where the collaborative partnership initiated with co-management became enshrined in the organizational structure of the Arctic Council. As section 2.4 describes, "International and other instruments increasingly recognize the rights of indigenous peoples to self-determination and representation in intergovernmental matters, and are evolving beyond issues of internal governance to external relations," culminating at the Arctic Council where, as noted in section 2.5, "Inuit are permanent participants at the Arctic Council with a direct and meaningful seat at discussion and negotiating tables."[55]

Objecting to the 2008 Exclusion of Inuit at Ilulissat

Directly challenging the Arctic 5 and their exclusion of the ICC at Ilulissat, section 2.6 argues,

> In spite of a recognition by the five coastal Arctic states (Norway, Denmark, Canada, USA and Russia) of the need to use international mechanisms and international law to resolve sovereignty disputes (see 2008 Ilulissat Declaration), these states, in their discussions of Arctic sovereignty, have not referenced existing international instruments that promote and protect the rights of indigenous peoples. They have also neglected to include Inuit in Arctic sovereignty discussions in a manner comparable to Arctic Council deliberations.[56]

This disappointment both precipitated the ICC's declaration on sovereignty and inspired the Inuit to use their declaration as a reminder of the tribe-state collaborative path not followed at Ilulissat, which they hoped would be restored. Section 3 of the declaration, "Inuit, the Arctic and Sovereignty: Looking Forward,"[57] thus starts out with a discussion of the collaborative "foundations of action." Section 3.1 notes, "The actions of Arctic peoples and states, the interactions between them, and the conduct of international relations must be anchored in the rule of law."[58] Section 3.2 adds, "The actions of Arctic peoples and states, the interactions between them, and the conduct of international relations must give primary respect to the need for global environmental security, the need for peaceful resolution of disputes, and the inextricable linkages between issues of sovereignty and sovereign rights in the Arctic and issues of self-determination."[59]

The remainder of section 3 focuses on tribe-state partnerships, under the subheading "Inuit as active partners," with section 3.3 asserting, "The inextricable linkages between issues of sovereignty and sovereign rights in the Arctic and Inuit self-determination and other rights require states to accept the presence and role of Inuit as partners in the conduct of international relations in the Arctic."[60] Section 3.4 adds, "A variety of other factors, ranging from unique Inuit knowledge of Arctic ecosystems to the need for appropriate emphasis on sustainability in the weighing of resource development proposals, provide practical advantages to conducting international relations in the Arctic in partnership with Inuit."[61] Section 3.5 further adds that "Inuit consent, expertise and perspectives are critical to progress on international issues involving the Arctic, such as global environmental security, sustainable development, militarization, commercial fishing, shipping, human health, and economic and social development."[62] Section 3.6 emphasizes the central importance of Inuit participation and partnership, arguing,

> As states increasingly focus on the Arctic and its resources, and as climate change continues to create easier access to the Arctic, Inuit inclusion as active partners is central to all national and international deliberations on Arctic sovereignty and related questions, such as who owns the Arctic, who has the right to traverse the Arctic, who has the right to develop the Arctic, and who will be responsible for the social and environmental impacts increasingly facing the Arctic. We have unique knowledge and experience to bring to these deliberations. The inclusion of Inuit as active partners in all future deliberations on Arctic sovereignty will benefit both the Inuit community and the international community.[63]

Section 3.7 further emphasizes the importance of Inuit as partners, owing in part to their extensive involvement in global, transnational, and Indigenous politics, making them, in essence, a bridge to the Arctic whose participation strengthens the cooperative foundations of the Arctic order:

> The extensive involvement of Inuit in global, trans-national and indigenous politics requires the building of new partnerships with states for the protection and promotion of indigenous economies, cultures and traditions. Partnerships must acknowledge that industrial development of the natural resource wealth of the Arctic can proceed only insofar as it enhances the economic and social well-being of Inuit and safeguards our environmental security.[64]

The rest of section 3 addresses the "need for global cooperation," specifying in section 3.8 that there's "a pressing need for enhanced international exchange and cooperation in relation to the Arctic, particularly in relation to the dynamics and impacts of climate change and sustainable economic and social development,"[65] and that "regional institutions that draw together Arctic states, states from outside the Arctic, and representatives of Arctic indigenous peoples can provide useful mechanisms for international exchange and cooperation"[66]—institutions like the Arctic Council. Environmental, climate, and health security concerns and the need to involve the Inuit in tackling these challenges bring section 3 to a close.

Section 4 turns back to the issue at hand, Arctic sovereignty, with section 4.1 reiterating Inuit disappointment with the Arctic 5 at Ilulissat for not going "far enough in affirming the rights Inuit have gained through international law, land claims and self-government processes."[67] This captures in a nutshell the thesis put forth throughout this manuscript that these processes of empowerment at the local, regional/territorial, national, and international levels have transformed the Inuit into an essential partner in the exercise of sovereignty within and between Arctic states. Owing to their steadfast commitment to cooperative interactions and collaborative decisionmaking, the Inuit are positioned as a key pillar of Arctic exceptionalism—not the only one, of course, since in northern Europe the tradition of Nordic cooperation provides us with another pillar, both being sine qua nons for the continuation of a cooperative Arctic region. Section 4.2 reminds the Arctic states of the Inuit view, not always fully embraced by the Arctic states but likewise neither fully dismissed, that the

> conduct of international relations in the Arctic and the resolution of international disputes in the Arctic are not the sole preserve of Arctic states or other states; they are also within the purview of the Arctic's indigenous peoples. The development of international institutions in the Arctic, such as multi-level governance systems and indigenous peoples' organizations, must transcend Arctic states' agendas on sovereignty and sovereign rights and the traditional monopoly claimed by states in the area of foreign affairs.[68]

And, as section 4.3 further states, the Inuit affirm that "issues of sovereignty and sovereign rights in the Arctic have become inextricably linked to issues of self-determination in the Arctic. Inuit and Arctic states must, therefore, work together closely and constructively to chart the future of the Arctic."[69]

The declaration thus concludes with a pledge of cooperation with the states, something of an olive leaf after rebutting the Arctic 5's exclusion of the Inuit at Ilulissat: "We, the Inuit of Inuit Nunaat, are committed to this Declaration and to working with Arctic states and others to build partnerships in which the rights, roles and responsibilities of Inuit are fully recognized and accommodated."[70] So rather than alienating or radicalizing the Inuit leadership, their exclusion was greeted as yet another teachable moment to engage with their sovereign counterparts, and to reinvigorate their collaborative relationship, in the hope of restoring the Inuit to their place at the diplomatic table in all matters related to Arctic sovereignty.

The Globalization of Resource Co-management and Collaboration

Two years later, the ICC issued its next declaration, one less sweeping and perhaps less theoretical than their 2009 declaration on sovereignty, but one every bit as committed to the collaborative relationship between the Inuit and the circumpolar states on matters of Arctic resource development, reasserting the principle of co-management and Inuit participation in resource development decisionmaking and planning. In the Circumpolar Inuit Declaration on Resource Development Principles in Inuit Nunaat,[71] the ICC reasserts in its preamble their recognition of "the Arctic's great resource wealth, the increasing global demand for the Arctic's minerals and hydrocarbons, the scope and depth of climate change and other environmental pressures and challenges facing the Arctic,"[72] the "core rights of Inuit as recognized in the United Nations Declaration on the Rights of Indigenous Peoples, as provided for in a variety of other legal and political instruments and mechanisms, including land rights settlement legislation, land claims agreements (treaties), and self-government, intergovernmental and constitutional arrangements, and as asserted in A Circumpolar Inuit Declaration on Sovereignty in the Arctic,"[73] and the "ingenuity, resilience and wisdom of previous generations of Inuit, confident of the ability of every generation of Inuit to adapt to change, and determined to provide for the material and cultural well-being of Inuit into the future."[74]

The Inuit thus declare a sweeping set of themes and issues connected to Arctic resources and the Inuit vision for balanced development to ensure that the renewable resource base upon which the Inuit have long depended, and the many nonrenewable resources increasingly sought by both states and post–land settlement Inuit corporations, are managed responsibly and sustainably. While the quest for and responsible stewardship of renewable resources reflect the long tradition of Inuit subsistence, the ICC recognizes that

> responsible non-renewable resource development can also make an important and durable contribution to the wellbeing of current and future generations of Inuit. Managed under Inuit Nunaat governance structures, non-renewable resource development can contribute to Inuit economic and social development through both private sector channels

(employment, incomes, businesses) and public sector channels (revenues from publicly owned lands, tax revenues, infrastructure).[75]

Toward a "Proper Balance"

Such development efforts must, they contend, strike a "proper balance," as the Inuit "desire resource development at a rate sufficient to provide durable and diversified economic growth, but constrained enough to forestall environmental degradation and an overwhelming influx of outside labour," and in a manner that "promote[s] economic diversification through contributions to education and other forms of social development, physical infrastructure, and non-extractive industries."[76] As such, ICC pledges that the "Inuit welcome the opportunity to work in full partnership with resource developers, governments and local communities in the sustainable development of resources of Inuit Nunaat, including related policy-making, to the long-lasting benefit of Inuit and with respect for baseline environmental and social responsibilities."[77]

As with their earlier declaration on sovereignty, their 2011 declaration on resource development is not a statement of Inuit unilateralism at all, but one of enduring commitment to collaboration. And as with their declaration on sovereignty, their declaration on resource development insists that the gains achieved by Inuit from participation in the management of the Arctic are respected and preserved as the foundational starting point for future resource development in the Arctic. With a nod to globalization, the ICC notes in the first section, on "Candour, Clarity and Transparency,"[78] that "The world's peoples and their social, cultural and economic systems are becoming more interconnected, the pace of change is accelerating, the challenges faced by the world are escalating in complexity, and the risks associated with human activities are of increasing significance."[79] Meaningful collaboration is therefore key to managing these increasing and accelerating interconnections and their associated challenges: "To prosper under these circumstances, the peoples and states of the world must conduct their relations cooperatively with candour, clarity and transparency—an approach in keeping with Inuit culture and custom."[80] Inuit culture and its alignment with these principles of collaboration thus strengthen the underlying fabric of cooperation across the Arctic. Engaging with Arctic states, and conveying these values, thus contributes to a strengthening of this foundation, whereas the converse weakens it (and weakens the underlying stability of the Arctic region, much the way that sovereignty itself, without aligning with Indigenous values, is weakened and not strengthened, contrary to the Westphalian mind-set). Thus the ICC asserts another teachable moment: "It is our desire to declare our key understandings, positions and intentions in relation to resource development, recognizing that doing so will benefit Inuit and the global community."[81]

The ICC notes that the principles put forth in their declaration on nonrenewables "are, in many ways, applicable to the use of renewable resources" owing to their being "inextricably linked,"[82] further binding states with their thirst for nonrenewables with the Inuit, who, before the age of land claims endowed them with corporate structures and investment capital, primarily sought renewables, first for subsistence and later for trade. However, since land claims were settled across Arctic North America, Inuit have increasingly sought nonrenewable resource opportunities on the corporate side of their postsettlement economies, along with their pursuit of renewables as part of the traditional economic systems protected by the Canadian land claims model.

Embedding Inuit Rights in International Law

Section 2 of the ICC declaration on resource development focuses on the UN Declaration on the Rights of Indigenous Peoples and asserts that resource development in the Inuit homeland "must be grounded" in it, as UNDRIP "recognizes the right of indigenous peoples to self-determination" under which "Inuit have the right to freely determine collectively our political, social, economic, and cultural development. Resource development in Inuit Nunaat directly engages our right to self-determination, and many other provisions of the UN Declaration."[83] Inuit rights, "including our right to self-determination, may be exercised in a practical way through governance structures that combine both Inuit and non-Inuit constituents. No matter what level or form of self-determination the Inuit of any particular region have achieved, resource development in Inuit Nunaat must proceed only with the free, prior, and informed consent of the Inuit of that region."[84] Moreover, "Private sector resource developers, and governments and public bodies charged with the public management of resource development, must all conduct themselves in concert with the UN Declaration,"[85] complete with openness and transparency. After embedding Inuit resource rights in international law via UNDRIP in section 2, section 3 embeds Inuit resource rights in their 2009 declaration on sovereignty, including its "many principles that are relevant to the governance and carrying out of resource development in Inuit

Nunaat, including the importance of the rule of law and recognition of the rights of Inuit as an Arctic indigenous people under both international and domestic law."[86]

Inuit as "Essential Partners" in the Arctic Order

The core of the resource declaration, section 4, turns to "Inuit as Partners in Policy Making and Decision Making," which is tied directly to the Inuit declaration on sovereignty: "Central to A Circumpolar Inuit Declaration on Sovereignty in the Arctic is the requirement that Inuit must be active and equal partners in policy-making and decision-making affecting Inuit Nunaat."[87] Noting the diversity of governing structures across the Arctic, section 4.2 states, "Partnerships with Inuit in relation to resource development will have different characteristics depending on the circumstances, but the spirit and substance of partnership must extend to both public sector governance and private sector enterprise," and section 4.3 adds these partnerships "must include the meaningful engagement and active participation of Inuit in local communities who are most directly affected by resource development in Inuit Nunaat."[88] Moreover, as section 4.4 notes, "Partnerships must draw upon the growing capacity and aspirations of Inuit businesses and enterprises through use of vehicles such as joint ventures, commercial mechanisms for facilitating equity participation, and the issuance of land and resource rights through licences, leases and similar instruments."[89] These indicate Inuit interests in not just preserving their subsistence traditions and the pristine ecosystems they depend upon, but participating as equity partners, contractors, and subcontractors with economic stakeholders in the development of nonrenewables as well.

Looking ahead to future clashes between regions, particularly on development projects that could affect neighboring areas, and recognizing the highly localized co-management systems established by most Arctic land claims as noted in the literature on streamlining efforts in the Yukon and Northwest Territories with some support in Ottawa, section 4.5 acknowledges "the need within Inuit Nunaat to create and implement inter-Inuit consultation mechanisms to ensure that approval of major resource development projects in one Inuit region, with major environmental and other implications for one or more adjacent Inuit regions, is accompanied by sufficient opportunity for an informed exchange of information and opinion between or among the Inuit regions."[90]

Global Environmental Security: A Convergence of Inuit Values and International Security

Especially pertinent to the transition from our study of the many levels of analysis in the globalizing Arctic, dating back by some measures to the Ice Age and by others to the early colonial expansion by chartered companies into Arctic and sub-Arctic North America, to the study and understanding of Arctic security in the contemporary world, shaped in many ways by this gradual and increasing inclusivity of Indigenous peoples in governance of the North, section 5 turns its attention to the issue of "Global Environmental Security."[91] As put forth in section 5.1, the ICC asserts that "Inuit and others—through their institutions and international instruments—have a shared responsibility to evaluate the risks and benefits of their actions through the prism of global environmental security,"[92] and in section 5.2 they further note, "Resource development in Inuit Nunaat must contribute to, and not detract from, global, national and regional efforts to curb greenhouse emissions and should always be seen through the reality of climate change."[93] Beyond this, the Inuit insist that as part of "their implementation of mechanisms for adaptation to climate change, states and the international community as a whole must commit to paying the cost of climate change adaptation measures and the upgrading of fuel-related infrastructure in Inuit Nunaat regions and communities,"[94] part of a long-standing effort that had earlier been expressed through litigation to ensure that the states that ultimately have caused global warming must pay the costs of mitigating the effects of global warming on the Inuit, whose culture is especially well-adapted to a colder world and thus at great risk (at least in the view of much of the Inuit leadership) to climate change. More practically, the ICC asserts in section 5.4 that "Resource development projects must not exacerbate the climate change–related stresses on the survival of Arctic wildlife," and more broadly, section 5.5 adds that to "minimize risk to global environmental security, the pace of resource development in the Arctic must be carefully considered."[95]

Healthy Communities in Healthy Environments with Healthy Economies

Section 6 focuses on the importance of maintaining "Healthy Communities in a Healthy Environment," and as section 6.1 asserts, "The physical and mental health of human communities and individuals cannot be separated from the health of the natural environment,"[96] and as a result, subsequent sections add that "Resource development proposals for Inuit Nunaat must be assessed holistically, placing human needs at the centre,"[97] and "Resource development in Inuit Nunaat must promote the physical and mental health of communities and individuals within Inuit Nunaat."[98] The ICC next extends their holistic interconnection of resource development from their broad conception of Arctic security to food security, asserting that "Resource development must enhance, not detract from, Inuit food security."[99] Beyond that a wide basket of human security issues together ensure a healthy community—the heart of Arctic security through an Inuit lens: "In a contemporary context, healthy communities in the Arctic require the establishment, maintenance and improvement of core infrastructure needs, including housing, education, health care and social service delivery infrastructure, and core transportation and communications networks that facilitate both public sector activities and private sector entrepreneurship."[100]

Section 7, on "Economic Self-Sufficiency and the Sustainable Development of Resources in Inuit Nunaat," notes, "Inuit seek to make use of the economic opportunities available through long-term development of the resources of Inuit Nunaat,"[101] and this requires that "resource development in Inuit Nunaat must be sustainable"[102] and "must serve the needs of Inuit today without compromising the ability of Inuit [to] meet their needs of tomorrow."[103] As a result, "The proponent of a resource development project bears the burden of demonstrating that the proposed development is sustainable."[104] Further, such a demonstration of sustainability must be based upon "the best available scientific and Inuit knowledge and standards must be determined and employed,"[105] and any "international standard-setting bodies must seek and secure direct and meaningful input from Inuit,"[106] while "national, regional and local bodies, such as offshore and land management regimes, must be designed and operated to be effective, transparent and accountable, thereby gaining and sustaining the confidence of the Inuit public at all times,"[107] with "demonstrated support of those communities directly affected by a resource development proposal."[108]

Section 8 further drills down on "Impact Assessment, Prevention and Mitigation," asserting in section 8.1 that "there is no free-standing or unqualified 'right' to proceed with non-renewable resource development in Inuit Nunaat,"[109] and any such "projects must be scrutinized by Inuit and proved to be in the best interests of Inuit and the wider public."[110] Looking to Inuit experience with land use planning and environmental assessments through their co-management systems, the declaration further asserts, "Land and offshore management regimes must include (a) long-term land use plans that set out ground rules for development applicable to specific projects, and (b) robust impact assessment processes to gauge the likely impacts of specific projects,"[111] and that "management, land use planning and impact assessment regimes must address the cumulative impacts of existing and potential projects and, where prudent, limit the number and scope of projects permitted."[112] Additionally, environmental "impact assessments covering broad geographic areas are important and necessary management tools, and their completion in advance of specific project proposals should be encouraged"[113] and "should examine all potential environmental, socio-economic and cultural impacts anticipated both during the project and after the project is completed or abandoned."[114] The rest of section 8 drills further down into these assessment processes, insisting on proper application of both the "precautionary principle" and the "polluter pays principle" during "all stages of project planning, assessment, implementation and reclamation,"[115] that habitat reclamation and recovery "must be thoroughly planned and fully funded in advance of and throughout project implementation,"[116] that "all development in Inuit Nunaat must adhere to the most developed and demanding environmental standards taking Arctic conditions fully into account," including ensuring that "preventing spills offshore and eliminating release of toxic substances to land and waters are paramount,"[117] and that "response to spills, contamination of lands or waters, and mining emergencies must meet the highest technological standards and be anchored in proven cleanup technologies with full Inuit participation."[118]

Section 9, on "Improving Inuit Living Standards and Expanding Inuit Governance," presents the Inuit expectation that "new resource development projects will contribute to an improvement in our material wellbeing,"[119] an expectation they explain as being "well-rooted in the fundamental features of relevant international indigenous and human rights laws and standards, in the underlying constitutional constructs and political values of the four Arctic States in which Inuit live, and in the application of fairness and reason."[120] Noting the complexity of the Inuit decisionmaking environment, with its many and diverse jurisdictions across four countries, section 9.2 observes how, "through a variety of mechanisms—land rights settlement legislation, land claims agreements (treaties), self-government arrangements, and intergovernmental and Constitutional provisions—Inuit have acquired critical means and levels of control over the governance of Inuit Nunaat. Many of these mechanisms provide for direct Inuit participation in specialized resource management bodies, including planning, project review, and regulatory bodies."[121] And, "While this trend is primarily a result of Inuit effort and determination, it has often been assisted and welcomed as healthy and normative by and within the four Arctic States."[122] "Accordingly, resource development projects must take into account the trend toward greater Inuit self-governance and, to the extent possible, advance it."[123] Moreover, the ICC asserts, "Public sector revenues derived from all phases of resource development should be distributed in a fair and visible way according to the following hierarchy of priorities: (1) providing security against unplanned or unintended environmental consequences, (2) compensating for negative community and regional impacts, (3) contributing to the improvement of community and regional living standards and overall well-being, and (4) contributing to the fiscal health and stability of institutions and mechanisms of Inuit governance," and that "Only after the legitimate needs of the Inuit of Inuit Nunaat are met, should public sector revenues contribute to the coffers of central state treasuries."[124] Moreover, "Inuit employment at all levels must be maximized in resource development activities in Inuit Nunaat, and "Inuit must derive direct and substantial employment income benefit from resource development projects."[125] And, with a hint at section 10 that shortly follows, "An Inuit education fund should be established in each of Canada, Greenland, Russia and the U.S.A. with public sector investments."[126]

Section 10 looks at "Promoting and Accommodating a Dynamic Inuit Culture" and roots such efforts in UNDRIP, which "acknowledges that indigenous peoples have the right to maintain, control, protect and develop their language, traditional knowledge and cultural heritage and expressions."[127] Noting that "Inuit culture is both well-rooted and dynamic," the declaration asserts that "Inuit are committed to ensuring that resource development projects must be planned and implemented in such a way as to support and enhance Inuit culture, rather than subvert or overwhelm it."[128] While "Inuit are committed to safe-guarding Inuit culture against excess adverse pressures and impacts"[129] of development, the ICC insists that "Governments, public bodies and private sector actors in Inuit Nunaat must share in these commitments."[130] Turning to these non-Inuit partners in resource development and northern co-management, the declaration closes with an invitation to continue their collaboration conditional on their respect for Inuit values: "We, the Inuit of Inuit Nunaat, are committed to the principles on resource development in Inuit Nunaat set out in this Declaration. Inuit invite—and are entitled to expect—all those who have or seek a role in the governance, management, development, or use of the resources of Inuit Nunaat to conduct themselves within the letter and spirit of this Declaration."[131]

Collaborative Versus Unitary Sovereignty: The Case for an Inuit Sovereign Restoration

Both ICC declarations are declarations of continuing collaboration, and though they are strongly voiced assertions of Inuit rights and values, they do not seek to topple the states system or erode state sovereignty but rather to strengthen the collaborative bond between the Inuit and the states that share their homeland. While insisting that states adhere to Inuit values and respect, they proffer continued cooperation and, thereby, continued political stability in the Arctic. Though critical of the Arctic 5's exclusion of the Inuit at Ilulissat, both declarations offer a map forward, restoring the spirit of Inuit inclusivity that gained traction during the age of arctic land claims amid the myriad co-management structures created (especially in Canada). The language of Inuit self-determination appears throughout, but it is contextualized by the historic collaboration with the state rather than in contradistinction to it.

This is very different from the declaration of Inuit nationhood asserted by Charles Etok Edwardsen Jr. in his many essays on this topic. His 1993 article, "Science and the Indigenous Arctic,"[132] is archived at the Fourth World Institute website. Curiously, the article includes Etok's phone number and address at the top; sadly, he passed away several years ago, otherwise it would be a delight to call him to discuss his views on Inuit sovereignty while writing this book. His article is, at heart, a rejection of Western science and its exclusion of traditional knowledge, and there has been much progress since the 1980s and 1990s, when Etok's ideas were articulated, so today a well-educated Indigenous community has its own scientists and researchers whose very participation science transforms it to better reflect Indigenous values, eroding the firewall that had hitherto existed between traditional knowledge and Western science. However, at the time Etok wrote, that firewall was still very much paramount. Etok thus writes, "We, the Inuit, cannot really know what a contemporary experiment means unless we understand what materials and what instruments and what sciences are involved in its design. This is why the growing edge of science is so inaccessible to our common experience in Alaska and our Circumpolar Inuit homeland."[133] But instead of continuing a discourse on Western science and its inaccessibility to the Inuit, he launches into one of his favorite and most discussed topics: Inuit nationhood.

As he declares, "The Inuit, of the Circumpolar Region, qualify as a nation state under international law. Therefore the Inuit of Canada, Denmark, United States, and Russia have met the criteria of Article I of the Montevideo Convention on the Rights and Duties of States. Our inherent rights to sovereignty as defined by longest peaceful existence have never been extinguished by the claims of discovery by the Spanish, Russians, British, Portuguese, Danes, Americans, nor Canadians."[134] (The language of the land claims treaties across Arctic North America contains clauses that do formally extinguish aboriginal title to the land in exchange for cash compensation, fee simple title to a portion of the lands, and a variety of governing and co-management structures that empower Indigenous people within the constitutional frameworks of their sovereign polities—but these clauses have been universally unpopular. The Inupiat, however, are the one native group in Alaska that voted against ANCSA, making them the one Inuit group in Arctic North America that rejected, among its many provisions, the extinguishment of aboriginal rights enshrined in ANCSA and all subsequent Arctic land claims.)

Etok roots Inuit nationhood in their

> tradition of the longest peaceful occupation, co-existence, territorial integrity and sovereignty of the Arctic since time immemorial. Based upon our self-determination and supported by International Law, we make this Declaration of Sovereignty which signifies Inuit Independence from all Anglo-European original or derivative states, and from any infringement of Inuit Sovereignty. And, therein lies the healing truth for the Inuit and for our homelands which have been threatened by quantitative science and selective enforcement of laws outside of the scope of the Constitutions of the United States, Canada, and Russia.

He thus calls on not only the restoration of an Inuit nation but on the accompanying threat posed by Western science—presumably in favor of a restoration of traditional knowledge as a guiding system for the Inuit nation.

Etok considers various legal precepts, including the implicit preservation of sovereignty of unrecognized regimes:

> Sovereign immunity for unrecognized regimes has been practiced by the United States Supreme Court and began with The Schooner

> Exchange vs. M'Faddon. In an opinion by Justice Marshall, the Supreme Court affirmed the dismissal of the libel because a warship "in the service of a foreign sovereign, with whom the government of the United States is at peace" should be exempt from U.S. jurisdiction. Marshall's analysis began with the premise that no sovereign would voluntarily subject itself to the jurisdiction of another. Marshall states that "all sovereigns have consented to a relaxation . . . of that absolute and complete jurisdiction within their respective territories which sovereignty confers."[135]

And so, his logic goes, the Inuit too retain their sovereignty, regardless of the will of the United States to ignore it. He continues,

> The act of state doctrine was adopted by the Supreme Court as a [principle] of judicial restraint to avoid the unseemliness and potential problems that might arise if the courts of one nation sat in judgment over a foreign sovereign. In the words of the *Restatement (Third) of Foreign Relations Law of the US*, "[i]n the absence of a treaty or other unambiguous agreement regarding controlling legal principles, courts in the United States will generally refrain from . . . sitting in judgment on . . . acts of a governmental character done by a foreign state within its own territory and applicable there."[136]

He thus looks back upon various efforts to abolish Inuit sovereignty without Inuit consent, such as with the Alaska purchase and later modernization of the Arctic maritime frontier shared with Russia, as illegal acts and, given their crushing effects on Inuit identity, as genocidal acts: "Most certainly the Treaty of Cession of 1867 and the Maritime Boundary Treaty of 1990 are flagrant examples of unconsented taking of indigenous lands in the Inuit Homeland. The United States government has been the sole benefactor of the largest illegal expropriation of Inuit Homelands."[137]

Later in his article, Etok criticizes the Alaska Land Claim, finding it to lack legitimacy:

> The State of Alaska does not exist in a vacuum of itself . . . [but] is subject to a compact, within the Statehood Act, between the United States and the indigenous Tribes of Alaska. The State of Alaska is not a sovereign of itself and, therefore, cannot be a signator[y] to any treaty by itself. This basic principle of federal law has been with us all this time. It is just that it is in the best interests of the State of Alaska to maintain this fraud. Read what the Bureau of Indian Affairs and the U.S. Fish and Wildlife Service published in 1976: "Furthermore, the Indian treaties were not a grant of rights to the Indians, but rather a

> grant of rights from them to the non-Indians, with the Indians reserving to themselves those rights not granted. The treaties specifically protect those reserved rights. These basic principles of Federal law, which undergird the decisions in Indian treaty rights cases, have been the subject of much misunderstanding and some non-Indians have found them difficult to accept."[138]

Etok further adds, "Someone may say that the Alaska Native Claims Settlement Act precludes treaty rights. This statement is based upon another fraudulent assumption, as it is important to realize that a treaty is not an act of Congress. The Alaska Native Claims Settlement Act as an act of congress was a forced incorporation of indigenous peoples, and can now be reexamined within the context of the United States Genocide Treaty Convention."[139] Etok believes America's commitments under the genocide convention plant a new, and welcome, seed for the eventual liberation of the Inuit from the many imposed and oppressive systems that, since the Alaska purchase, have illegally offset the unsurrendered sovereignty of the Inuit. Etok has a particular disagreement with one scientist, whom he believes dismissed Inupiat knowledge of bowhead stocks, thereby threatening Inupiat subsistence—not an uncommon phenomenon in the early history of co-management when scientists were still inexperienced in their efforts to incorporate traditional knowledge into their analyses and often quite hostile to practitioners of subsistence and their field expertise, a situation that has since changed dramatically in favor of Indigenous knowledge systems.

In closing, Etok argues that the extension of American sovereignty to the Arctic lacks consent from Alaska's Indigenous peoples and is thus unlawful and illegitimate:

> The legal basis of the United States within the territory of Alaska is not derived from within the United States Constitution. First, there was never a treaty nor consensual relationship between Czarist Russia and the indigenous tribes of Alaska. Secondly, the Treaty of Cessions of 1867 between Russia and the United States is not a transfer of sovereignty nor a secession of lands from the indigenous population. Therefore, the United States government is an occupational force within Alaska.[140]

Etok thus contends:

> This occupation can now be challenged under international law, particularly in that the United States and Russia cannot manufacture, between themselves, sovereignty which they have not acquired nor

> can ever acquire either on March 30, 1867 nor through the ratification of the Maritime Boundary Treaty of September 16, 1991. For the last 250 years in Alaska, Russia and the United States, have been outside of civilized international law and outside of their own constitutions. The United States and Russia have been the benefactors of this unconstitutional occupation and have enjoyed one trillion dollars of ill-gotten gains.[141]

Etok's contention of unextinguished Indigenous sovereignty cannot be entirely dismissed, as uncomfortable as it made more pragmatic Inupiat leaders like Eben Hopson, who worked within, and pushed the boundaries of, the existing constitutional system—maximizing Inupiat rights as much as possible. Etok's logic, if not his rhetoric, may in fact undergird the ICC's quite assertive response to their exclusion at Ilulissat and their concerted effort to restore the collaborative system that they have so diligently nurtured since the 1970s. Though the ICC is not (at least yet) propelled toward Etok's vision of formal, sovereign independence, it does seem willing to hint that it may reconsider its pragmatic approach should Arctic states not cooperate with continued Inuit pragmatism and commitment to mutual collaboration.

Embracing Arctic Collaboration: Benefits for All

The logic of Etok's view is compelling, not just for the Inupiat but for all the isolated Indigenous peoples of the Arctic for whom state expansion and the assertion of their sovereignty over them was never truly accompanied by the establishment of genuine Westphalian sovereignty. Indeed, lines on many maps changed, but until the age of Arctic land claims, there was no associated institutional development or modernization, with the exception perhaps of the quasi-sovereign trading posts of the HBC and RAC, which did not truly undermine traditional sovereignty but rather integrated these sovereignties into an alliance system that retained their overlapping multiplicity in parallel with the emergence of the new "company-state" of the chartered companies. Only in northern regions with significant influxes of settlers did new and competing systems of governance arise that truly reflected a new state sovereignty—but in the otherwise unsettled or lightly settled parts of the Arctic, Etok's contention of an unextinguished Indigenous sovereignty cannot be entirely dismissed, and may in fact undergird the ICC's quite muscular response to their exclusion at Ilulissat and their concerted effort to restore the collaborative system that they have so diligently nurtured since the 1970s.

It seems the Inuit have succeeded, since even as US Arctic policy stepped back from a universal commitment to Arctic collaboration in 2016 after a long run (since 1994), it has maintained its commitment to engage and cooperate with Indigenous peoples in the Arctic. A similar continuity is evident in Canadian policy and in Denmark's as well, though Russia, as always, has continued to assert a more conventional, centralized, and unitary Westphalian approach even as its Arctic policy increasingly embraces Indigenous issues. Inuit efforts to restore the collaborative path they have embraced since Eben Hopson and his generation of Inuit leaders pursued a pragmatic and balanced approach to sovereignty in the Arctic have largely found reciprocity among their Arctic states. Even Arctic states without Indigenous populations, like Iceland, strongly assert an enduring commitment to Indigenous inclusivity and collaboration, with the structure of the Arctic Council enshrining what holds as a consensus between the Indigenous peoples and the states of the Arctic on collaborative sovereignty and the continuing necessity for Indigenous participation and consent.

While Etok's vision of a restoration of Inuit nationhood has not caught on in mainland North America, it is very much evident in Greenland's incremental pursuit of independence through negotiation with the Danes, occasionally punctuated by referenda. And his logic remains intertwined with the pragmatic vision of his counterparts who favor a more collaborative approach, since in the event the Arctic states do step back from their commitment to collaboration with the Inuit (as Fenge has observed above), Etok's vision offers a powerful reminder of an alternative path, one that would challenge the long and mutual embrace of Arctic exceptionalism by Indigenous and state stakeholders for generations.

Greenlandic Independence and Arctic Exceptionalism: Toward a Westphalian Inuit Polity?

To the surprise of many during the summer of 2019, the global media and diplomatic community went into a frenzy after the *Wall Street Journal* published an article about thenPresident Trump's keen interest in purchasing Greenland from Denmark, generating worldwide headlines comparable to those that greeted Secretary of State William H. Seward when word leaked out of his 1867 secret treaty with Russia to purchase its ailing Alaskan colony, a move widely ridiculed as "Seward's Folly" (but which proved to be enormously prescient). News of Trump's

Sewardian interest in Greenland generated an immediate critical reaction in both Greenland, where a movement for increased autonomy and a gradual, incremental evolution toward sovereign independence has had majority support for many years, and Denmark, leading to a brief display of diplomatic sparks between Denmark and the United States. As Greenland's foreign minister Ane Lone Bagger put it, "We are open for business, but we're not for sale." Prudently, Greenland's leaders, while vehemently opposed to the idea floated by Trump, nonetheless embraced the immediate (and sustained) rise in attention his proposal elicited, and in the months that followed enjoyed multiple benefits associated with rekindled US interest in the world's largest island, including fast-tracking the reopening of a US consulate in Nuuk for the first time since 1953.

At 2.13 million square kilometers, Greenland is equal in size to the combined areas of the world's next three largest islands: New Guinea (785,753 square km), Borneo (748,168 square km), and Madagascar (587,041 square km). It occupies a strategic location along the northeastern flank of North America similar to Alaska's position in the far northwest, with a comparable geostrategic importance for hemispheric security, recognized during World War II, again in the Cold War, and now once again as the polar thaw invites increased global interest in the Arctic. While Greenland has long been a colony of Denmark, its formal governing status has evolved in recent years from outright colonial governance to more collaborative home-rule governance in 1979, to increasingly robust and meaningful self-rule in 2009, in the wake of its 2008 referendum on autonomy and independence that garnered overwhelming support of Greenland's electorate (75.54 percent), with a path toward peaceful secession mutually endorsed by both colony and colonizer.

Roots of Greenlandic Autonomy: The Circumpolar Inuit Rights Movement

A key driver of this movement for increasing autonomy has been the steady empowerment of the island's majority Inuit population—part of a wider, circumpolar movement for Inuit rights spearheaded by the ICC. This movement includes, and for many has been defined by, securing the protection of Inuit land rights through various mechanisms, such as the land claims process in Alaska and the Canadian Arctic, which has formally transferred land of approximately one-tenth the land in Arctic North America to the Inuit along with a variety of co-management tools to protect those lands and its recourses (in contrast to the Russian Arctic, where in the absence of gaining land title there has instead been a more limited use of joint venture economic development projects, occasionally augmented by the creation of national parks). In addition to regaining and formalizing land rights, the Inuit rights movements have sought, and successfully strengthened, the preservation and revitalization of Inuit culture and language, along with the increasing empowerment of Inuit through greater self-governing powers, with notable achievements in both Alaska and the Canadian Arctic in addition to Greenland.

The movement for autonomy in Greenland, and the collaborative path toward eventual independence with the support of Denmark, is both part of this circumpolar movement and distinct from it, as noted by Hannes Gerhardt in 2011. It takes inspiration from the pioneering gains of the Arctic's evolving experience with Inuit land claims, starting with ANCSA in 1971, which jump-started the process of Inuit-state reconciliation that continued through subsequent revisions until 2005. But at the same time, the constitutional and historical context for the Inuit rights movement in Greenland is markedly different from on the mainland of Arctic North America, with a colonial system largely in place albeit with much increased autonomy. It is this asymmetry of constitutional and historical contexts that has propelled Greenland on what Gerhardt describes as a Westphalian trajectory, toward the emergence of what would be the first truly sovereign majority Inuit state.[142]

ANCSA was the first land claim to transform the political geography of the North, and while it had many structural flaws and imperfections, it laid a foundation from which Arctic land claims continued to evolve. Each new iteration provided the Inuit with greater powers, increasingly augmenting Inuit self-government, first through the integration of co-management with a land settlement and later—in the 2005 Labrador Inuit (Nunatsiavut) Land Claim—embracing ethnically defined Inuit self-government.[143] Outside of Greenland, these iterations have been constitutionally subordinated to the sovereign states governing the Arctic region, with Inuit autonomy defined either municipally, regionally, or territorially but always subordinated to constitutional supremacy of the sovereign state itself. But in the case of Greenland, there is for the first time a process in which sovereign independence is a distinct possibility, as mutually recognized by both Greenlanders and the Danish state.

How this movement toward independence evolves and the diverse constitutional forms the emergent sovereign Greenlandic state may potentially take have generally not been discussed in great detail in the academic literature or press, apart from being the logical conclusion to the incremental approach to expanded Greenlandic autonomy that has taken place so far. Though limited to the island of Greenland itself, this does not mean that Greenland's independence will remain confined by its present geography, or that over time we won't see other manifestations of Inuit sovereignty and configurations of Inuit states. Because the future of Greenland and the ultimate extent of Inuit sovereignty asserted are of such great importance to the stability of the high North Atlantic region and to North American security, it is vital that we consider all possible models and outcomes.

Climate Consequences: Energizing the Inuit Rights Movement

The dramatic and volatile effects of the polar thaw are adding urgency to the contemporary circumpolar Inuit rights movement, bringing global attention to a region that has been largely neglected since the Cold War. What was once something of a niche field understudied by IR and strategic studies scholars has become a topic of global attention from the lay public to the highest levels of government, so much so that numerous non-Arctic states have their own Arctic policies, and non-Arctic observer states now outnumber the Arctic member states on the Arctic Council. Growing global interest in the Arctic brings along new diplomatic challenges. Most recently China's assertions of a special "near-Arctic" status aligned with its Polar Silk Road initiative was noted in the 2019 US Coast Guard Arctic Strategic Outlook, along with other US policy papers and strategy documents, as particularly concerning.

The results of these competing interests in the near as well as more distant future are exceedingly difficult to predict, so considering a wide range of scenarios is essential. For instance, a determined China could develop regional alliances and dependencies through strategic capital infusions to the sovereign island states of the high North Atlantic, which owing to their small populations remain vulnerable to rapid demographic upheavals from a small number of development projects staffed by overseas contractors—resulting in a potential stealth invasion of the region. Iceland, with a population around 364,000 and a long sovereign experience, is far more resilient to such an external demographic threat, though were Iceland to break from NATO and pursue a nonaligned future, its vulnerability could increase. Greenland, while part of the

Kingdom of Denmark, likewise remains embedded in a solid alliance architecture, but with only 56,000 people could, once it becomes independent, become highly vulnerable to external pressures, whether economic, demographic, or even military.

With Russia's renewed appetite for foreign intervention (following its annexation of Crimea, incursion in eastern Ukraine, and military interventions from Libya to Syria culminating in its invasion of Ukraine in 2022) and the renewed specter of a clash with NATO over the small Baltic states, or potentially the nonaligned northern European states, scenarios of extreme instability in the high North Atlantic can also be envisioned. In short, global interest in the Arctic introduces new risks that could in time threaten the hemispheric security of North America. This follows a long period that with few exceptions was marked by steadiness and predictability; the pre-thaw Arctic region was more a strategic buffer in world politics that—even at the height of the Cold War—was defined foremost by its stability. The movement for Inuit rights emerged during this period of calm, and the incrementally increasing empowerment of the Inuit proceeded with the same stability during a half century of policy innovation that began with ANCSA, culminated with the formation of the Nunavut Territory in 1999, and achieved its conclusion on the mainland of Arctic North America with the passage of the Labrador Inuit (Nunatsiavut) Land Claim in 2005.

Were the strategic and geographic landscape to remain stable and largely free of uncertainty, the emergence of a sovereign and independent Greenland would likely continue the current mutually amicable autonomy process and its multidecade incrementalism and not upend regional geopolitics. But because of the dynamic uncertainties of the polar thaw and the return of Westphalian state competition to the Arctic region in recent years, the potential independence of Greenland becomes instead a strategic wildcard requiring close study and proactive engagement to ensure a future sovereign Greenland maintains the close, collaborative, and friendly relationship with the United States and the West, optimally as part of NATO, that it currently enjoys as a constituent component of the Kingdom of Denmark.[144]

Alternate Models for a Post-Denmark Greenland

While most conversations regarding Greenland's constitutional evolution, at least those prior to the Trump White House's surprise overture to purchase the island from Denmark, consider Greenland's continuing movement for increasing autonomy strictly within the context of peaceful, and negotiated, Danish constitutional politics, there is good reason to consider alternative scenarios not widely discussed—including scenarios of a broader Inuit secession from the states that now assert constitutional authority over Inuit lands.

The Inuit homeland has, in contrast to the more complex and diverse sub-Arctic region, a cohesive ethnopolitical culture, and while it has until now deftly adapted its aspirations for the many different sovereign polities whose borders intersect the Inuit homeland, it is conceivable that as the prospect of formal independence grows for Greenlandic Inuit, the appetite for enhanced autonomy elsewhere, such as in neighboring Nunavut, northern Labrador (Nunatsiavut), Canada's Western Arctic (the Inuvialuit settlement region), and Alaska's North Slope, may grow in lockstep. It is imperative to thus ask: What if Nunavut and/or other Inuit regions join Greenland in a coordinated secession movement? Or, in the face of successful lobbying by either party or a groundswell of support at home, what if Greenland decides to change sovereigns and join either Canada or the United States? While these scenarios may seem unlikely, prior to 1989 one did not hear comparable discussions about a collapse of the Warsaw Pact, breakup of Yugoslavia, or implosion of the Soviet empire before events rapidly spiraled out of control, radically transforming the constitutional fabric of Eastern Europe. With the prospect of an Inuit state amicably emerging from the negotiated constitutional dialogue between Greenland and Denmark, the implications of this profound and catalyzing transformation must not be overlooked. As the famed RAND nuclear strategist Herman Kahn reminded us during the Cold War, it is now again essential to think about the unthinkable. Below are some of what Kahn might dub relevant "metaphors and scenarios" to consider.

Joining Canada and Its Constitutional Protection of Indigenous Rights

Were Greenland to change sovereigns rather than seek formal independence, there is some logic to the notion of selecting Canada as its new constitutional partner. The Nunavut model of territorial self-governance aligned with a land claim treaty has many potential benefits well-suited to its needs and would be a refreshing break from its current colonial governing structure. Moreover, residents of northern Greenland are relatively new arrivals from Canada, one of the last stages of the modern Inuit migration across the top of the world, and tied to Canada's Eastern Arctic by language and kinship. Joining Canada in modern times is not without precedent: The province of Newfoundland and Labrador did so as late as 1949, around the time the United States was contemplating its own acquisition of Greenland. It brought under the Canadian constitution not only the island of Newfoundland but the northern coastal territory of Labrador, home to several thousand Inuit who serendipitously are among the most recent beneficiaries of a comprehensive land claim treaty with Ottawa, presenting a logical path for Greenland to follow. Imagining how such a scenario could unfold will require much further study and would depend, in part, on the emergence of a transnational movement for Inuit unification that as of now has not taken root in either Greenland or Nunavut. Should such a movement arise, and for it to succeed, Denmark would have to agree to Greenland's departure and to Canada's expansion (as would Canada), unless it were to occur in two separate steps sufficiently paced to forestall Denmark's opposition—a scenario that may seem extremely unlikely today but that, in a situation of war in northern Europe, might become more feasible.

Indeed, before President Trump's unsolicited overture, Canada seemed the more likely alternative sovereign partner for Greenland, given the advances achieved by Canadian Inuit through their dynamic mixture of comprehensive land claims, robust co-management, and increasingly powerful self-government processes as well as the long, close collaboration between Greenlandic and Canadian Inuit at the ICC. In recent years, Inuit leaders have expressed much dissatisfaction with Ottawa's commitment to fulfilling the promises of Nunavut, turning to the courts in frustration on multiple occasions and in 2006 calling for the moral intervention of famed jurist Thomas Berger, to many Canadian natives the conscience of Canada. More recently, native rail blockages across Canada have reinflamed long-simmering tensions between the Indigenous and settler communities in Canada, straining reconciliation efforts of the Trudeau government, a situation that will likely be even further inflamed by the recent discovery of a mass grave at a residential school in Kamloops, British Columbia. Amid such circumstances, one can no longer presume that Canada, by default, would be the only logical choice for Greenlandic Inuit should they seek to change sovereigns.

Joining the United States: Not Necessarily as Illogical or Ahistorical as Many Think

Indeed, Trump's unexpected interest in Greenland—despite the initially critical response to his surprise overture to purchase the island—provides an alternative to joining Canada or remaining part of Denmark, with Alaska's transition from territory to state serving as a model for Greenland's future. Such a possibility, of Greenland joining the United States in a constitutional union not unlike that of Alaska, has been considered at the highest levels of the US government before, particularly in the immediate aftermath of World War II, during which Greenland was a US protectorate and its strategic significance in the coming Cold War was keenly appreciated. Such a tectonic shift in North America's sovereign political geography is uncommon now; it has been a century and a half since a change of similar magnitude directly affected the United States, when Alaska was purchased from Russia in 1867. But it's more frequent, and recent, than many would think, such as with the aforementioned entry of Newfoundland and Labrador to Canada's confederation.

And while political leaders in both Greenland and Copenhagen quickly insisted that Trump's idea was without merit, and inconsistent with their own step-by-step process of decolonization underway, Greenlandic officials did subsequently embrace renewed US diplomatic and economic interest in what had largely been an overlooked Cold War outpost rediscovered amid the dynamic flux and strategic uncertainty of the polar thaw. The consequent and significant rise in high-profile official delegations to and from Greenland included an accelerated reopening of a US consulate in Nuuk, the first since 1953. In addition, high-profile visits by leading members of the US strategic and diplomatic community culminated in the May 14, 2021, visit by Secretary of State Tony Blinken—notably via Copenhagen, and with Denmark's official blessing—who was greeted upon arrival on the island by Denmark's foreign minister and both Greenland's premier and foreign minister, in contrast to the unilateral nature of the Trump initiative.[145]

Supporters of President Trump's Greenland purchase initiative, few as they were, noted the increasing strategic importance of Greenland in a thawing Arctic. They also understood it was a backdoor approach to recognizing the strategic implications of climate change in an administration that did not formally or publicly acknowledge global warming. The economic integration of Arctic resources with the global economy is not only a contemporary phenomenon accelerated by climate change but a historic one dating back centuries. During the gold rush era, the mineral potential of the Far North was equally recognized for its strategic economic value, leading to a brief demographic imbalance in the Yukon Territory, with more Americans in the Klondike than Canadians and Ottawa rightly concerned there could be instability and potentially a secession risk. In the twentieth century, with the advent of air power, the strategic military value of the region was recognized for its own sake, while its energy and mineral resources continued to be highly sought after by all of the Arctic states, driving a new wave of northern resource development.

During World War II, thousands of US soldiers were stationed across the North, building the strategic Alaska-Canada (Alcan) Highway, the lesser-known Canol Road, and a slew of air bases, as well as protecting the vital Northwest Staging Route ferrying lend-lease aircraft to the eastern front, where the Nazi military onslaught was brought to a halt, at great sacrifice by then US ally Russia. At the same time, the US Coast Guard defended Greenland, which came under direct US military protection after Denmark fell to the Germans, from becoming a North American beachhead for further Nazi advances—indeed, the specter of Greenland falling and the Gulf of Saint Lawrence becoming vulnerable to Nazi conquest concerned war planners until the machinery of the Nazi state was decisively demolished. It wasn't long after World War II came to an end that President Truman floated the idea of purchasing Greenland from Denmark for $100 million, an idea that *Time* magazine endorsed for its strategic wisdom the very next year, with widespread encouragement from war planners who recognized Greenland's strategic prominence for the postwar world.

During the Cold War, the Distant Early Warning (DEW) Line was built across the Inuit homeland, protecting the North American continent from strategic bomber assault. The Thule air base augmented US air power in the region, and a US consulate in Nuuk helped to reinforce the American presence. The waters off both Greenland and Iceland, the famed Greenland-Iceland-UK (G-I-UK) gap, came to play a central role during the US-Soviet confrontation and in US forward maritime strategy near the Cold War's end. Soon after the Cold War ended, columnist and foreign affairs expert Walter Russell Mead proposed in all seriousness, in a July 1992 column in the *LA Times*, to purchase Siberia from Boris Yeltsin's Russia for $1–$2 trillion.

While the outright purchase of such a large portion of the globe is now uncommon, and due to this relative infrequency is widely perceived to be better suited to the world of yesteryear, it wasn't all that long ago that large-scale shifts in borders were more the norm and less the exception. And in today's world so much is in flux—let's not forget that it was less than a decade ago that Crimea quickly fell to Moscow's expansionist ambitions, experiencing a rapid annexation by a resurgent Russia[146] and changing hands largely without bloodshed (in contrast to the subsequent contest for eastern, and later all of, Ukraine), suggesting that such tectonic shifts in political geography do remain possible, and in some cases might even contribute to geopolitical stability. It is thus conceivable that in a future world, Greenland's union with the United States could again be envisioned.

It is no secret that Russian President Vladimir Putin has been fortifying his vast Arctic territories. Mothballed military bases unused since the Cold War are undergoing strategic refurbishment on a scale comparable to Beijing's fortifications of contested islands in the South China Sea, gaining increasing strategic attention and media coverage as both Russia and China seek to expand their global military influence and to secure their most proximate island chains.[147] There is every reason to expect the United States to do much the same, by strengthening its own strategic architecture and refortifying its own proximate island chains along its ramparts, from the Aleutian Islands guarding the Bering Strait, to the Canadian Arctic archipelago standing watch over North America's northern flank, to the islands that anchor the G-I-UK gap. Rekindled US interest in Greenlandic security and the return of US forces to Iceland (now on a rotating basis) reflect this strategic reawakening, as does its long-delayed icebreaker modernization program, the Polar Security Cutter Program. Thinking ahead to a more fluid geopolitical world, and toward the protection of the more isolated island outposts in the Far North that remain vulnerable in such a world, is a prudent exercise given the perceived risks associated with both increasing state rivalry in the Arctic and Arctic climate change.

Exchanging Colonial Sovereigns: An Alternative to Independence

To imagine a hypothetical world in which Greenland might become a US territory, on a path toward statehood not unlike that which Alaska followed, is to imagine a world in which North America is more secure and united than it is today. Greenlandic Inuit, who suffer from a long legacy of neglect and whose colonial experience, despite recent gains in autonomy, has not been entirely positive, particularly in the smaller and more remote villages lacking basic infrastructure and economic opportunities, could indeed stand to benefit in multiple ways.

First and foremost, the defense of Greenland in time of war would be strengthened by its constitutional integration into the US polity, much the way Alaska's has been since its purchase, and this alone could deter war from ever taking place. That the legacy of Russian colonialism, which under the RAC was brutal and exploitative, could be gradually reversed in Alaska over time—where particularly since the ANCSA of 1971 and ANILCA of 1980 gains by Alaska natives have been notable, even if still a work in progress—is illustrative of the changes we can expect in such a hypothetical future.

While Greenland's gradual process of increasing autonomy and decolonization under Danish rule would be upended by such a change of sovereigns, a US purchase could be a win-win for both Greenlanders and the United States. There would have to be consent to such an exchange of sovereigns, to be sure, and that would require a referendum to be held. But, once the people of Greenland make their choice, if they choose to join the United States, it would be a historically transformative event on the scale of the Alaska purchase and with the same long-term potential strategic and economic benefits. It may not be likely, and at the moment it is not under consideration by the United States, the Danes, or the Greenlanders. Indeed, it may well have been just a brief flirtation and not a long-term strategic commitment of the president's, and thus different from, say, Secretary of State William H. Seward's embrace of Alaska as a critical component of US expansion. But it is no less a scenario worthy of study.

Because the globalized world is vastly different from the colonial world, however, it would require additional parameters to be met than required for the Alaska purchase. The people of Greenland, who are majority Inuit, would have to also welcome a change in sovereigns and find US policies and investment, in addition to its military protection, attractive enough to forgo independence. Because securing and sustaining independence with a population of 56,000 is quite a difficult challenge, a formal sovereign association with the United States might, in some circumstances, appeal to the people of Greenland. Just as the earlier mentioned scenario of Greenland joining Canada would require Greenlanders to find merit in the Nunavut model of regional self-governance, for Greenland to become part of the United States, Greenlanders would have to find merit in the Alaska model, with its combination of multilevel governance, settled land claims, still-evolving structures for co-management between the various levels of governance (federal, state, tribal, and municipal), and robust natural resource development experience.

Indeed, were Greenlanders to one day consider joining the United States or Canada in lieu of pursuing independence, their choice may be between resource development (more robust in Alaska, with its strong state government) versus cultural and environmental protections (more robust in Arctic Canada, owing to the absence of a powerful state level of government, more readily enabling an alignment of Indigenous and federal interests), the very same fault line that divides contemporary Alaska. Issues of future settlement by non-Inuit would also surely be an issue: Alaska today is less than 15 percent native (and over 85 percent settler), while Nunavut remains nearly 85 percent native (and less than 15 percent settler), as if an inverse mirror image of one another. Greenland is, at present, closer to 90 percent native, so it is more aligned with Nunavut in terms of its ethnography. On the other hand, Canada, by virtue of its proximity to the United States, is largely buffered from the risks and dangers of world politics and unlikely to face an external threat to its sovereignty. Greenland, however, is quite exposed, on the outer edge of the North American continent, and in waters that are not only increasingly strategic but also potentially contested. It faces existential risks as a polity much the way Iceland does, perhaps more so because of its smaller demography and more expansive geography. If alliance membership alone does not guarantee its independence, perhaps a closer constitutional relationship, such as territorial status and a path toward statehood, could one day prove to be an appealing option to the people of Greenland.

Dual-Secession Models: Strength Through Confederation

Often, when Greenlandic independence is considered critically, its small population and vast territorial breadth (and its even more vast exclusive economic zone offshore) are presumed to leave the island in an inevitably unviable position, particularly should its independence ever be contested by force—suggesting there is at least some strategic and economic logic to a change of sovereigns rather than a bid for fragile sovereign independence. A mere change of sovereigns may fail to find domestic support among Greenlanders; even so, going it alone need not be the only path left to pursue. Indeed, Greenland may in the face of so many obstacles to sovereign viability choose to remain part of the Kingdom of Denmark by mutual consent, and thus decelerate its march toward independence and continue to refine its relationship, and expand its autonomy over time, with the same incrementalism as in the past.

Alternatively, it is also possible that the movement for independence in Greenland could inspire a similar movement in Canada—particularly Nunavut, which has close historical, cultural, and familial ties to Greenland, with its north settled by Canadian Inuit just over a century ago. In addition, its close geographical proximity to Greenland, where Inuit already govern at the territorial level but face continuing implementation resistance from Ottawa, results in the theoretical potential for a transnational movement for dual secession from both Canada and Denmark. While Canadian Inuit have gained much in native rights and self-governing powers, more than in any other Arctic region, their communities remain traumatized, with multiple indicators of societal collapse including epidemic levels of suicide, substance abuse, and community violence. Should Ottawa remain unwilling or unable to fulfill the aspirations of Inuit for meaningful autonomy and the many material comforts enjoyed by southern Canadians (ranging from health security and community safety to educational opportunity), one can envision a movement emerging in Nunavut that aligns with Greenland's independence movement, where seeds for dual secession may find fertile soil.

Dual secession would be complex to manage, posing even greater risks to the United States and its NATO partners' security, comparable in scale to the Quebec independence movement farther south, which over the years has threatened to destabilize the northeast of North America, coming perilously close to a Quebec secession in 1995. While much progress has been made toward healing the enduring French-English divide in Canada, it is not inconceivable that the movement could reintensify, and if it did, the fate of Nunavut to its north and Nunatsiavut to its east could be profoundly affected—either by overt Quebec military expansion to secure these vulnerable, majority Inuit flanks, or covert efforts to destabilize these regions to prevent them from serving as staging grounds from which Anglo Canada could destabilize Quebec's energy-rich north. Further, in the event Quebec one day secedes from Canada, Inuit to Quebec's north could seize the opportunity to make a similar bid for independence—and in so doing, may find many reasons to partner with Greenland in its quest. Such a scenario, while clearly improbable at the moment, remains worthy of future study, since the role Nunavut and Nunatsiavut might play in a post-Quebec Canadian confederation, and their likelihood of being caught in the whirlwind of chaos precipitated by a Quebec secession, could be of much consequence.

Joining Iceland and/or the Faroes: Dual Secession with a Twist— a Confederation of Former High North Atlantic Colonies

A similar dual-secession movement might emerge to Greenland's east, where a separate movement for a gradual and mutual secession from Denmark has emerged in the Faroe Islands, modeled in part on the successful independence movement by Iceland a century ago. If the movements for Faroese and Greenlandic independence were to merge into a single North Atlantic independence movement, one can envision the potential confederation of the Faroes, Greenland, and Iceland—a union of former Danish colonies that share a colonial heritage and many postcolonial synergies. This scenario would create, in essence, a greater Icelandic sovereign entity, leveraging Iceland's diplomatic, economic, and political strengths along with its central strategic role as a strategic hub between Europe and North America within the NATO alliance. The demographic diversity of such a union, united by its shared heritage of remote subsistence marine resource harvesting communities, whether seal, whale, or cod, would lay the foundation for a fascinating polity enriched by its own regional variation and distinctiveness.

In contrast to a dual Nunavut-Greenland secession, with all its risks and complexities, such an island confederation of former Danish colonies could offer the region much promise of stability and cohesion, since the two new secessions would be equally embraced by both island colonies and their mutual colonizing sovereign, and their constitutional union with Iceland would further bring solace to their mutual alliance partners concerned with the emergence of new exclusive economic zones in newly independent microstates lacking the self-defense, search and rescue, or monitoring capacity for enforcement. Reykjavik could serve as a mentor to both Greenland and the Faroes, offering independence through confederation, much the way the union of the Malay peninsula with most of northern Borneo did after World War II—harnessing the movement for independence and away from colonialism while avoiding the specter of balkanization and instability. Originally, Singapore, the economic engine of that region, was set to be part of the new Malaysian state, along with Sarawak and Sabah, and had Singapore stayed in Malaysia, its role would likely mirror that which Reykjavik could play in an Iceland–Greenland–Faroe Islands confederation—as an engine of economic growth, a model for effective, efficient, and democratic self-governance, and a center for education and training.

While a constitutional union with the United States was, from the moment the Trump White House proposed it, highly controversial and widely lampooned, a confederation of former North Atlantic colonies seems less likely to face as much criticism, even if not actively under consideration now. Iceland, independent now for over seventy-five years and autonomous for more than a century, can thus serve as a mentor to its neighbors as they follow down the path that it earlier blazed, with the full blessing of their mutual former colonizer and a close relationship with Copenhagen after independence. Iceland's sovereign independence is a fixture of the Arctic's political geography that contributes not only to the region's collaborative balance between East and West but also to its dynamic balancing of the interests of small and large states, as well as between hard power and soft power approaches to international relations and regional security.

So while each movement for independence in the high North Atlantic is generally viewed as sui generis, viewing them in tandem and in sequence raises an interesting possibility of confederation after independence. All three, once sovereign, will emerge on the world stage as relatively vulnerable microstates with populations that would be hard pressed to secure independence in wartime on their own, but despite their asymmetries in scale, their strategic geography, infrastructure, and resource potential position them as important future alliance partners. With a shared security challenge and a common history of colonization, and with strong cultural, maritime, trade, and diplomatic ties, there could be numerous benefits to forging a common union, one highly decentralized to ensure maximal achievement of sovereign aspirations but not necessarily independent of one another. Just as there is a logic in Greenland uniting with the Inuit of Canada's eastern and high Arctic, one can as logically imagine a North Atlantic union. Whether the will for such a union ever emerges or could overcome centrifugal forces pulling the three insular polities apart remains to be seen, but the theoretical potential is no less intriguing.

Village Sovereignty, Multivillage Secession, and the Return of the Sovereign "Polis"?

There is one other secession model to consider, one more complex than the models of dual secession contemplated above, and that is a secessionary cascade in the more isolated and remote villages that share a strong tradition of independence and survival against adversity, harsh climate, and limitations in resources and their accessibility. Whether that means cutting ties with both Ottawa and Iqaluit on one side of Baffin Bay, and with Copenhagen and Nuuk on the other, or just breaking free of Nuuk and Copenhagen, the result is the same: a Balkanization process in favor of localized forms of microsovereignty. One can even envision a restoration of a security relationship between pro-Copenhagen remote communities and Denmark even as they sever their ties to what some perceive as a neocolonial governing structure in Nuuk, no better than the more distant sovereign Nuuk aspires to replace.

Such yearnings have been felt across the Arctic and sub-Arctic for generations and appear from time to time in different places, such as a 1992 "study" the Inupiat leadership told the Inuit Circumpolar Conference (as it was then called) they would sponsor to investigate their potential to secede from Alaska and form not an independent sovereign polity but a "51st state" in the US union, a research endeavor that catalyzed a formal response by the state of Alaska to forestall. After the Alaska land claim was enacted in 1971, standing up a new system of native corporations and in time a new generation of corporate leaders, a tribal sovereignty movement erupted across much of rural Alaska in search of a return to village autonomy and federal protection from not only state-level governance but from the assimilating pressures of modernization and globalization. Alaska's constitution allows for the formation of municipal boroughs for Alaska's distinctive natural regions, such as on the North Slope where the Inupiat reside, providing a path for municipal governance to embody the sovereign aspirations of a unified region—but in 1992 that seemed far too limited against a state government in Alaska that opposed greater protections of Inupiat subsistence rights. Across the border, efforts to construct a Western Arctic Regional Municipality to jointly govern Inuvialuit and Gwich'in communities shared a similar aspiration, albeit one that ultimately fell short of implementation despite coming close more than once, for a more localized version of sovereignty and autonomy.

Thomas Berger, when heading up the Alaska Native Review Commission in the mid-1980s, called his series of hearings a "village journey," which ultimately called for a retribalization of Alaska lands, away from corporations and corporate values, toward more traditional governance. Not long after, in 1993, the US Department of the Interior recognized the more than 200 Alaska native villages as tribes under federal law. It is possible that the people of Greenland, along with their counterparts in the Canadian and Alaskan Arctic, may choose to pursue a similar model, one that would lead to a new conversation about what sovereignty means and how it can be nurtured at the local level. The fact that the city-state of Singapore, known affectionately as the "little red dot" on maps of peninsular Southeast Asia, has achieved such a vision, building a modern state with its own sovereign form at the municipal level, is proof there is no limit to what can be achieved within the city gates. In the Arctic region each village is in its own way a country of its own, so why not assert a new sovereign form that embodies the strength of the village as a unit?

Seizing the Moment: Asserting a Stronger Commitment to, and Presence in, the Arctic

Given the widespread attention and curiosity that accompanied the critical response to the Trump White House's Greenland initiative, even in the absence of forward movement on the plan, the White House's renewed and continuing interest in the Arctic and its increasing commitment to engagement and forward presence in the region have nonetheless been positively reinforced in the many months since. This surely has not escaped the attention of principal US rivals Beijing and Moscow, or of its friends in Greenland and across the lightly settled and strategically vulnerable Atlantic Arctic.

Leveraging this moment by extending the US Arctic presence through greater economic, diplomatic, and military engagement with the region and its people can achieve many of the very same benefits of an outright sovereign accession of Greenland but without either its risk or controversy. As Greenland considers its many options in its transformation from colony to increasingly sovereign polity, continued engagement and support of the people of Greenland, no matter what sovereign model they choose, will go far to ensure the Arctic remains secure and free. Indeed, these many potential constitutional configurations illustrate potential trajectories along which Greenland may evolve on its journey of decolonization toward sovereign restoration, in many forms—from union with either Canada or the United States, to confederation between Kalaallit Nunaat (the homeland of Greenland Inuit) and Inuit Nunangat (the homeland of Canadian Inuit), to a splintering of Kalaallit Nunaat into smaller, more localized polities, to the formal reemergence of Kalaallit Nunaat as a sovereign state. These paths to achieving Greenland's sovereign aspiration are thus manifold, though the most likely sovereign form to emerge in the Inuit homeland is an independent Kalaallit Nunaat with its present geography, achieved through a peaceful, negotiated secession from the Kingdom of Denmark with continued NATO membership and close bilateral ties to the United States, Canada, and Iceland, along with new economic ties to trading partners in Asia and Europe, and close diplomatic ties to the host of multinational and intergovernmental fora to which it presently belongs as part of the Kingdom of Denmark.

Indeed, Greenland is already asserting a more autonomous diplomatic and strategic voice in world politics. Greenland has been strengthening its diplomatic presence on the world stage, with overseas diplomatic representations now in four countries, starting with the United States in 2014, Iceland in 2018, China in 2021, and as part of the Permanent Representation of Denmark to NATO (DANATO) in Belgium in 2023. Greenland has plans for diplomatic representation in Canada in the near future, along with plans to take the lead for the Kingdom of Denmark at the Arctic Council. The United States, Iceland, China, NATO, Canada, and the Arctic Council all represent key stakeholders for Greenland, with overlapping alignments of their own. The United States, Iceland, and Canada are fellow Arctic Council members and fellow NATO members, reinforcing their shared western alignment, while China is an observer to the Arctic Council and is already an important economic partner for Greenland as it seeks new markets in quest of economic self-sufficiency. Relationships with all will be essential to Greenland's future when it becomes an independent state. That Greenland is laying such a solid diplomatic foundation now will go far to ensure future success upon its sovereign restoration. In addition to expanding its diplomatic footprint, presently as part of the Kingdom of Denmark (and, importantly, the Arctic component of the kingdom), Greenland has recently presented its very first Arctic strategy, with a comprehensive and nuanced vision for a peaceful and inclusive Arctic with a decidedly western strategic tilt and a globalized economic vision. Just as the ICC creatively asserted its own distinct vision of collaborative Arctic sovereignty and resource development over a decade ago, thus aligning Inuit and state interests in a mutually beneficial collaborative relationship, Greenland is now doing much the same, reinforcing its ambition to achieve sovereign restoration in the Westphalian world by leveraging its unique position at the crossroads of the Indigenous and Westphalian worlds.

In February 2024, Greenland released its new Arctic strategy, *Greenland in the World: Nothing About Us Without Us*. It articulates a vision of Greenland as a cooperative if largely Western-aligned member of the Arctic North American region, playing an increasingly important role not only in achieving its full economic potential but also in matters of state and defense, reminiscent of the ICC's earlier declarations on sovereignty and resource development, which did not declare outright Inuit sovereignty but instead articulated a similar vision of collaborative sovereignty as an essential partner on all matters of strategic importance to the Arctic, captured so well by Greenland's subtitle.[148]

The strategy's thirteen sections include the following eleven topics (in addition to its introduction and closing remarks), which would all be important factors in Greenland's sovereign restoration: the Arctic Council; climate and the ocean; the United States; Iceland; Canada; international trade; connectivity; a newly proposed Arctic North American forum; East Asia; multilateral cooperation and international political agendas (including the Arctic, ICC, Nordic cooperation, United Nations, European Union, other organizations); and security and defense policy. As I've argued in these pages and elsewhere over the years, the spirit of co-management that has come to define the Indigenous interface with Arctic states manifests itself throughout Greenland's Arctic strategy, wherein Greenland's proposed collaboration with multiple states as partners is outlined in detail, and its catchy subtitle of "nothing about us without us" could likewise be a nutshell definition for co-management as it evolves toward what Shadian has dubbed a post-Westphalian sovereignty. As Greenland's Arctic strategy eloquently describes: "History has shown us that it is necessary to have a seat at the table if Greenland is to ensure that its voice is heard. Greenland aims to play an active role in the world's main multilateral bodies wherever possible and, more importantly, whenever it is relevant. This is an inherent consequence of *Nothing about us without us*."[149]

As the strategy states in section 1, "Greenland's foreign, security and defense strategy is an expression of our vision and desire for more collaboration with the world around us—not just because this is the course that we intend to pursue, but because it is necessary. Greenland is seeking greater international cooperation and trade because our development and progress require it."[150] It further elaborates:

> The strategy aims to lay down the overall framework and guidelines for Greenland's relations and policies with other countries to safeguard the security and defend the interests of Greenland and the Greenlandic people, to define our objectives, and to foster cooperation for the benefit of our country and our partners. Our foreign policy must support and strengthen our foreign trade and bolster our progress towards achieving a self-sustaining economy to enhance our self-determination, defend our interests, and increase our independence as stipulated in the Act on Greenland Self-Government and under international law.[151]

On the independent spirit that infuses its Arctic strategy, the introduction further states: "With the right to self-determination and the goal of independence, our country and people aim to increase their cooperation with other countries. It is important for us as responsible citizens of the world, in our own name, to have the courage to take a stand on issues and events around the world."[152]

Among the "shared values that underpin Greenland's approach to relations with other countries" are its commitment that "democracy and human rights are at the core of all relations"; "Greenland and the Arctic is an area of low tension"; "improving the lives and livelihoods of the Greenlandic people is of key importance"; "all relations are based on the premise that Greenland and the Greenlandic people constitute an independent people and nation"; "all relations must be solution oriented"; and "we stand in solidarity and work hand in hand with other Arctic communities."[153] Moreover, the strategy affirms that "our country is aware that we are part of the Kingdom of Denmark, but we also strive for independence. We want to make a difference in the world, and we have something to contribute to the world." And within the Kingdom of Denmark, it is Greenland's aim that "all member countries must cooperate with respect for their differences and with equal rights and conditions."[154] Within the Indigenous world, the strategy affirms that "Inuit in Greenland will continue to serve as a role model for other indigenous peoples, and we can champion their rights."[155]

As the strategy further explains in section 13:

> *Greenland in the World—Nothing About Us Without Us* is an ambitious strategy that demonstrates that Greenland is insisting on a seat at the table. Greenland has something to contribute, and it is important that Greenland's voice is heard and that we safeguard our interests and articulate our values. At the same time, this is a strategy that sets out our objectives for developing ties with our neighboring countries, partners and the international community.[156]

While an Arctic strategy, it presents not only an Arctic vision but a global one: "In addition to pursuing our collaborations with countries farther south, as an Arctic nation we also look to the East and the West, including our closest neighbors, and intend to develop our North-to North ties as a source of inspiration and collaborative opportunities."[157] Thus, Greenland plans to

> utilize its foreign policy and diverse collaborative relationships to convey to the outside world our culture and our values, and to promote peace and democracy, international law and the right to self-determination, all the while providing an understanding of the conditions in the Arctic and the unique and valuable interplay between the peoples of the region. Our climate is changing, and the ice is getting thinner.

> We have proud peoples in the Arctic who continue to live and adapt—and we seek greater cooperation on today's terms. We will not allow ourselves to be hemmed in by outdated mindsets and structures, which give us no control over the course of developments. Nothing about us without us.[158]

Indeed, the strategy is a bold declaration of decolonized thinking about Greenland's place in the world, one where it is increasingly independent when it comes to matters of state and statecraft. A peaceful and cooperative Arctic remains the goal, but with Greenland's increased engagement and agency as a NATO ally and partner. This is all to occur while fostering the formation of a new Arctic North America Forum to unite the predominantly indigenous regions of Arctic North America (Alaska, Yukon, NWT, Nunavut, Nunavik, and Greenland); while pursuing continued cooperation at the international and regional level through the UN, EU, Arctic Council, Nordic cooperation councils (Nordic Council of Ministers, Nordic Council, and West Nordic Council), and Inuit Circumpolar Council (ICC); while maintaining close bilateral ties with neighboring allied states, foremost of which is the United States for its military power and prominence in NATO, but also Canada and Iceland; and while nurturing its evolving relationship with China and other East Asian trading states. This reveals a complex, balanced, pragmatic, thoughtful, and independent approach to statecraft, illustrating Greenland's maturation as an emergent sovereign polity, ready not only for its seat at the table but its place among the world's family of nations.

Notes

1. Such as Wilson, Alcantara, and Rodon, *Nested Federalism and Inuit Governance in the Canadian Arctic*, illustrating the domestic; and illustrating the international, Shibata, Zou, Sellheim, and Scopelliti, eds., *Emerging Legal Orders in the Arctic*; and Takahashi, ed., *The Influence of Sub-state Actors on National Security Using Military Bases to Forge Autonomy*, which includes Greenland as a case study (the editor's specialty) as well as additional Pacific basin case studies including Okinawa that extend this literature from the Arctic to the Pacific and beyond.

2. See Armitage, ed., *Adaptive Co-management*, as a prime example, in addition to the more recent edited volume, Parlee and Caine, eds., *When the Caribou Do Not Come*. For a glimpse of co-management's recent emergence as a governing paradigm in the Russian Arctic, see Parlato, Fondahl, Filippova, and Savvinova, "The Evolution of Forming 'Territories of Traditional Nature Use' in the Sakha Republic (Iakutiia)," 1–27. Though there is still reason for some caution and skepticism regarding the potential for co-management to transform governance in the Russian

Arctic to the extent seen in Arctic North America, as explained by Lauren Kaljur, "Why Russia's Indigenous People Are Wary of National Parks," *The New Humanitarian*, May 9, 2017.

3. Among the most recent of this crop of publications: Burke, *Diplomacy and the Arctic Council*; Nord, *Changing Arctic*; English, *Ice and Water*; in addition to Keil and Knecht, eds., *Governing Arctic Change*.

4. For a discussion of this structural duality of IPOs, see Barry Scott Zellen, "A 'Fourth Image' at the Interface of Tribe and State: Tribal Homelands, Expanding States, and the Foundation of International Order from the Western Arctic to Borneo," presented at the 14th International Borneo Research Council Conference, August 6, 2018. Additionally, the theoretical contours of the "fourth image" are described in my earlier work, "Rethinking the Subcomponents of World Order," and "Clan, the State, and War," 20–21.

5. As studied in Zellen, *Breaking the Ice*.

6. See the following works from Jessica Shadian, "Reconceptualizing Sovereignty Through Indigenous Autonomy: A Case Study of Arctic Governance and the Inuit Circumpolar Conference" (PhD diss., Department of Political Science and International Relations, University of Delaware, 2006); "Remaking Arctic Governance: The Construction of an Arctic Inuit Polity,": 249–259; "From States to Polities: Reconceptualizing Sovereignty Through Inuit Governance," 1–26.

7. Rodon and Therrien, "Resource Development and Land Claims Settlement," 119–131.

8. Ibid.

9. Nuttall, "Aboriginal Participation," 617–634.

10. Notzke, "New Perspective," 187–209.

11. Armitage, Berkes, Dale, Kocho-Schellenberg, and Patton, "Co-management and the Co-production of Knowledge," 995–996; also see Dale and Armitage, "Marine Mammal Co-management," 440.

12. Nuttall, "Aboriginal Participation."

13. Armitage et al., "Co-management and the Co-production of Knowledge."

14. Claudia Notzke, "A New Perspective."

15. Armitage, Berkes, and Doubleday, eds., *Adaptive Co-Management*.

16. Penikett, "Unfinished Journey?," 1127–1156.

17. Shadian, "Reconceptualizing Sovereignty"; Shadian, "Remaking Arctic Governance"; Shadian, "From States to Polities."

18. Gerhardt, "Inuit and Sovereignty," 6–14.

19. Fenge, "Inuit and the Nunavut Land Claims Agreement."

20. Njord Wegge, "The Political Order in the Arctic: Power Structures, Regimes and Influence," *Polar Record* 47, no. 2 (2011): 165–176.

21. Shadian, "Remaking Arctic Governance"; Shadian, "Reconceptualizing Sovereignty."

22. Shadian, "Remaking Arctic Governance," 255; Shadian, "Reconceptualizing Sovereignty," 388.

23. Shadian, "Remaking Arctic Governance," 257.

24. Thierry Rodon, "Inuit Diplomacy: Reframing the Arctic Spaces and Narratives," in *The Internationalization of Indigenous Rights: UNDRIP in the Canadian Context* (Centre for International Governance Innovation, 2014), 18.

25. Gerhardt, "Inuit and Sovereignty."

26. Fenge, "Inuit and the Nunavut Land Claims Agreement."

27. Application of "shared space" to co-management from Snook, Cunsolo, and Dale, "Conceptual Representation of the 'Shared Space' of Co-management," 52–56.

28. I first discussed the comparison of the colonial-era partitioning of Borneo and the Arctic in Zellen, "'Fourth Image.'"
29. Shadian, "Remaking Arctic Governance."
30. Rodon, "Inuit Diplomacy."
31. Wilson, Alcantara, and Rodon, "Inuit Diplomacy in the Circumpolar North."
32. Gerhardt, "Inuit and Sovereignty."
33. Snook, Cunsolo, and Dale, "Conceptual Representation of the 'Shared Space' of Co-management."
34. See Erin Hanson, University of British Columbia, Indigenous Foundations web page, https://indigenousfoundations.arts.ubc.ca/un_declaration_on_the_rights_of_indigenous_peoples.
35. I first discussed this in Barry Scott Zellen, "Cold Front: Hillary, Ottawa, and the Inuit: A Year After the Inuit Re-assert Their Sovereignty, Washington Takes Their Side," *Journal of Military and Strategic Studies* 12, no. 3 (Spring 2010).
36. Andrew Clark, "Mervyn King Goes Dog Sledding but All Avoid Seal Meat at G7 Summit in Canada," *The Guardian*, February 8, 2010. Also see Andrew Clark, "Finance Ministers Promised a Meal of Seal at G7 Summit in Iqaluit," *The Guardian*, February 3, 2010.
37. Inuit Circumpolar Council (ICC), A Circumpolar Inuit Declaration on Arctic Sovereignty, 2009.
38. Ibid., sec. 1.3.
39. Ibid., sec. 1.4.
40. Ibid.
41. Ibid.
42. Ibid., sec. 1.5.
43. Ibid.
44. Ibid., sec. 1.6.
45. Ibid., sec. 1.7.
46. Ibid.
47. Ibid., sec. 1.8.
48. Ibid., sec. 2.1.
49. Ibid., sec. 2.
50. Ibid., sec. 2.1.
51. Ibid., sec. 2.1.
52. Ibid., sec 2.2.
53. Ibid., sec. 2.3.
54. Ibid.
55. Ibid., sec. 2.5.
56. Ibid., sec. 2.6.
57. Ibid., sec. 3.
58. Ibid., sec. 3.1.
59. Ibid., sec. 3.2.
60. Ibid., sec. 3.3.
61. Ibid., sec. 3.4.
62. Ibid., sec. 3.5.
63. Ibid., sec. 3.6.
64. Ibid., sec. 3.7.
65. Ibid., sec. 3.8.
66. Ibid.
67. Ibid., sec. 4.1.
68. Ibid., sec. 4.2.

69. Ibid., sec. 4.3.
70. Ibid.
71. ICC, Circumpolar Inuit Declaration on Resource Development, 2011.
72. Ibid.
73. Ibid.
74. Ibid.
75. Ibid.
76. Ibid.
77. Ibid.
78. Ibid., sec. 1.
79. Ibid., sec. 1.1.
80. Ibid., sec. 1.2.
81. Ibid., sec. 1.3.
82. Ibid., sec. 1.4.
83. Ibid., sec. 2.1–2.2.
84. Ibid., sec. 2.3.
85. Ibid., sec. 2.4.
86. Ibid., sec. 3.2.
87. Ibid., sec. 4.1.
88. Ibid., sec. 4.2–4.3.
89. Ibid., sec. 4.4.
90. Ibid., sec. 4.5.
91. Ibid.
92. Ibid., sec. 5.1.
93. Ibid., sec. 5.2.
94. Ibid., sec. 5.3.
95. Ibid., sec. 5.4–5.5.
96. Ibid., sec. 6.1.
97. Ibid., sec. 6.2.
98. Ibid., sec. 6.3.
99. Ibid., sec. 6.4.
100. Ibid., sec. 6.5.
101. Ibid., sec. 7.1.
102. Ibid., sec. 7.2.
103. Ibid.
104. Ibid.
105. Ibid., sec. 7.4.
106. Ibid.
107. Ibid., sec. 7.5.
108. Ibid., sec. 7.6.
109. Ibid., sec. 8.1.
110. Ibid.
111. Ibid., sec. 8.2.
112. Ibid., sec. 8.3.
113. Ibid., sec. 8.4.
114. Ibid., sec. 8.5.
115. Ibid., sec. 8.6.
116. Ibid., sec. 8.7.
117. Ibid., sec. 8.8–8.9.
118. Ibid., sec. 8.10.
119. Ibid.

120. Ibid., sec. 9.1.
121. Ibid., sec. 9.2.
122. Ibid., sec. 9.3.
123. Ibid., sec. 9.4.
124. Ibid., sec. 9.5.
125. Ibid., sec. 9.6–9.7.
126. Ibid., sec. 9.7.
127. Ibid., sec. 10.1.
128. Ibid., sec. 10.2.
129. Ibid., sec. 10.3.
130. Ibid., sec. 10.4.
131. Ibid.
132. Charles Etok Edwardsen Jr., "Science and the Indigenous Arctic: Legal Basis for Inuit Sovereignty in the Circumpolar Region," Fourth World Documentation Project, June 15, 1993.
133. Ibid.
134. Ibid.
135. Ibid.
136. Ibid.
137. Ibid.
138. Ibid.
139. Ibid.
140. Ibid.
141. Ibid.
142. Gerhardt, "Inuit and Sovereignty."
143. See Zellen, *Breaking the Ice*, and Zellen, *On Thin Ice*, for more details.
144. For an overview of Arctic geopolitics in the post–Cold War era, including the era of the polar thaw, see Zellen, *Arctic Doom, Arctic Boom*; for more on the evolution of the land claims model and its contribution to Arctic stability and security, see Zellen, "Stability and Security in a Post-Arctic World." For a discussion of the complex diplomatic environment of the contemporary Arctic, through the lens of the US-China strategic competition, see Zellen, "China and the 'Near-Arctic,'" *Stars and Stripes*, August 8, 2019; Zellen, "China and the 'Near-Arctic': An Opportunity Lost Over 150 Years Ago," *Georgetown Journal of International Affairs*, September 5, 2019; and Zellen, "Missed Opportunity," 26–28. For a discussion of Inuit involvement in Arctic diplomacy, please see Zellen, "Cold Front: Hillary, Ottawa, and the Inuit."
145. Ned Price, "Press Statement: Secretary Blinken's Travel to Denmark, Iceland, and Greenland," US Department of State, May 14, 2021.
146. It may also be noted that Moscow's Crimean annexation took place only sixty years after the Soviet Union had transferred control over Crimea from Russia to Ukraine in 1954, albeit within the overarching Soviet constitutional framework, so not a formal transfer of sovereignty. But Ukraine's independence from the Soviet Union in 1992 would cement the loss of Crimea from Russian sovereign dominion, planting the seed for the armed conflict that eventually erupted in 2014 and intensified in 2022. Mark Kramer, "Why Did Russia Give Away Crimea Sixty Years Ago?" *Hindsight Up Front | Ukraine*, Cold War International History Project (CWIHP) e-Dossier No. 47, *WilsonCenter.org*, https://www.wilsoncenter.org/publication/why-did-russia-give-away-crimea-sixty-years-ago.
147. For instance, see Andrew E. Kramer, "In the Russian Arctic, the First Stirrings of a Very Cold War," *New York Times*, May 22, 2021.

148. Government of Greenland Ministry for Statehood and Foreign Affairs, *Greenland in the World: Nothing about us without us—Greenland's Foreign, Security and Defense Policy 2024-2033—an Arctic Strategy* (Nuuk, Greenland: Naalakkersuisut, February 21, 2024).
149. Ibid., Section 11: Multilateral Cooperation and International Political Agendas, 37.
150. Ibid., Section 1: Introduction, 7.
151. Ibid., Section 1: Introduction, 7.
152. Ibid., Section 1: Introduction, 7.
153. Ibid., Section 1: Introduction, 7.
154. Ibid., Section 1: Introduction, 7.
155. Ibid., Section 1: Introduction, 7.
156. Ibid., Section 13: Closing Remarks, 47.
157. Ibid., Section 1: Introduction, 8.
158. Ibid., Section 1: Introduction, 8.

5

The Rise and Collapse of Consensus

Central to the post–Cold War consensus on Arctic cooperation between Arctic states and between those states and the Indigenous polities of the region has been the Arctic Council, which turned twenty-five in 2021. By 2022, just as its second quarter century got underway, it faced its first deep existential crisis, catalyzed by the global suspension of collaboration with Russia, the largest of the Arctic states, after it invaded Ukraine.

Balancing Tribe and State: The Ottawa Declaration and the Birth of the Arctic Council

The Ottawa Declaration, formally called the Declaration on the Establishment of the Arctic Council, was promulgated by the eight Arctic states—Canada, Denmark, Finland, Iceland, Norway, the Russian Federation, Sweden, and the United States—in 1996. In it they affirmed their "commitment to the well-being of the inhabitants of the Arctic, including recognition of the special relationship and unique contributions to the Arctic of indigenous people and their communities,"[1] the "sustainable development in the Arctic region, including economic and social development, improved health conditions and cultural wellbeing," and the "protection of the Arctic environment, including the health of Arctic ecosystems, maintenance or biodiversity in the Arctic region and conservation and sustainable use of natural resources," while recognizing "the contributions of the Arctic Environmental Protection Strategy to these commitments," "the traditional knowledge of the indigenous people of the Arctic and their communities and taking note of its importance and that of Arctic science and research to the collective understanding of the circumpolar Arctic," and "the valuable contribution and support of the Inuit Circumpolar Conference, Saami Council, and the Association of the Indigenous Minorities of the North, Siberia, and the Far East of the Russian Federation in the development of the Arctic Council."[2] The Arctic states also expressed their mutual desire "to provide a means for promoting cooperative activities to address Arctic issues requiring circumpolar cooperation, and to ensure full consultation with and the full involvement of indigenous people and their communities and other inhabitants of the Arctic in such activities," and "to provide for regular intergovernmental consideration of and consultation on Arctic issues."[3]

They thereby declared the establishment of the Arctic Council as a

> high level forum to: (a) provide a means for promoting cooperation, coordination and interaction among the Arctic States, with the involvement of the Arctic indigenous communities and other Arctic inhabitants on common Arctic issues, in particular issues of sustainable development and environmental protection in the Arctic; (b) oversee and coordinate the programs established under the AEPS on the Arctic Monitoring and Assessment Program (AMAP); Conservation of Arctic Flora and Fauna (CAFF); Protection of the Arctic Marine Environment (PAME); and Emergency Prevention, Preparedness and Response (EPPR); (c) adopt terms of reference for, and oversee and coordinate a sustainable development program; [and] (d) disseminate information, encourage education and promote interest in Arctic related issues.[4]

In addition to the member states, it also designated in paragraph 2 that the "Inuit Circumpolar Conference, the Saami Council and the Association of Indigenous Minorities of the North, Siberia and the Far East of the Russian Federation are Permanent Participants in the Arctic Council" and that "permanent participation equally is open to other Arctic organizations of indigenous peoples with majority Arctic indigenous constituency, representing: (a) a single indigenous people resident in more than one Arctic State; or (b) more than one Arctic indigenous people resident in a single Arctic state."[5] This inclusive definition made possible the variation in ethnic composition of the Permanent Participant groups as well as the variation in scale, from representing as few as several thousand constituents from a single tribal group to over a quarter million constituents from over forty tribal groups, with Permanent Participants ranging in order of magnitude from four digits to six. Once the council determines "that such an organization has met this criterion," the Ottawa Declaration caps the total number of Permanent Participant groups to "at any time . . . less than the number of members," and since the Arctic states number eight, there can be no more than seven Permanent Participants, one more than the present six.[6] While Permanent Participants lack the agency of member states to implement Arctic policies, which remain the prerogative of states, many Permanent Participants or components thereof have achieved formal governing powers within their home states owing to an impressive variety of institutional powers, whether constitutional, legislative, or co-managerial. Indeed, at the local and regional levels, their powers often are represented by a corresponding level of governance. This further fulfills the aspiration of the Ottawa Declaration that the "category of Permanent Participation is created to provide for active participation and full consultation with the Arctic indigenous representatives within the Arctic Council."[7] In addition to the Arctic states and the Permanent Participants, the Ottawa Declaration also recognized three categories of observers that include "non-Arctic states; inter-governmental and inter-parliamentary organizations, global and regional; and non-governmental organizations that the Council determines can contribute to its work."[8] In short, the council was forged to reflect the very "mosaic of cooperation" described by pioneering Arctic scholar and noted regime theorist Oran R. Young[9]—a mosaic that at once captures the unifying spirit of multilateral and multilevel cooperation in the modern Arctic, as reflected in the policies of not only the Indigenous organizations operating at both sub- and transnational levels (illustrated by our discussion of Inuit policy above) but also the Arctic states and the non-Arctic states with Arctic interests and ambitions. We will consider those below, starting with an in-depth retrospective examination of America's long embrace of a multilevel, multilateral Arctic.

America's Long Embrace of a Multilevel, Multilateral Arctic

An examination of US Arctic policy from the Clinton to the Biden administrations shows a largely unbroken chain of official documents committing the United States to collaborative, multilateral, and multilevel Arctic engagement in partnership with Arctic Indigenous peoples and states—initially including Russia, with which the United States first aligned its interests in the immediate post–Cold War era, an alignment that endured for three decades (1991–2021). The Clinton administration's Presidential Decision Directive/NSC-26 on Arctic policy issued on June 9, 1994, was reaffirmed in the George W. Bush administration's Arctic policy update, National Security Presidential Directive (NSPD-66), on January 9, 2009, and again in the Obama administration's National Strategy for the Arctic Region issued on May 10, 2013 (NSAR 2013), and the Biden Administration's NSAR update in October 2022.[10] This enduring partisanship experienced only a handful of setbacks, most of which—such as after Russia's annexation of Crimea or President Trump's overture to purchase Greenland and his administration's departure from the Arctic Council's otherwise united consensus regarding climate change—were brief in duration; the present collapse in consensus is the first to be protracted. Indeed, over the two decades spanning the collapse of the Soviet Union and reemergence of a sovereign Russian state, the two decades of the global war on terror and well into the acceleration of a polar thaw, regardless of ruling political party, US Arctic strategy remained at heart collaborative and multilateral.

In the years that have followed the release of NSAR 2013, global interest in the Arctic has greatly intensified. In response the Arctic Council opened its doors to a wide range of non-Arctic states as official observers, with many opportunities to participate in Arctic Council business, particularly at the working group level, where distinctions between members and observers is less salient. The increasing complexity of world politics and of Arctic international relations in particular, the recent assertion of interests in the Arctic by non-Arctic states, and the strategic reappraisal of Russian military ambitions that followed its Crimea annexation have precipitated a reassessment of US strategic policy with regard to the Arctic region and the range of inclusivity of Arctic exceptionalism, as viewed through the lens of alliance solidarity and hemispheric security interests.

This reassessment was first articulated in the December 2016 Report to Congress on Strategy to Protect United States National Security Interests in the Arctic Region, and reaffirmed by the updated United States Coast Guard Arctic Strategic Outlook, released in April 2019, in which dual concerns with both Russia's and China's Arctic ambitions contrast with the collaborative spirit that infused preceding strategy documents. Importantly, Arctic collaboration and cooperation is still a guiding principle—but it is framed in prudence more by alliance cooperation and joint military partnerships than by the presumed universality and inevitability of Arctic cooperation.

The cooperative foundation undergirding Arctic exceptionalism for more than a generation has thus been shaken, but so far it has withstood these new competitive pressures. Enduring structural changes to Arctic governance, particularly the rise of co-management as a paradigm for joint governance between natives and the state and the gradual internationalization of co-management from a governing principle to a diplomatic principle, have much to do with the durability of Arctic exceptionalism amid the recent rise in Westphalian state competition across the Arctic region.

More recent US policy updates reflect a growing perception in government circles of the reemergence and reintensification of state competition in the Arctic, a region hitherto perceived to be a sea of calm at the top of the world. These tensions leapt forth from the pages of policy and strategy documents to cause diplomatic fireworks at the May 2019 Arctic Council ministerial meeting in Rovaniemi. But even there, public comments from the US secretary of state on the return of Westphalian state competition to the Arctic, made outside the council chambers both for the benefit of his domestic audience and for news headlines to convey the newly reassessed US position, did not prevent the secretary's continued collaboration with fellow council members behind closed doors during the very same council session, nor did it disrupt ongoing collaboration at the working group level, though it did prevent the issuance of the traditional joint consensus statement at the meeting's end—perhaps an affirmation of the importance of consensus to the council rather than its rejection.

The ministerial did not conclude with a consensus statement due to new disagreements within the council on the causal origins and global nature of the polar thaw, which had previously enjoyed consensus on the council. This collapse in the climate consensus, with roots outside the Arctic firmly planted in a US policy reversal, was the culprit, and not a collapse in the council's mutual embrace of and commitment to consensus governance. Indeed, widely quoted comments by the US secretary of state rejecting China's claim of a special "near-Arctic" status and calling for a transparent, rules-based order in the Arctic have been warmly (if more discreetly) embraced by several other council members.

Moreover, internal divisions within the council membership, such as those between the Arctic 5 (the five states littoral to the Arctic basin—Russia, the United States, Canada, Denmark/Greenland, and Norway) and the three other council member states (Iceland, Sweden, and Finland) that flared up a decade ago at Ilulissat and the potential rift between Russia and the other seven council member states after its Crimea annexation are potentially more divisive than disagreements over climate change—and thus a far graver threat to the council's collaborative dynamic. And yet, the council's collaborative dynamic has held. Indeed, the Arctic was the one front during the Cold War that remained free of armed conflict. It was recognized not only by the Inuit as a natural zone of peace but also by the last premier of the Soviet Union for this very same virtue—a view that was later embraced by the authors of US Arctic policy after the Soviet collapse and that defined and continued to guide US policy until quite recently when Arctic cooperation was reimagined through an alliance-centric lens.

Arctic sovereignty, as we see in Chapter 4 above, has become de facto a collaborative venture between Indigenous peoples and the state. This collaboration and the many systems of co-management that have emerged as a result secure the Arctic region and ensure its stability, undergirding what has become known as Arctic exceptionalism. This is as evident in US Arctic policy as it is in Canadian policy. While the latter has gained wide recognition for its whole-of-government approach to reconciliation between Indigenous peoples and non-Indigenous peoples and its embrace of an inherent and constitutional right to aboriginal self-governance, the former has a long, albeit complicated, history of tribal recognition that has improved over time as policies of assimilation lost favor to policies of inclusion. It was, after all, Alaska that pioneered the comprehensive land claim model that was later embraced and, in many ways, enhanced by Canada, breaking cleanly from the longer tradition of the reservation model that took root in both countries and laying the collaborative foundation stone that transformed and modernized Arctic North America's political economy.

Canada's Arctic and Northern Policy Framework: Setting a New Gold Standard for Indigenous Inclusion

From its initial announcement in 2017 to its formal release in September 2019, Canadian Arctic policy has been redefined through the lens of co-management, yielding the Arctic's most collaborative and inclusive Arctic policy approach. Ottawa's Arctic and Northern Policy Framework (ANPF) establishes what former Inuit Circumpolar Council chair,

Dr. Dalee Sambo Dorough, describes as a "new gold standard" for inclusive Arctic policy.[11] As Dorough explains,

> Canada's September 10th release of their Arctic and northern policy framework is an extraordinary and unprecedented expression of co-development of policy accompanied by what appears to be a promise of Co implementation in favor of the inherent rights holders of the region, the Inuit. Indeed, the Government of Canada has raised the bar for all other states. I regard this Arctic policy as the new gold standard because of its clear recognition of Inuit rights.

In a foreword introducing the new framework, Canada's minister of Crown-Indigenous Relations and Northern Affairs explained,

> The Arctic and Northern Policy Framework is a profound change of direction for the Government of Canada. For too long, Canada's Arctic and northern residents, especially Indigenous people, have not had access to the same services, opportunities, and standards of living as those enjoyed by other Canadians. There are longstanding inequalities in transportation, energy, communications, employment, community infrastructure, health and education. While almost all past governments have put forward northern strategies, none closed these gaps for the people of the North, or created a lasting legacy of sustainable economic development. . . . Co-developing the new framework became a bold opportunity to shape and direct change in the region by collaborating with governments, northerners and Indigenous governments and organizations. Consultation was not enough to meet the challenges and harness emerging opportunities in the Arctic and North. In a significant shift, the federal government, Indigenous peoples, Inuit, First Nations and Métis, 6 territorial and provincial governments (Yukon, Northwest Territories, Nunavut, Newfoundland and Labrador, Quebec, and Manitoba) contributed to this framework together.[12]

The ANPF is so infused with Indigenous participation that "indigenous" is mentioned 187 times in the policy's text. This marks an important evolution to co-management in national Arctic policy, as it will now be developed through a multilevel consultative process with the peoples of the Arctic region. Peter Kikkert and P. Whitney Lackenbauer describe the framework's arrival:

> After four years of development, the document appeared on the Crown-Indigenous Relations and Northern Affairs website. It included no photos, maps, or even a downloadable pdf—just a wave of words,

> over 17,000 in the main chapter alone. The single infographic that accompanied the framework's release captured its main "highlights": that a "whole-of-government, co-development" process that created the framework involved the three territorial governments, over 25 Indigenous organizations, as well as three provincial governments. This collaborative process represents the "profound change of direction" that the Government of Canada highlights in the opening sentence of the ANPF.[13]

Ottawa's "emphasis on collaborative governance recognizes that when Ottawa has defined problems facing the North incorrectly or has set the wrong priorities, with little consultation from Northerners, policy responses have been shortsighted and ineffective."[14] Despite the generally positive reception of the framework's codevelopment process, Kikkert and Lackenbauer point out that it "appeared with no budget, timelines, or clear plan to address the wide array of challenges and issues identified. Critics have labelled the framework a 'half-baked' and 'chaotic mess' that simply lists well-known issues and gives 'lip service to addressing the problems,' while providing no 'concrete' plan for action."[15] Indeed, the framework "highlights many well-known issues that Northerners have identified for years, including climate change impacts, food insecurity, poverty, health inequalities, and housing shortages," and this is "useful in reinforcing common understandings of these problems with those most affected, reiterating the importance of these issues to the general Canadian public, and setting priorities for federal policy."[16] The framework's "expressed objective, however, is to provide a 'roadmap' to achieve the 'shared vision' co-developed by the groups involved in the process. If this is a roadmap, it is one with few clear directions—a map that identifies hazards, problems, and opportunities, but does little to illuminate how the federal government will work practically with its partners to navigate the complex terrain around myriad Arctic policy priorities and seemingly intractable political dilemmas."[17] Moreover, "Canada's 'new' Arctic and Northern Policy Framework offers few substantive approaches or mechanisms to meet them"; the "absence of a coherent strategy embedded in the ANPF speaks to the complicated process of co-developing policies across a wide and disparate region. Many different voices need to be taken into account."[18]

Among these voices is that of the Inuit, and "throughout the co-development of the ANPF," Canadian Prime Minister Trudeau "emphasized its 'Inuit-specific component, created in partnership with Inuit' that would take Inuit Nunangat as the primary lens through which to view policies focused on Inuit."[19] But as with the other voices, whose priorities vary, there remain many questions left unanswered: "How will Ottawa operationalize its focus on Inuit Nunangat moving forward? Will it support the re-drawing of Canada's political boundaries if self-determining Inuit decide that this is what they want? How will Inuit Nunangat, as a political jurisdiction, interact with the current roles and responsibilities of public territorial and provincial governments?"[20] While much remains unknown, the framework's ambition sets it apart from the Arctic policies of other states, as Dorough noted above. For comparative purposes, if we examine the frequency and context for the use of the term "Indigenous" (as well as "tribal" and "native") in the Arctic policies and strategies of the United States, we find their presence to be persistent but at a much lower order of magnitude, indicative perhaps of a less inclusive Arctic policy development process in the United States as compared to this "new gold standard" set by Canada in its current approach to Arctic policy.

For instance, in the 2019 US Coast Guard Arctic Strategic Outlook "indigenous" is mentioned twice, [21] "tribal" seven times,[22] and "native" seventeen times,[23] even though the Coast Guard is arguably the US service with the closest working relationship to the Indigenous community, and with a native liaison officer serving as a cultural bridge between the two communities in its Anchorage district office. Further, when we turn to the July 2020 US Air Force Arctic Strategy we find less terminological presence of these words: "indigenous" is mentioned three times, while "tribal" and "native" aren't mentioned at all.[24] While these US policy mentions are only a fraction of those in Canada's ANPF—a far cry from what Canada's Indigenous peoples have achieved—their presence indicates at least an awareness of and commitment to Indigenous peoples and their participation in policy implementation, if not its formulation. Though still at vastly different scales, one can find evidence of a convergence when it comes to Indigenous inclusion in Arctic policies of both the United States and Canada.

Interestingly, we also see Indigenous participation prominently featured in the Arctic policies of the United States' strategic rivals, including its current Arctic opponent, Russia. Russia's July 2020 Arctic Policy Foundations to 2035 has fifteen mentions of "indigenous."[25] Additionally, "native" is mentioned twice.[26] This compares favorably with China's own 2018 Arctic White Paper in which "indigenous" is mentioned seven times, while "tribal" and "native" aren't mentioned at all.[27] Intriguingly, we find the exact same number of mentions in Japan's 2015 Arctic Policy, where "indigenous" is mentioned seven times, and "tribal" and "native" are not mentioned at all.[28]

Just as Canada's Arctic and Northern Policy Framework elevates the principles and collaborative dynamics of co-management from the local and regional levels to policy formulation at the national level, the Arctic Council has since its formation internationalized co-management principles as a result of its inclusion of the Permanent Participants. They represent the interests of Arctic Indigenous peoples, bringing not just these terms but awareness of and commitment to their inclusion to the entire circumpolar family of nations as well as non-Arctic observer states with increasingly active Arctic policies, embedding this inclusivity into the numerous international agreements that have emerged under the auspices of the Arctic Council.

Thus it is no coincidence that we see such a convergence on Indigenous inclusion and collaboration in the Arctic policies of not only the council member states but the observer states as well. While Arctic governance beyond the North American continent can vary greatly in form and structure from the North American paradigm of collaborative co-management, there has nonetheless been a convergence on the matter of Indigenous participation and inclusion, at least at the rhetorical level—resulting in increasing Indigenous participation, albeit with great variation in the actual empowerment of and quality of participation by Indigenous peoples.

Arctic Exceptionalism Navigates Intensifying State Rivalry

Despite Russia's military, diplomatic, and economic resurgence, China's continued rise and aspiration to become a more prominent Arctic influencer, and US recognition in recent years of the growing challenge presented by a Russia-China alignment in the Arctic—creating what University of Calgary political scientist Rob Huebert eloquently described as a "New Arctic Strategic Triangle Environment" (NASTE),[29] discussed further below—the depth of commitment to Arctic collaboration by Arctic states and Indigenous peoples (until Russia invaded Ukraine in 2022) had not only not been meaningfully undermined or threatened by a renewal of Westphalian state competition but appeared to be gaining momentum and crossing new borders into states with a far less demonstrated track record of Indigenous collaboration. And while the setbacks to Arctic cooperation after Russia invaded Ukraine led many to conclude Arctic exceptionalism was dead, it remains far too early to come to such a definitive conclusion.

One reason for optimism is that state rivalry, even amid the widely recognized condition of Arctic exceptionalism, never truly went away:

it was there before Russia's annexation of Crimea galvanized a recommitment to the protection of national interests and sovereignty in the Arctic, and it was there after the Trump administration established new policies rejecting the otherwise universal consensus on climate change and emphasizing US interests over and above transnational collaborative interests, and one can hope this remains so despite the unprecedented isolation of Moscow since it invaded Ukraine in 2022. Indeed, conflicts between the Arctic states have instead been successfully counterbalanced by mutual collaborative values with deep historical roots reaching back long before the Arctic Council was formed.

That is why the United States and Canada embraced mutual collaboration on multiple fronts even while disagreeing, as friends and neighbors, on issues including their joint boundary in the petroleum-rich Beaufort Sea and the legal status of the Northwest Passage—neither a trivial matter. And Norway and Russia likewise have found many reasons to cooperate, even with one firmly embedded in the NATO alliance and the other NATO's principal adversary and increasingly perceived as a reemergent threat to the West. Within the council there have been many tensions before, including, as noted above, those between the five littoral Arctic states with shores along the Arctic basin and the three other Arctic states, as seen at Ilulissat in 2008. A mutual commitment to collaboration does not simply erase the existence of competing national interests; it just manages this competition within a structure defined by mutual commitment to collaboration rather than conflict—a commitment reflected in many founding Arctic Council members' historic embrace of co-management, which likewise managed the competing interests of tribe and state within a domestic collaborative mechanism.

As explained at the start of this chapter, while the council's member states and Indigenous Permanent Participants reflect the original parties to the co-management process that took root in Arctic North America after land claims were settled, recent years have witnessed a proliferation in the number of observer states from well outside the Arctic that now participate in council meetings and on its working groups. Even though those new voices don't have the organizational status or influence of the founding members or Indigenous Permanent Participants, the increasingly diverse voices at the council and increasingly complex interactions that take place both within the working groups and on the sidelines of its ministerials may have effects outside the council. Some of these new stakeholders have taken a collaborative approach with their own Indigenous peoples, such as Singapore, which has recently embraced a reassessment of its precolonial history and restored its place in the national narrative. Japan has resumed progress on its recognition of Ainu indigeneity, the restoration of Ainu rights, and the revitalization of Ainu identity in recent years. Thus these observers appear well disposed to embrace the co-management paradigm in their Arctic relations while embarking on their own domestic journeys toward greater Indigenous inclusivity, co-management, and collaboration, while others do not, such as China, which continues to oppress non-Han minorities from occupied Tibet to Xinjiang, and which does not formally recognize its Indigenous peoples, though it does celebrate the diversity of its many minority cultures while also articulating its respect for Arctic Indigenous peoples in its Arctic policy.[30]

Indeed, China's courtship of Indigenous partners at the Arctic Council and in its bilateral relationships with Indigenous polities such as Greenland may further acculturate Beijing's diplomatic officials to the council's paradigm of diplomatic collaboration between states and Indigenous peoples and increase their sensitivity to and awareness of Indigenous issues—lessons they can bring home that in time may percolate down into domestic policy reforms. Such a dynamic may in fact be underway in Singapore and Japan, though causality has yet to be documented linking their experiences as council observers to their domestic policy reforms on Indigenous inclusion and diversity. By embracing co-management and navigating both the complex administrative landscape of the post–land claims Arctic and the collaborative institutional environment of the council where Indigenous peoples' organizations and states jointly manage nonsecurity dimensions of Arctic international relations with humility and respect, non-Arctic states can find enthusiastic partners in villages from one end of the Arctic to the other, and through the mutuality of collaboration enjoy the reciprocity of what's been called the "shared space" of co-management.[31] This can in turn build trust and smooth relations not only with Indigenous communities and local levels of governance but also with the national agencies and departments that participate in the co-management process. It's important to remember that Arctic co-management has been around for over half a century now, with a robust and evolving system of interlocking institutions of multilevel governance that increasingly unite the levels of analysis—from the local to the international level across the state-tribe interface via an Indigenous, transnational bridge, becoming an emergent paradigm for collaborative northern governance—and that stakeholders representing the increasingly harmonized interests of Indigenous and state are deeply and mutually committed to the co-management process and the reconciliation of competing interests through mutual respect.

The state of Arctic exceptionalism may not be an inherent, default condition of Arctic international relations, but instead the by-product of the Arctic's distinct historical, strategic, and geographical context and evolving governance systems. That condition could erode as historical trends, the strategic environment, and the region's insular geography transform as a result of climate change, which may introduce myriad new stressors to disrupt the Arctic calm.

The 2019 Arctic Council Ministerial: Competitive Tensions Undermine Consensus

As mentioned above, the 2019 Rovaniemi Arctic Council ministerial meeting was shaken by the abrupt policy reversal of the Trump administration. Indeed, as the Biden administration came to power in 2021, it quickly put into reverse many of the previous administration's policies regarding climate change and Arctic environmental protection, while maintaining the more nested vision of Arctic cooperation articulated by the preceding administration, framed by a general distrust of China and Russia in global affairs. But the tone did quickly shift away from open confrontation to a more collaborative rhetoric. As Betsy Baker, an international lawyer in Alaska, told Michele Kelemen on National Public Radio's *Weekend Edition Saturday* on the eve of Secretary of State Antony Blinken's first council ministerial in Iceland in May 2021, "The Biden administration has the opportunity to say, this is a new start. We're returning to what's been the longstanding recipe for success in the Arctic Council, which is to focus on what we have in common and not on the tensions that divide us."[32]

Because the council is so deeply collaborative at its core, with its members, Permanent Participants, and observers, including both state and nonstate entities, finding many opportunities to work together on the council's many substantive working groups tackling the region's many challenges regardless of their relative formal powers within the council, like any collaborative governing body, disagreements are quite natural and are better resolved through ongoing discussion rather than pressures for an immediate (and less enduring) agreement.

As more and more states from outside the Arctic region join the council as observers, new and diverse interests in the Arctic will find a voice in these discussions and transform what was an easier dialogue between fellow Arctic states with a common heritage and shared geopolitical frontier into a more dynamic, complex, and at times more difficult conversation. But a conversation it will be, and it's at this table where old Arctic powers like the United States and Russia found common ground for nearly a quarter century in spite of the previous three-quarters of a century marked by ideological and geopolitical rivalry, suggesting that the introduction of new voices, from non-Arctic states with increasing interest in Arctic economic opportunities, need not jeopardize the underlying spirit of collaboration that has guided the Arctic Council since its inception. That being said, many Arctic strategies and policies of the stakeholders have shifted in tone in recent years to more nationalist and regionalist views, but intriguingly, there has also been a notable rise in attention to Arctic Indigenous peoples and Indigenous-state relations from a diverse range of stakeholders.[33] As mentioned above, Russia's Arctic Policy Foundations to 2035 (released in July 2020) mentions the term "indigenous" fifteen times, and "native" twice. Additionally, "protecting the environment in the Arctic, preserving the native lands and traditional way of life of indigenous peoples residing in the Arctic zone of the Russian Federation" is considered to be part of the "primary national interests of the Russian Federation in the Arctic."[34] And perhaps most relevant with regard to the rise of the co-management paradigm in Russia is the inclusion among Russia's "primary objectives for the development of international cooperation" in the Arctic the aim of "providing assistance to indigenous minorities in building cross-border cooperation, cultural contacts, and establishing economic activities with other minority and ethnic groups residing outside the Russian Federation. In addition, facilitating the contribution of indigenous minorities to international cooperation on ethnocultural development between the states and in accordance with international agreements of the Russian Federation."[35]

A similar though less often mentioned commitment to Indigenous engagement and collaboration is evident in the January 2018 Arctic White Paper by the State Council Information Office of the People's Republic of China, in which the term "Indigenous" is mentioned seven times. It states, "To protect the Arctic, China will actively respond to climate change in the Arctic, protect its unique natural environment and ecological system, promote its own climatic, environmental and ecological resilience, and respect its diverse social culture and the historical traditions of the indigenous peoples."[36] Central to China's Arctic policy is its commitment to the mutuality of respect that defines and permeates Arctic domestic and international politics:

> "Respect" is the key basis for China's participation in Arctic affairs. Respect should be reciprocal. It means all States should abide by international treaties such as the UN Charter and the UNCLOS [UN Convention on the Law of the Sea], as well as general international law. They should respect the sovereignty, sovereign rights, and jurisdiction enjoyed by the Arctic States in this region, respect the tradition and culture of the indigenous peoples, as well as respect the rights and freedom of non-Arctic States to carry out activities in this region in accordance with the law, and respect the overall interests of the international community in the Arctic.[37]

Inherent in Beijing's commitment to mutual respect is its capacity to ensure a "win-win" outcome, which is ultimately what Arctic exceptionalism is all about: "'Win-win result' is the value pursuit of China's participation in Arctic affairs. It means all stakeholders in this area should pursue mutual benefit and common progress in all fields of activities. Such cooperation should ensure that the benefits are shared by both Arctic and non-Arctic States as well as by nonstate entities, and should accommodate the interests of local residents including the indigenous peoples."[38]

These expressions of support for Indigenous collaboration and the harmonization of Indigenous and state interests in the Arctic by such a wide range of states is not oxymoronic; increasingly, Arctic states of all types recognize that the stability of the Arctic region and their ability to assert sovereignty over the region depends upon the consent of Indigenous Arctic peoples, just as non-Arctic states of all types with interests in the region recognize a comparably important role for Indigenous Arctic peoples as partners and legitimizers of their Arctic aspirations.

Russia's Arctic Ambitions: Defensive or Destabilizing?

Russian ambitions for its sector of the Arctic have until very recently focused almost exclusively on the economic development of its vast, resource-rich interior and uniquely shallow and seasonally ice-covered continental shelf—which Moscow smartly wants all the world to recognize as Russian waters,[39] and which article 234 of the UN Convention on the Law of the Sea largely supports by providing that

> coastal States have the right to adopt and enforce non-discriminatory laws and regulations for the prevention, reduction and control of marine pollution from vessels in ice-covered areas within the limits of

> the exclusive economic zone, where particularly severe climatic conditions and the presence of ice covering such areas for most of the year create obstructions or exceptional hazards to navigation, and pollution of the marine environment could cause major harm to or irreversible disturbance of the ecological balance.

While Russia's farther offshore extended continental shelf claims are ambitious, reaching across the pole into waters also claimed by Canada, Denmark, and Norway, its hope is that science—upon which all extended continental shelf claims are based—will support Russia's expansive aspirations.

Moscow's controversial but headline-grabbing 2007 diplomatic stunt beneath the North Pole was perceived as a grab for the polar seabed, but may be better understood as an assertion that there is a Russian side of the Arctic.[40] Moscow may, as Rob Huebert intriguingly suggests, very likely welcome the selection of the North Pole as the boundary point as it was during the Cold War, though in the end, the UN's Commission on the Limits of the Continental Shelf and the International Seabed Authority may, once all the claims are filed and adjudicated, find that Canada's, Denmark's, or Norway's extended shelf extends past the pole into what Moscow views as "its" side, depending in part on what everyone can prove as their own continental shelf extensions.[41] By the same token, science may also reveal that Russia's extended shelf extends into what many in the West perceive as "our" side of the Arctic. As Huebert explained in *Up Here* magazine in 2007,

> Russia's claim to the North Pole would give them an advantage. "The North Pole is not the geographical centre between Russia, Alaska, Greenland and Canada; it's in fact further in towards the Russian coast. So claiming it would give them an advantage." Still, Huebert says the Russians won't be able to claim the entire region to exploit as it sees fit. "My guess is we'll see a complete division of the Arctic Ocean—except for two very clear depressions that are not part of a continental shelf," Huebert says. "Everyone would have a sector, like the Mediterranean or the North Sea."[42]

The primary Arctic tension—other than that between its Indigenous peoples and the broad group of southerners who assert sovereign claim to the high North—has been over offshore boundaries, and here the main fault line remains between Russia on the one hand and the West (Canada, the United States, and its European allies) on the other, even as political tensions thaw between neighbors that share histories of rivalry. Before the war in Ukraine galvanized a diplomatic alignment of the entire Western world against Moscow, Russia and Norway had settled their long Arctic maritime boundary dispute, while the United States and Canada began cooperating more closely than ever, even without agreeing fully on their Beaufort Sea boundary dispute or the status of the Northwest Passage. And the rhetorically muscular dispute between Ottawa and the Danes over Hans Island (which seemed mostly for domestic play on both sides) was quietly settled in June 2022 when both countries agreed to split the diminutive island roughly into two sectors.[43] Though in April 2010 Russia and its Cold War rival Norway buried the hatchet and resolved their long-simmering disagreements over their offshore boundary line, easing the way forward to the joint development of the bountiful offshore petroleum resources in the hitherto contested waters, we should not presume that it will always be smooth sailing ahead.[44] Economic collaboration can yield to nationalist rivalries and even war between trading partners. In the end, the millennial East-West rivalry may well eventually resurface, much as autopilot switches on during inclement weather, and this reinforces the notion that the Arctic as a region, and a potential theater of conflict, fits logically into United States European Command's (EUCOM's) area of operations and its continuing mission of securing Europe from external threats.[45]

Russian interests in the vast Eurasian Arctic are largely defined by its exploitation and development of the enormous natural resource wealth both along and beneath its northern shores, and its rehabilitation of the Northern Sea Route, which during the immediate post–Cold War era lay largely abandoned (particularly off the shores of Eastern Siberia) but which has lately enjoyed Moscow's recommitment to bringing its vast treasure chest of northern resources to market.[46] With its extensive, shallow, and increasingly accessible Arctic continental shelf chock full of petroleum resources in exploitable quantities, Russia has much to gain from an Arctic thaw. But the strategic importance of this natural resource wealth to Russia's economic resurgence also provides ample motivation for Moscow to ensure an adequate defense of its northern domain. It can no longer count on nature to defend its northern flank with a "great wall" of ice, and this could result in rising security tensions along the old East-West fault line.[47] Just as Canadians have a powerful emotional attachment to their northern frontier, Russians view their Arctic lands and seas as an extension of their heartland—which for them has been and remains their key to survival, militarily and economically. The intensity of this attachment and the strategic importance of the heartland to Russian geopolitics, which saved the Russian nation from Napoleon's armies as it did from Hitler's, combine to define a vital national interest for Moscow. This means that Russia, more than the other littoral Arctic states, is more inclined to fully utilize its Arctic assets—even though the post-Soviet economic collapse led to a decade-long abandonment of much of its centrally subsidized megaprojects in the vast and now rusting Russian Arctic, as well as its maritime infrastructure along its Northern Sea Route. But in recent years, with higher commodity prices changing the calculus, Moscow has reversed course, and there is now a growing commitment to a fuller utilization of its Arctic resources, and a growing awareness that Russia's destiny, and a critical source of its future wealth, is tied to its fate in the North.[48]

Already there has been a restoration of Arctic naval, land, and air exercises to show the world that Moscow is serious about its Arctic ambitions, though these may be viewed as largely defensive in nature thus far.[49] Along its borders, where the defensive nature of its regional military deployments could appear to be more menacing, this has led to a reemergence of historic tensions with its neighbors, the first stirrings of which followed Moscow's smackdown of Georgia, as symbolic an act as its North Pole flag planting, with greater muscularity. That caught the attention of its many neighbors, particularly the former Soviet satellite states, who united in their critique of the reawakened bear (and whose concerns have been greatly magnified by Moscow's full-scale invasion of Ukraine in 2022). After Georgia, there could be little doubt that Russia would aggressively defend its Arctic interests if Moscow felt they were threatened, a conclusion magnified manifold by the Ukraine war.[50]

Still raw for many Russian nationalists is Russia's loss of empire—beginning with its 1867 sale of Alaska to the United States, which many in Russia still feel was nothing short of wholesale theft, a transaction whose history remains clouded by distrust. The Russian-American Company was shuttered by Moscow after decades of sacrifice and investment by its explorers, who risked much to explore and colonize the high North Pacific, leaving many Russians perplexed by the abandonment of Alaska.[51]

Some Russian ultranationalists, such as the infamous Vladimir V. Zhirinovsky, include a still-Russian Alaska on their national maps, though this may be largely symbolic and not necessarily a reflection of their true military ambitions, and in today's Siberia, "Alaska Is Ours" billboards have been reported widely in the media.[52] In our own time, with the Soviet collapse, Russia became even smaller and more vulnerable with the loss of its empire. Its remaining Arctic lands and seas are thus perceived to be especially highly valued as a sacred and inseparable part of Mother Russia—a key to its future, and one of the last sources of pride and hope that it has left. With more heavy icebreakers than all of its neighbors combined, Russia might well emerge the predominant military power in the high North one day; it already possesses one of the most developed Arctic resource economies and has rapidly restored to operation its Northern Sea Route. While behind its Arctic neighbors on issues of Indigenous rights recognition and collaboration, it has made progress on protecting Indigenous lands through park status not unlike the way most of northeast Greenland is park land, as is much of Alaska (primarily the federal part)—even if it is not as enthusiastic about developing a political space for Indigenous cultural and political autonomy from the state as its Western peers. It is therefore in many ways on the right track toward parity with the other Arctic states on these matters, thereby strengthening the fabric of Arctic exceptionalism. But there is still a long way to go.

As Lackenbauer and Dean have noted, "Russia has been the most determined Arctic player for nearly a century. As such, its own sense of 'Arctic exceptionalism' flows from a conviction that only it 'has the necessary experience and knowledge to contribute to the economic and social development of the region and to the protection of its ecosystem.'"[53] They further observe, "Russia has declared that it intends to transform the Arctic into its 'foremost strategic base for natural resources' and that dramatically expanding shipping along the Northern Sea Route (NSR) is a top priority; indeed, President Putin called in August 2019 for annual shipments to reach 80 million tons by 2024,"[54] but on the other hand, critics worry "Russia's military modernization programs in the Arctic represent an aggressive buildup aimed at regional domination, while others point to 'dual-use' and 'soft security' applications that pose no threat to regional stability."[55] Lackenbauer and Dean further suggest,

> It is certain that revisionist moves that undermine Arctic state sovereignty or sovereign rights would have disproportionately negative impacts on Russia, thus making military confrontation in the region unlikely on the grounds of Russian national self-interest. As Katarzyna Zysk astutely observed: "One of the region's biggest assets as a promising site for energy exploration and maritime transportation is stability. . . . Given the economic importance of the Arctic to Russia it is likely that leaders will avoid actions that might undermine the region's long-term stability and security."[56]

Lackenbauer and Dean see in Russia's approach dynamics born of "domestic politics and national self-interest more than any ideological commitment to 'Arctic exceptionalism' rooted in post–Cold War internationalism," which they further clarify by adding, "Rather than seeing this as normatively distasteful, this form of exceptionalism might encourage maintaining regional institutions and relationships, even with Putin['s] [*sic*] belief that liberalism has 'become obsolete,'" as had been reported by BBC in 2019.[57]

Though Russia was at the table at Ilulissat in 2008 and pledged to support international law and the UN Convention on the Law of the Sea mechanism,[58] one must wonder what Moscow would do if the world community sided with Canada or Denmark in terms of continental shelf extensions at Russia's expense. And though Moscow has resolved its border dispute with Norway, a welcome sign of a more collaborative Russia, sentiments and political winds can change. On the other hand, the Arctic, just as Gorbachev proposed in the 1980s,[59] could become a compelling testing ground for a new relationship between Russia and the West, and perhaps—if cooperation trumps competition over time—a path toward eventual NATO membership. But if competition trumps cooperation instead, the Arctic may become one of the first regions in which a newly assertive Russia confronts the West. This fundamental ambiguity regarding Moscow and its approach to Arctic exceptionalism remains unsettled, forcing NATO to proceed with caution on matters relating to Arctic cooperation with nonalliance partners, and reinforcing the need for a continued alliance-centric approach that views Russia in the Arctic through the lens of renewed Atlanticism, as described by Lackenbauer and Dean, among others.

Lackenbauer and Dean write that

> part of the post–Cold War euphoria that allowed proponents of the liberal institutionalist interpretation of Arctic exceptionalism to conceptualize the region as an "exceptional space" flowed from the rapid collapse of the Russian military and the apparent absence of any regional military competition in the Yeltsin era after 1991. By 2007, however, an increasingly confident Russia led by President Vladimir Putin was rebuilding its armed forces with oil and gas revenues, resuming strategic bomber flights in the Arctic, and mounting regional naval operations. Coupled with Russia's invasion of Georgia in 2008 and its increasingly apparent "diplomatic opposition to Western interests," some commentators chastised what they saw as naïve idealists in the West clinging to "Arctic exceptionalism" when Russia was indicating its intention to return to coercive politics and even unilaterally demarcate and defend its Arctic borders.[60]

They further explain,

> Through a Russian strategic lens, the Arctic, North Atlantic, and North Pacific constitute a single operational zone in which to counter U.S. and NATO strategic forces. For the Russian Northern Fleet and strategic bomber forces the Arctic region is a "bastion" of deterrence and defense or a thoroughfare to project power—all to maintain global strategic balance. In the Western sector of the Russian Arctic, land and air forces stand ready against NATO (particularly Norwegian) capabilities, while the conventional component of the Northern Fleet protects Russia's economic interests in the Barents Sea and offers support/auxiliary services to nuclear forces. The Northern Fleet and the Murmansk Command of the Border Guards also protect the Northern Sea Route (NSR) and the Arctic Ocean coastline, while the Pacific Fleet and the Petropavlovsk-Kamchatsky Command of the Border Guards control the Bering Sea, Bering Strait, and access to the Chukchi Sea. Thus, although one lens leads Russia to view its Arctic as a distinct domestic space that needs to be defended and protected from external encroachment, another sees it as a core element in its broader geostrategic map of the world.[61]

Confronted by this fundamental ambiguity, NATO must be prepared to defend its interests and its members' sovereignty from Russia, even while (until March 2022, when such efforts were suspended by the West in protest of Russia's aggression in Ukraine) its Arctic members cooperated with Russia as fellow Arctic Council member states, a delicate diplomatic and strategic dance that required great agility as times and perceptions change, one that establishes an inherent ambiguity in Arctic exceptionalism regarding which actors in the Arctic system to include, and which to exclude, within the zone of cooperation when dealing with various Arctic issues.

That's one more reason why the US Department of Defense's EUCOM has found itself invariably drawn into the increasingly salient and ever challenging mission of securing the Arctic. If you look at which countries are Arctic nations, the *coastal* nations include Russia, Norway, Denmark/Greenland, Iceland (though its territory is nearly all sub-Arctic, with the exception of diminutive Grimsey Island, which straddles the Arctic Circle), Canada, and the United States; the *noncoastal* Arctic states include Finland and Sweden. Most are European, and the non-European Arctic states are NATO members with close historical, cultural, and strategic links to Europe. Only Russia's sparsely populated Far East, Alaska's equally sparsely settled southern coasts, and Canada's far western province of British Columbia abut Pacific waters.

Increasingly, transatlantic relations and the security of the West, and the continuing integration of the economies of the industrialized Far East with those of the West, will depend upon ensuring the security of the Arctic—suggesting that EUCOM may be the right command, in the right place, to play a key role in Arctic defense efforts, particularly as NATO's Nordic expansion approaches completion, further intertwining the fates of Europe and the Arctic (while a China-Russia alignment reintroduced a Cold War Arctic bifurcated between a democratic West and an autocratic East. EUCOM—like the Arctic—enjoys intimate proximity to Russia that ensures their fates will remain tied together for the years that lie immediately ahead. Proximity to an awakening Russian bear and experience in taming its more aggressive instincts will be an important key to a secure and peaceful north. While it can always be hoped that the bear can be enticed to join the West as a friend and partner, one must always be prepared for its more aggressive instincts to return, as we have seen increasingly between 2014 and 2022. EUCOM, whose mission has been to defend the West from the darkest days of the Cold War through the glorious transformation to the post–Cold War era, has the experience to do both as the world again bifurcates.

Until April 2011, three commands shared some responsibility for the defense of the Arctic; as Associated Press correspondent Dan Elliot observed, "Previously, that responsibility was shared by the U.S. Northern, Pacific and European commands."[62] Soon thereafter, this division was rebalanced, with NORTHCOM and EUCOM dividing the responsibility of the defense of the Arctic and INDOPACOM (Indo-Pacific Command) seemingly left out of the mix. *Stars and Stripes* reporter John Vandiver noted that "U.S. Northern Command's area of responsibility was expanded earlier this year to include the North Pole and the Bering Strait," while "U.S. European Command's area was extended to include the water space of the Laptev and Eastern Siberian seas north of Russia. While NORTHCOM will be the lead advocate for Arctic issues within the Defense Department, EUCOM will manage military relationships with other Arctic nations in Europe," who with the exception of the sole Eurasian Arctic state, Russia, are located in Europe or North America.[63] Vandiver noted both NORTHCOM and EUCOM "officials have launched a review of the assets that will be required in the region in the years ahead," and that "Col. Daniel Neuffer, the lead officer for Arctic issues at EUCOM, said the review will look at the Arctic from a long-term perspective," further citing Neuffer as saying, "What capabilities will we need 30 years from now? That's the assessment we're going through. . . . I think for Russian sustained growth, they will continue to need to harvest more natural resources. But nobody wants a conflict, because you can't extract anything if you're ducking bullets. In the Arctic, I think, cooler heads will probably prevail."[64] Even within the military command defending the Arctic, the logic of Arctic exceptionalism and the imperatives of Arctic resource development were paramount in the years between the Soviet collapse and Russia's military resurgence (roughly 1991–2011, or 1991–2021, depending on what year you pinpoint the Russian bear's reawakening).

If "cooler heads" don't prevail, a more inclusive command structure will invariably prove necessary, one that draws not just on EUCOM's expertise containing the Russian bear, and NORTHCOM's expertise securing North America's high North, but also on INDOPACOM for its expertise defending the high North Pacific, containing China's naval expansion, and its long and important legacy securing US Pacific frontiers during the Cold War era and into the new, more chaotic, post–Cold War world.

With China also on a historic rise, even if its ambitions remain largely economic, the prospect of Beijing aspiring for greater geostrategic recognition as a great power, perhaps even a superpower, whose reach extends far beyond the South China Sea into the global maritime commons that include the Arctic basin, cannot be safely discounted. Tokyo's aspirations were similarly overlooked in the years that preceded World War II. Japan's assault not only on Pearl Harbor but on Alaska's outer Aleutian Islands was foreseen and predicted by noted air theorist Billy Mitchell more than a decade before it took place (and, just as he presciently predicted, on a Sunday morning when we were least prepared to face such a daring assault),[65] but his wise counsel was ignored at great peril and ultimately very high cost in blood and treasure. It is to this shattered calm, and the resulting "war in the fog," that we next turn.

Arctic Exceptionalism and Atlantic Security

In the years since World War II, there has been much discussion of whether the Arctic is best perceived as part of the Asia Pacific security system, the Euro-Atlantic security system, or the North America security system, and given the Arctic's interconnections to all three regions, the best answer may well be "all of the above," with relative regional importance contextualized to match the current strategic environment (now, after Obama-era efforts to foster an Indo-Pacific pivot, tilting once again toward EUCOM). Whether INDOPACOM, NORTHCOM, or EUCOM is the command best suited to manage Arctic security challenges will depend on the details of the moment. Indeed, in this historical moment, with Ukraine under assault by Russia and the democratic Arctic states united in their efforts to both contain and isolate Russia, a case can be made for EUCOM's continuing suitability as the primary regional command for the defense of the Arctic, since the most salient external risk to the region, according to NATO risk assessments, emanates from Moscow in the wake of its military resurgence. This should not come as a surprise, since a bolder, resource-enriched, and diplomatically resurgent Russia has been on the rise under Putin's leadership for over a decade.

Indeed, Russia's symbolic 2007 polar flag-planting on the deep sea floor made international headlines and provided notice to the world that Russia was prepared to draw a "line in the ice," and to strongly defend its northern national interests—with some critics suggesting the media was overhyping the risk; fifteen years later, the risk is clearly no longer hype or media sensationalism, but perhaps validation of the media's initial prescience.[66] Geography also sides with the relevance of EUCOM to Arctic security, since Russia owns by far the largest sector of Arctic coast—and by quirk of geography, the shallowest and most resource-accessible Arctic continental shelf—so as it thaws, Russia will have greater access to a greater share of the Arctic's long-hidden offshore resource wealth than any other Arctic state, and will thus have much reason for a strong defense of its northern onshore and offshore domain, illustrated by its Arctic military modernization efforts in recent years.[67] Recent history also supports viewing the Arctic as part of EUCOM's area of operations, as the longest recent conflict in Arctic waters was not the relatively brief battle for the Aleutians but the much longer Battle of the Atlantic. Later, the US Navy's 1986 Maritime Strategy at the Cold War's end perceived the Arctic's undersea domain as a vital forward theater to contain then–Soviet Russia's fleet in its home waters, before it could menace North America.[68]

While the latest Pentagon Arctic strategy dates from 2019 with an updated version expected in early 2024, the White House updated its National Strategy for the Arctic Region as recently as October 2022. Iris Ferguson, deputy assistant secretary of defense for Arctic and Global Resilience, told the 2023 Arctic Circle Assembly, as reported in *High North News*, that the United States was "developing our new Arctic strategy, which will likely be launched in early 2024. I want to make it clear that the DoD intends to maintain a stable and peaceful Arctic region. We will absolutely protect our interests but also contribute to stability."[69] Reframing Arctic exceptionalism within an alliance-centric lens, Ferguson explained the United States will "work closely with our allies and partners while developing the strategy. It is critical that we have alignment with them, and a strategic focus is on how we collectively can use joint presence to ensure stability. It is also crucial that we align with our combatant commands. So a major coordination process is now taking place, but we aim to conclude it early next year." As *High North News* noted, "The Arctic is covered by three American combatant commands—one Northern, one European, and one for the Indo-Pacific region."[70] Describing the dynamics that will influence this alignment with the combatant commands, Ferguson explained,

> Recently, various significant geopolitical changes have occurred in the Arctic region. One of the major changes is NATO's expansion, with Finland as a new member and Sweden close behind. It is essential to address what this entails for the alliance's regional plans and how we conduct deterrence and develop capabilities. . . . Some of our competitors also conduct more military activity in the region. In addition, we see increasing alignment between them—particularly off the coast of Alaska. There is no decline in activity in the Russian Arctic despite the war in Ukraine. This says something about the significance of the region. We must be mindful of what is happening and assess whether we have the right capabilities to deter the threats.[71]

The rapid linkage of Russia's aggression in Ukraine to Arctic cooperation, evident in the West's decision to de facto boycott Arctic Council activities for the duration of the war, as well as the pivot by neutral Arctic Council Nordic states Finland and Sweden to pending NATO membership, more recently illustrates this point. For these reasons, the key to a secure Arctic remains tied to the fate of Europe and the ambitions of its largest state: Russia. Arctic waters also came into play during the six-year Battle of the Atlantic from 1939 to 1945, considered by many to be the longest continuous military campaign of World War II.

Efforts to assert command of the seas, especially vital to ensure Britain's survival as an independent country, but also important for resupply efforts of our wartime allies including Soviet Russia, and German efforts to deny North Atlantic waters to us, resulted in an ongoing naval clash between Allied and Axis sea power.[72] There were a total of 78 Arctic convoys that resupplied the northern ports of Arkhangelsk and Murmansk under the protective escort of the US Navy, Royal Navy, and Royal Canadian Navy—enabling some 1,400 ships to deliver lend-lease supplies to the Soviet Union. Ever since the long Battle of the Atlantic, the high North Atlantic and Arctic waters have been viewed through the lens of the Atlantic alliance, and as essential to the stability of the North Atlantic.

It was Greenland's vulnerability to external aggression that brought US military power to the island, a year after Denmark was invaded and occupied by Nazi Germany on April 9, 1940. After Denmark had fallen, the Germans eyed Greenland as their first stage of an invasion route of mainland North America via the Gulf of St. Lawrence through to upper Canada along the Great Lakes—much the way Britain did during the War of 1812. The vulnerability of Greenland resulted in the United States extending defense protection on behalf of the Danish government in exile, which continued after the war through the entire Cold War era as Soviet naval power grew. Had the Germans gained possession of Greenland, it could have put their fleet in striking distance of Newfoundland, enabling a two-pronged attack of strategic British territories. Guaranteeing Greenlandic security was viewed as a necessity to ensure the independence of Britain. If Britain had fallen, keeping the Germans out of Greenland, Newfoundland, and inevitably Canada would have been harder—and America's northeast maritime and land frontiers would have been highly vulnerable, much as the Aleutians proved in the face of aggressive use of Japanese naval power.

According to a statement from the US Department of State issued on April 10, 1941, one day after the United States and Denmark entered into a defense agreement for Greenland, "During the summer of 1940 German activity on the eastern coast of Greenland became apparent," when "three ships proceeding from Norwegian territory under German occupation arrived off the coast of Greenland," and then "in the late fall of 1940, air reconnaissance appeared over East Greenland under circumstances making it plain that there had been continued activity in that region."[73] And on March 21, 1941, "a German bomber flew over the eastern coast of Greenland and on the following day another German war plane likewise reconnoitered the same territory. Under these circumstances it appeared that further steps for the defense of Greenland were necessary to bring Greenland within the system of hemispheric defense envisaged by the Act of Habana."[74] So on April 9, 1941, an agreement was made "between the Secretary of State, acting on behalf of the Government of the United States of America, and the Danish Minister, Henrik de Kauffmann, acting on behalf of His Majesty the King of Denmark in his capacity as sovereign of Greenland," granting "to the United States the right to locate and construct airplane landing fields and facilities for the defense of Greenland and for the defense of the American Continent"—but only "after explicitly recognizing the Danish sovereignty over Greenland."[75] The agreement acknowledged that "as a result of the present European war there is danger that Greenland may be converted into a point of aggression against nations of the American Continent, and accept[ed] the responsibility on behalf of the United States of assisting Greenland in the maintenance of its present status."[76] The United States asserted it had "no thought in mind save that of assuring the safety of Greenland and the rest of the American Continent, and Greenland's continuance under Danish sovereignty."[77]

Early in the Cold War, a new external threat to Greenland and to North America arose, not from the decisively defeated and now divided Germany but from the former wartime partner, the Soviet Union. On April 27, 1951, a new treaty was signed, the "Defense of Greenland: Agreement Between the United States and the Kingdom of Denmark."[78] Article I of the 1951 treaty affirmed that both countries, "in order to promote stability and well-being in the North Atlantic Treaty area by uniting their efforts for collective defense and for the preservation of peace and security and for the development of their collective capacity to resist armed attack, will each take such measures as are necessary or appropriate to carry out expeditiously their respective and joint responsibilities in Greenland, in accordance with NATO plans."[79] This treaty remained in place for more than half a century. On August 10, 2004, Denmark, Greenland, and the United States updated their 1951 defense agreement, when "after two years of negotiations, all three parties—the U.S. on one side, and Denmark/Greenland on the other—reached consensus on the terms of the treaty. The United States was granted permission to upgrade Greenland's Thule Radar Station as part of the American Missile Defense (MS) program. The agreement itself implicitly recognized former Danish colony Greenland as an equal partner with influence over its own foreign affairs."[80] Among the most notable changes in the treaty's language was the emphasis on "partnership with Greenland," the inclusion of Greenland as a party to the treaty, and the evident spirit of equality among these three parties. According to Greenland's minister for foreign affairs, Josef Motzfeldt, "For us at home, this date marks the day that Greenland took a decisive step toward equality and responsibility on par with other countries of the world, and away from the indignity and indifference of the colonial era. By entering this agreement complex, Greenland has taken an active step toward increased foreign policy independence."[81] Colin Powell, then serving as the US secretary of state, echoed Motzfeldt's sentiment, adding that "it is important to demonstrate that Greenland is a full-fledged member of this partnership. And the best way of showing that is by being on hand today."[82]

"Whoever Holds Iceland Holds the World"

It was not just Greenland and its security that were vital to the defense of the West. As important was Iceland. One could modify Billy Mitchell's well-known geopolitical maxim on Alaska from the 1930s—"I believe that, in the future, he who holds Alaska will hold the world, and I think it is the most important strategic place in the world"[83]—and apply it to the high North Atlantic, at least with regard to European and North American security. In this case, he who holds Greenland *and* Iceland seems destined to command the North Atlantic. Indeed, as recounted by the US Marine Corps Historical Division,

> "It has been said," wrote Winston Churchill, "Whoever possesses Iceland holds a pistol firmly pointed at England, America, and Canada." At the time of which he wrote, the "pointed pistol" threatened most immediately the British lifeline: The northern convoy route between Great Britain and the Western Hemisphere, upon which the island kingdom was dependent for most of the materials to sustain its war effort as well as much that was needed for its very subsistence. Iceland perched on the flank of these shipping lanes, which were under heavy attack by German submarines. Hostile air and naval bases on the island would almost certainly render the northern route unusable, and put pressure, perhaps untolerable pressure, on the longer and more vulnerable southern route.[84]

During the Cold War, novelist Tom Clancy imagined Iceland becoming the strategic pivot in a future conventional battle for the North Atlantic between NATO and the Soviet Union, a conflict that seemed all but impossible to imagine in the post–Cold War era prior to the Russian invasion of Ukraine, when Clancy's vivid and imaginative scenario became ominously salient once again. The role of the G-I-UK gap throughout the Cold War, for both Soviet and NATO naval strategy, was indeed central—though ultimately untested by war in contrast to Clancy's fictional imaginings. Ironically, it was Clancy's conception of Iceland as a strategic pivot that influenced President Ronald Reagan on the eve of the almost history-changing but in the end unsuccessful Reykjavik Summit with Soviet Premier Mikhail Gorbachev. Gorbachev's bold proposal to rid the world of nuclear weapons was rebuffed because it would require a mutual commitment not to develop a strategic missile defense such as envisioned by Reagan's Strategic Defense Initiative, Reagan's cherished dream known to many as Star Wars. As recounted in the December 8, 1986, edition of *Time*,

> The phrase "Reagan is not a detail man" is a mantra among Reaganites and suggests that he sees the big picture, that "details" are for smaller minds. Yet such detachment can prove dangerous. In preparation for the Iceland summit, Reagan did not study the history and nuances of America's arms-control strategies; instead he practiced ways to sell Gorbachev on SDI. To get himself into the right frame of mind, he read Tom Clancy's *Red Storm Rising,* a potboiler about a non-nuclear war between NATO and the Soviet bloc.[85]

The Maritime Strategy of 1986 likewise recognized the Arctic and the high North Atlantic as important areas for forward operations to contain the projection of Soviet naval power. Critics feared it would destabilize deterrence but in the end it helped reassure Europe that Soviet power was far less potent than Moscow wanted people to believe. And in terms of economic potential, the commercial and strategic sea lanes of the North Atlantic, the vast North Sea oil fields, and the bountiful fisheries in the high North Atlantic almost led to alliance members Iceland and the United Kingdom coming to blows during their "Cod Wars" of the 1950s–1970s and placed Iceland's NATO membership and access to the strategic air base at Keflavik in jeopardy. Three distinct "wars" included the deployment of British warships to defend the British fishing fleet from Iceland's use of net cutters, and that resulted in some ramming of vessels as well as some warning shots fired from September to November 1958, September 1972 to November 1973, and November 1975 to May 1976,[86] illustrating the strategic-economic importance of the high North Atlantic as a bridge connecting Europe and North America.

As the Arctic thaws, North Atlantic fisheries,[87] natural resource extraction efforts,[88] and sea lanes will edge farther north into Arctic seas,[89] eventually facilitating the emergence of an Asia–Europe–North America sea bridge some predict will be a modern-day silk road[90]—but the fundamental strategic relationship will remain the same. Consequently, it may continue to make sense to view the Arctic, as it becomes increasingly navigable and economically integrated, as an extension of the North Atlantic—since in addition to its historical linkages to the Euro-Atlantic community, the Arctic basin is only semi-enclosed, with its opening flowing into the North Atlantic, while in the Pacific it encounters a physical barrier, with only the narrow and shallow Bering Strait connecting the two. With an Arctic thaw, the Northeast Asian trading states will find a shorter and quicker direct route to markets in Europe and North America, but because of the narrowness (85 km) and shallowness (55 meters feet on average) of the Bering Strait,[91] they will not find as ready an opportunity to expand naval influence into a sea still dominated by NATO and Russia.

The longer-term potential of trans-Arctic shipping, via increased usage of the Northwest Passage and the Northern Sea Route, while promising, has a long way to go before being viable. The Koreans, Chinese, and Japanese are eyeing shorter and safe shipping lanes to Europe over the top, and the Koreans have taken the lead with regard to commissioning a new generation of ice-hardened tankers, though the Russians still dominate when it comes to heavy icebreakers.[92] While connecting Northeast Asian markets to Europe through an Arctic maritime bridge is compelling, there will always be winter, and with winter, new ice will form in the Arctic basin, limiting the year-round viability of such sea routes, so it is unlikely that we will see the center of gravity tip entirely toward the Pacific, particularly given the extraordinarily close and enduring transatlantic relationships that have been forged across centuries of trade, wartime and peacetime alliances, and the much less united strategic environment in Northeast Asia. As transpolar shipping becomes more frequent, however, we may find reason for INDOPACOM and EUCOM to conduct joint operations in the Arctic, perhaps formalizing the current regional command overlaps into a new, cross-regional subunified command, not unlike the new US CYBERCOM that is subordinate to STRATCOM but that takes ownership of the distinct and emergent defense challenges of the information domain.

Even with Asian states eyeing Arctic trade routes, the North Atlantic still features prominently in most of their plans, most notably as the end destination for their marine exports or the starting point for their imports. Iceland could well become a primary transshipment hub for Asian cargo ships, positioning the high North Atlantic to remain of critical strategic importance. That may be one reason why Moscow was first to step up with an offer of neighborly assistance to bail out Iceland when its economy collapsed.[93] Perhaps it hoped to nudge Iceland a bit out of the Western camp and help Moscow expand its influence in the high North Atlantic, counterbalancing the Scandinavian states that share maritime borders with Russia and that have historically contained its naval influence. As Konstantin Rozhnov reported on BBC,

> When Iceland announced it was seeking a $6bn (£4bn) loan from Russia to help rescue its crisis-ravaged economy, some in the NATO alliance, of which Iceland is a member, took fright. They suspected that Russia was acting to further its geopolitical interests in the region in the guise of a white knight. Reports of Russia seeking—or even securing—rights to Icelandic fisheries, energy and metal sectors, as well as in tourism, poured fuel on these fears. Russia has denied any political interest in its dealing with crisis-hit Iceland, but even some Russian media outlets have expressed scepticism, publishing caustic headlines such as "Ready to buy Iceland for good money." "If Russia becomes the country which saves the Icelandic economy, Russia could also end up securing an extended level of power in the North Atlantic," BarentsObserver website said in an editorial.[94]

This was not the first time that Iceland looked east instead of west when in need. During its third Cod War with Britain in 1976, Iceland's government had sought to acquire US Asheville class gunboats, but when its effort was thwarted by the US government, it considered Soviet Mirka class frigates as an alternative.

Sovereign Aspirations: Greenland as a Future Strategic Pivot?

Beyond Iceland, if Greenland were to pursue an unfriendly secession from Denmark and ended up hostile to Western interests, Moscow may find yet another friendly island state open to courtship, and that would certainly favor its strategic position, putting pressure on the West and its command of the high North Atlantic. For the moment, its independence movement is a friendly one, with Denmark's blessing—but that could always change if the cost/benefit calculation of Danish sovereignty over Greenland is reassessed in light of the global thaw.[95] Going forward, the United States and its NATO allies might be wise to cultivate warmer relations with all the microstates and territories of the high North Atlantic and Arctic. Alaska and Iceland have especially close political ties, so this could be a good foundation, leveraging the warm relations state leaders in Alaska have fostered with Iceland's government. Fostering a closer diplomatic relationship with Greenland is also under consideration, with its eventual independence anticipated. Consider a November 2007 State Department cable that observed,

> Greenland is on a clear track toward independence, which could come more quickly than most outside the Kingdom of Denmark realize. . . .

> With Greenlandic independence glinting on the horizon, the U.S. has a unique opportunity to shape the circumstances in which an independent nation may emerge. We have real security and growing economic interests in Greenland, for which existing Joint and Permanent Committee mechanisms . . . may no longer be sufficient. American commercial investments, our continuing strategic military presence, and new high-level scientific and political interest in Greenland argue for establishing a small and seasonal American Presence Post in Greenland's capital as soon as practicable.[96]

The cable discusses the "High Stakes for the U.S. in Greenland," and argues that the "time is now to begin investing in a flexible, low-cost, official U.S. presence in Greenland" that "would allow us to advance our strategic and commercial agenda directly and to shape the image of the U.S. in Greenland as never before. For now, we can offer Greenland an American perspective. Down the road, we must be prepared for the day when we welcome a new and independent neighbor, one that will be a true partner within the transatlantic community of the 21st century."[97]

Like Iceland, Greenland could well be the key to a stable Arctic; no one at this stage can predict with accuracy where the sentiment and loyalties of an independent Greenland will lie. If the festering tensions between Europe and Canada's Inuit are any indication, there's much need for some fence-mending. Embracing the Inuit and their seal-hunting traditions would also go far to reduce tensions between them and the Europeans who oppose seal hunting and the fur trade generally, despite their long history of fur empires, which, ironically, fostered their economic colonization of much of North America.[98] More concerted confidence-building measures could help to ensure that the interests of the Inuit and of the modern states that jointly assert sovereignty over their homeland remain aligned.

The Inuit may be few in number, but they control many local economic and political levers, and their interests are now fully backed by Ottawa—their partner in land claims, self-government, and northern development.[99] Resolving lingering tensions between Europe and the Inuit is a necessary step to ensure the security of the high North, as greater issues are now in play that could affect the destiny of nations more than one people's views of another's dietary preferences. It wouldn't take much diplomatic savvy for the Russian bear to seize the opportunity and break bread with the Inuit over tasty slabs of whale and seal meat, hoping to forever drive a wedge between the people of the Arctic and the European states whose security will increasingly be tied to the fate of the Arctic. Secretary Clinton's overture to the Inuit a decade ago was thus a well-timed and diplomatically preemptive move to ensure the West didn't lose the North on her watch,[100] as was the even more contentious bid by President Trump a decade later to purchase Greenland from the Danes, a proposal that caused a brief diplomatic rift with fellow NATO partner Denmark and that was widely criticized in Greenland,[101] yet reflected a similar realization at the highest levels of the US government of the importance of the Inuit to the security of not only the Arctic but to the United States, the Western Hemisphere, and the world.

Polar Silk Roads from World War II to the Present

It's not just the Atlantic gateway to the Arctic that was engulfed by World War II as technology and war united to end North America's isolation from both Europe and Asia. A generation before China's rise and restored ambition to become a preeminent global power, rewiring the world's networks and trade connections into a contemporary silk road interconnecting the Pacific, Indian, Atlantic, and Arctic Oceans, Japan tried much the same—but instead of forging its empire through economic interconnections, it sought to do so by force.

Japan's ambitious but ultimately tragic efforts ended just over seventy-five years ago with its August 15, 1945, acceptance of surrender. The emperor of Japan's very first radio broadcast to his nation, announcing the surrender, would never be forgotten—and it was one that almost never happened, as militarists tried in vain to seize the recording. But far less often evoked than the war's dramatic finale is the beginning of the end of the Pacific War. The liberation of the outer Aleutians reversed Japan's breathtaking imperial expansion in the preceding years—a campaign described aptly by Ira Wintermute in August 1943 as "war in the fog," as much for the impact of the region's infamous weather as for its many frictional consequences of the fog of war.[102] From the restoration of US sovereignty over the island of Kiska in Alaska's Aleutian Islands on August 15, 1943, to the emperor's dramatic surrender announcement two years later, a determined, bloody, and successful island-hopping campaign transpired from the frigid subarctic waters of the Aleutians to the tropical waters of the South Pacific. The historic importance of the war in the Aleutians cannot be overstated; what remains puzzling to this day, however, is the widely held view that the war in the Aleutians has become one of America's "forgotten wars," a perception so often repeated that without a determined effort by historians and students of war to "unforget" this war, it may well become a self-fulfilling prophecy. With growing global geostrategic interest in the Arctic and sub-Arctic regions, fueled in part by the accelerating polar thaw and its concomitant opening of new sea lanes running through the Aleutian gateway into the narrow Bering Strait, the lessons of the Aleutians war are evermore essential for remembrance.

From Infamy to Ennui

In early June 1942, the imperial Japanese navy launched carrier raids on Dutch Harbor, Alaska, and in the days that followed, invaded and occupied the islands of Attu and Kiska in Alaska's outer Aleutians, 847 and 671 miles, respectively, to the west of Dutch (itself another 796 miles from Anchorage). The bombardment of Dutch was believed at the time to be Tokyo's opening move in a wider campaign to establish a foothold on the Alaska mainland from which to expand and project power deeper into the Pacific Northwest, one successfully thwarted by prescient US efforts to fortify the region near Dutch in the months that followed the attack on Pearl Harbor.[103] More recent scholarship suggests that Japan aimed less for a beachhead on North America than to distract war planners with a feint in the hope of dividing the US fleet—with their ultimate goal being victory at Midway. Whether Tokyo's grander ambitions for war in Alaska were thwarted by an unexpectedly vigorous defense at Dutch, or it intended from the get-go to clear and hold only the outer Aleutians, its year-long occupation of Attu and fourteen-month-long occupation of Kiska became the first and only Japanese occupations of North American territory during the war.

While it was our only loss of North American soil to imperial Japan, its invasion and occupation of the outer Aleutians that followed the bombing of Dutch were one of multiple assaults upon remote Pacific US insular territories, and the second upon an incorporated US territory. The Aleutians invasion was preceded that previous December by near simultaneous assaults by Japan on the unincorporated US territories of the Philippines and Guam, and the incorporated territory of Hawaii. While each of these Pacific insular possessions was of strategic naval importance, and essential to maintenance of sea control in the Pacific, they are not all remembered equally today. Indeed, the assault on Pearl Harbor remains hallowed as a "day of infamy," widely discussed, debated, and dramatized in multiple Hollywood films—and by universal account, passionately avenged. The war in the Aleutians has, in comparison, been largely overlooked, less widely taught to American students, and all but ignored by Hollywood. While the assault on Alaska has yielded a far less fervent public remembrance in the Lower 48 than that on Hawaii, it does not stand alone for the relative dispassion of its place in our historical memory. It has been described as one of the United States' many "forgotten wars" dating all the way back to the War of 1812 and forward to the Korean conflict—and likely to include several theaters of the long-running global war on terror, including the war in Afghanistan, which one can argue has become mostly forgotten even while still underway.

A Partitioned Pacific?

When viewed through the lens of grand strategy, Japan's multiple assaults on Pacific US insular territories from Guam to Kiska may well have been calibrated by war planners in Tokyo to avoid a direct strike on the US mainland and physically occupy territories only on the far side of the antemeridian. By hitting the United States outside its constitutional core, much the way it attacked colonial possessions in British Malaya, French Indochina, and the Dutch East Indies, it could potentially gain territories that distant colonial powers might be prepared to surrender without a wholehearted fight, given the military pressures they felt upon their homelands by advancing German forces.[104] At Pearl Harbor, this gambit ultimately failed, precipitating US entry into global war with a passion that had failed to muster prior to this direct assault. Tokyo may have viewed US Hawaiian possessions as distant colonial territories on par with Guam and the Philippines, of importance to US forces but not sacred to the American people—but the United States did not.

In Alaska, however, Tokyo's gambit largely succeeded, at least for a time—Japan ultimately held the outer Aleutians for just over a year. Kiska, invaded on June 6, followed by Attu on June 7, 1942, remained Japanese-held until Attu was liberated on May 30, 1943. After a tough slog across treacherous, fog-shrouded muskeg by a joint US-Canadian force, over 500 Allied troops were dead, more than 1,000 were injured, and up to 2,000 more suffered nonbattle injuries, the majority from exposure; nearly the entire Japanese garrison of more than 2,000 were killed with only 28 or 29 survivors. Operations to liberate Kiska did not commence until August 15, providing sufficient time for preparation after the challenges encountered on Attu, with the island declared secure on August 24.[105] Ironically, its entire Japanese garrison of over 5,000 men had been evacuated under the cover of fog on July 28, but the US counterinvasion, despite the absence of opposing forces, nonetheless resulted in over 300 killed due to friendly fire, accidents, and the loss of the destroyer USS *Abner Read* from a mine strike—making Wintermute's aforementioned "war in the fog" especially pertinent.

Some historians have argued that control of the outer Aleutians was perceived by Japan to be critical to protect its main islands from US aerial assault, and thus to prevent the United States from gaining a foothold in the Kuril Islands in Japan's north, after the Doolittle Raiders struck Tokyo in the aftermath of the attack on Pearl Harbor. But renowned Alaskan military historian John Haile Cloe suggests this explanation for Japan's Aleutians invasion may be overstated, and notes the operation was planned well before the Doolittle Raiders struck, and that Japanese strategists knew the raiders did not come from the north.[106] Other historians, including Brian Garfield in his seminal 1969 book *The Thousand-Mile War: World War II in Alaska and the Aleutians* (which was a Pulitzer finalist in history that year), have argued that Tokyo's primary objective was to divide the US Pacific fleet (and its commanders' strategic attention) with a northern feint synchronized with its attack on Midway. But the loss of Attu and Kiska was not so significant to overtax US war planning or waging capabilities, or to materially undermine US efforts at Midway. Indeed, Tokyo's June 1942 assault of Alaska, bringing war even closer to the US mainland, was greeted with a relative shrug—which several observers have argued was because many details of the bombing and subsequent invasion and occupation were censored from American awareness.[107] The bombing of Dutch Harbor did only inflict modest damage, particularly when compared to the steep losses suffered at Pearl Harbor. Furthermore, by the time of the invasion of the outer Aleutians, the Pacific was already fully embroiled in war of such scale and intensity that the limited destructiveness of the assault on Alaska may have contributed to its relatively dispassionate reception. It was also overshadowed by the far more kinetic and costly Guadalcanal campaign in the Solomons. US and Canadian war planners responded to Japan's audacity—and its ability to hold territories once under the American flag—planning accordingly for the inevitable counteroffensive and accelerating efforts already underway to fortify Alaska through a program of rapid military modernization and development. That included building the Alaska-Canada (Alcan) Highway across the far northwest, roughly following the Northwest Staging Route, which more tightly integrated Alaska with the Lower 48 through modern infrastructure and transportation systems, and erecting the Canol Pipeline across the rugged Yukon to the oil fields of Norman Wells. At an operational level, one can conclude with certainty that the war in the Aleutians was in no way overlooked or forgotten by Alaska's defenders.

The Aleutians have long been recognized for their strategic potential. Geographers refer to the Aleutians as the North Pacific stepping stones or the stepping stones of giants, a concept that appealed to theorists of air and sea power who recognized the strategic importance of the region to the security of North America and Northeast Asia well before Japan's assault.[108] But even if Tokyo believed the Aleutians were "stepping stones" to Japan, most historians of the Aleutians war, including John Haile Cloe and Galen Roger Perras, believe that in reality, they proved to be little more than "stepping stones to nowhere," as Perras has described them.[109] Instead of precipitating an impassioned response like the earlier attack on Pearl Harbor, the United States demonstrated its resolve through restraint, waiting and preparing for one of its first efforts to reclaim Japanese-held islands across the vast Pacific. To retake these islands in such forbidding terrain required much planning and time. Weather and geography initially favored the defenders, with Aleutian geopolitics enhancing the old Clausewitzian precept that defense is superior to offense by three to one.

A War by Many Names: From "Williwaw" to "Forgotten" War

When the US counterstrike ultimately commenced, it was known by many names. One was the popular phrase the "Williwaw War," named for the Aleutians' sudden violent gusts of wind accompanied by rain, snow, and fog—where the greatest foe was the frostbite on ill-attired forces lacking proper winter clothing for the engagement.[110] In the foreword to the 1995 edition of Garfield's influential *Thousand-Mile War*, Terrence Cole recounts one B-24 bomber pilot described it as "the weirdest war ever waged," which Cole further describes as a "three-sided battle among the United States, the Japanese empire, and a force that proved to be more powerful than either Washington or Tokyo: the weather."[111] This sentiment was reflected by Major Ira F. Wintermute in his August 1943 "War in the Fog" chronicling how a "solid front of fog" forced his Liberator to ditch in frigid seas on his mission to Kiska.[112] While the Clausewitzian insights captured by Wintermute's "War in the Fog" are compelling, by far the most common motif for the Aleutians conflict is as a "forgotten war."

Indeed, Stan Cohen captured this in his *The Forgotten War: A Pictorial History of World War II in Alaska and Northwestern Canada*, as did Fern Chandonnet in *Alaska at War: The Forgotten War Remembered*, John Haile Cloe in *Attu: The Forgotten Battle*, and Lt. General David Huntoon in his 1988 monograph *The Aleutians—Lessons from a*

Forgotten Campaign (which the author further notes was waged in "a forgotten and lesser known region within the forgotten and unknown territory")—all describing a military engagement that took place largely outside the imagination and awareness of the American public.[113]

Military censorship may have fostered a general disconnect between the isolated Alaskan frontier and the US public, a case several historians and lay authors have made. However, that view has generated some criticism for overstating a news blackout that was not only porous but, like elsewhere in the war zone, much less a blackout than a subtler "blue-pencilling." Effective blue-pencilling by censors could help dull, but never fully erase, our collective memory of the long Allied preparations to retake the islands and protect the Alaska mainland, the fiercely opposed counterattack on Attu, and the unopposed Canadian-US counterassault on Kiska with its tragic toll in friendly fire casualties as Canadian and US forces approached each other from opposite sides of the island, each thinking it was taking fire from the enemy when in fact the only enemy remaining on the island was the fog of war.[114]

Despite taking place beyond the horizon of many Americans' awareness, the Aleutians war yielded much insight into the enemy and proved an important field test of strategy and tactics essential to the liberation of so many more Pacific islands. It resulted in the recovery of the "Akutan Zero," which crashed on Akutan island and was fully restored, and repeatedly tested, revealing the aircraft's few vulnerabilities that could be exploited in battle, thus "cracking the Zero mystery."[115] It revealed the desperate fury of Japanese Banzai charges, the first of which took place at the end of the Battle of Attu, one of the largest such attacks of the war, as described in detail by Rob Szasz, translator and historian of Japanese tactics.[116] Finally, the Aleutians war also witnessed what naval historian Samuel Eliot Morison described as the longest daylight surface naval battle of the Pacific War, and the last to take place between large surface ships without participation of aircraft or subs: the battle of the Komandorski Islands on March 26, 1943—a battle that "lasted without a break for three and a half hours of daylight."[117]

Unforgetting Forgotten Wars

The war in the Aleutians is not, of course, our only "forgotten war." Many conflicts are similarly considered to be forgotten, going as far back as the War of 1812, which ended without a clear-cut victory after three years of hostilities. Another three-year-long conflict, the Korean War, also ended in a stalemate—"wedged between America's biggest war, World War II, and Vietnam, our [then] longest war, Korea sometimes gets overlooked."[118] In similar ways, our two-decade-long war in Afghanistan—usurping Vietnam as our longest while matching Korea for its lack of decisiveness—is also a forgotten war, distinct perhaps for becoming forgotten by the US populace while still underway. This is also the case, to some degree at least, for the "fiasco" of the Iraq War, as famously described by Thomas E. Ricks.[119]

The war in the Aleutians is unlike these other less vividly remembered conflicts in that it brought us an important early victory over Japan and can be considered the beginning of our successful rollback of Japanese military power from the Pacific.[120] Illustrating its relatively forgotten legacy in the public imagination, the Aleutians war remains little-taught today in most of the Lower 48, and many Americans don't know Japan occupied the outer Aleutians during World War II—or that it took such an effort by US and Canadian soldiers, sailors, and airmen to regain this remote territory. The silence that so many Aleutians veterans brought home from their war may have contributed to our collective forgetting. As *New York Times* reporter Sarah Ferguson describes in "The Unknown War," "American soldiers stationed in the remote, windswept archipelago stretching for 1,100 miles west of the Alaskan mainland fought a secret war against Japanese invaders. Driven mad by the cold, the isolation and the 'dim, colorless light,' some were taken away in straitjackets, gazing into the distance, 'unfocused, unspeaking.' The rest of the country didn't even know they were there."[121]

As we approach the eighth decade since the Aleutians were invaded, enough time has now passed to lift this persistent fog in our collective memory. Much can be learned from revisiting our forgotten wars, and from the perspectives of participants on all sides. For the Aleutians war this includes antagonists in the conflict in addition to Allied partners and Indigenous communities caught in the crossfire, for whom it remains an unforgettable event. The Unangan of the outer Aleutians were displaced from their home communities to Hokkaido by the Japanese and to evacuation camps in southeast Alaska by US forces—resulting in protracted suffering and the creation of "lost villages" that remain uninhabited today. Our process of unforgetting can begin with a determined effort to understand their experience of loss and displacement from their homeland.[122]

Understanding the nature of remote, peripheral conflicts is especially relevant in today's increasingly fractured world, since future wars are expected to be much the same as our current and past "wars in the shadows." When looking back upon our history, forgotten wars may in fact be quite common—and need not always be remote, peripheral engagements fought far from home, as exemplified by the Aleutians war as well as by the War of 1812, both fought on North American soil. If remoteness is not the cause of our collective amnesia, then perhaps they became forgotten wars because they did not go well—either perceived to be poorly fought or somehow failed to culminate in a decisive victory, yielding ambiguous outcomes. Indeed, the War of 1812, Korea, and even our twenty-year struggle in Afghanistan are all wars that yielded stalemates—with the postwar contours of peace not appreciably differing from the status quo ante. But the Aleutians war did bring us a decisive and illustrative victory, just not an easy one. Perhaps because the tactical victory achieved on Attu and Kiska came two long and bloody years before Japan's surrender, its victory became overshadowed by the continuing struggle.

With the Arctic region now emerging from remote periphery to strategic crossroads as its ice cover declines, the Aleutians may yet fulfill their strategic potential as "stepping stones of giants," imagined long ago. In the warmer and more interconnected world now emerging, the Aleutians may indeed become critical stepping stones in multiple directions. Amid the rekindling of strategic interest in the Aleutians and adjacent waters it is now imperative that we begin the process of "unforgetting" this forgotten war, as there is much of importance that we can learn from our renewed efforts of remembrance.

From Japan's Twentieth-Century Aleutian Conquest to China's Twenty-First-Century Polar Silk Road: Historical Memory and Reimagined Arctic Futures

Ever since its strategic retreat from outer Manchuria near the end of the Qing dynasty, China formally ceded its claims to what is now the Russian Far East, leaving Japan as Asia's preeminent near-Arctic state. Now Beijing, concomitant with its rise as a great power, is eager to reclaim its earlier near-Arctic status.[123] Indeed, much attention has been devoted to China's Arctic ambitions, particularly since Beijing's white paper on Arctic policy was released on January 26, 2018, which asserted that "China is an important stakeholder in Arctic affairs. Geographically, China is a 'Near-Arctic State,' one of the continental States that are closest to the Arctic Circle." While Beijing's assertion drew widespread attention both within and beyond the Arctic policy world, not everyone embraced China's claim of near-Arctic statehood. This became evident in the recent comments from then secretary of state Mike Pompeo in

Rovaniemi at the last Arctic Council ministerial in May 2019. There, Pompeo dismissed the entire concept as pure fantasy: "Beijing claims to be a 'Near-Arctic State,' yet the shortest distance between China and the Arctic is 900 miles. There are only Arctic States and Non-Arctic States. No third category exists, and claiming otherwise entitles China to exactly nothing."[124] Secretary Pompeo did not dismiss China's interest in the Arctic or its ambition to engage with the Arctic diplomatically or economically. Indeed, transparent Chinese investment would always be welcome, but not—Pompeo added—debt-trap diplomacy of the sort that has overwhelmed governments from Sri Lanka to Malaysia, which became "ensnared by debt and corruption," a fate Pompeo wanted Arctic nations and vulnerable Indigenous communities to be spared.

A look at the map does indeed throw some cold water on Beijing's claim as a "Near-Arctic State." Beijing lies 3,468 miles from the North Pole but is only 2,747 miles from the equator. This suggests that China is more accurately a near-equatorial state than a near-Arctic one. But importantly, the world's highest point, the summit of Mount Everest at 29,035 feet, falls within the jurisdiction of China, as do four of the world's highest lakes. Comparable to India, which due to its own proud Himalayan heritage has long considered itself a polar power (the Himalaya being described as Earth's "third pole"), this qualifies China as much as India as a nation with inherent polar and cryospheric interests, if not necessarily "near-Arctic" by virtue of its latitudinal coordinates. But even these noteworthy claims of extreme geographical prowess were only attained by Beijing's brutal conquest of the independent kingdom of Tibet, a country that still suffers under the yoke of Chinese occupation, raising troubling questions on par with the sort currently being pondered with regard to Russia after its Ukraine invasion. Moreover, as the late neorealist IR luminary Kenneth Waltz once told me, polarity is inherently dyadic; so even if the Himalayas are often described as a third pole, it cannot really be so. Just as Pompeo argued there are only Arctic states and non-Arctic states, there are only two poles—north and south—but no third pole. As Waltz explained, there's simply no such thing as either monopolarity or multipolarity, despite the best efforts of many revisionist IR theorists to so argue when the bipolar order of the Cold War collapsed. If there can be no third pole, then China like India may be a Himalayan power, but this does not provide entry to the Arctic club.

Ironically, China may well have been not only a near-Arctic state but Asia's preeminent near-Arctic power had it retained sovereign control of the far northeastern corner of the Qing empire, which came firmly under Manchu dominion by the seventeenth century. It was recognized as such by its regional rival, Russia, in the 1689 Treaty of Nerchinsk, with the Manchurian border reaching as far north as the 56th parallel and as far east as the Sea of Japan across from Sakhalin, which was already a Chinese tributary since the preceding Ming era. This historical case is perhaps China's strongest for near-Arctic status, but one that Beijing, curiously, does not make publicly—but which perhaps it should. In the mid-nineteenth century, the Russian Empire expanded into outer Manchuria, and these territorial gains were formally conceded by the Qing dynasty in the 1858 Treaty of Aigun and the 1860 Treaty of Peking, resulting in today's Russo-Chinese border that wraps around Manchuria, much the way the trans-Siberian railroad hooks to the south as it approaches its eastern terminus in Vladivostok.

Had China not recognized Russia's imperial expansion onto Manchurian lands, and instead successfully contested Russia's sovereignty over these lands, it would be a near-Arctic state today. Even a vocal assertion by Beijing of traditional sovereign dominion over these lands would help convey a sense of urgency to its near-Arctic aspirations, much the way it has done in the South China Sea. But as Beijing fortifies contested islands to its southeast while reiterating its impassioned (even if not recognized as valid) claims, it remains silent on its long-surrendered northeastern lands, leaving Russia the undisputed sovereign of Eurasia's far northeast. Indeed, around the same time that China retreated from outer Manchuria, its neighbor Japan began to expand north to Hokkaido, in part to prevent further Russian expansion toward Japan and later to crush the last stronghold of Tokugawa loyalists who rose up in rebellion in 1868. Japan's sovereignty formally expanded to Hokkaido the next year and initiated a broader colonial expansion that later reached Sakhalin, the Korean Peninsula, and Manchuria itself. Two years earlier, the United States, via the Treaty of Purchase (1867), had gained possession of Alaska from Russia; Canada finalized its purchase of Rupert's Land from the Hudson's Bay Company in 1870, just three years after Canada's confederation as an independent state in 1867. The sovereign boundaries of much of today's Arctic were thus established during a relatively brief period of northward state expansion in both North America and Eurasia, but because of China's southward retreat during this same period, it remains, as Secretary Pompeo has pointed out, a non-Arctic and not a near-Arctic state.

But is there really no such thing as a near-Arctic state as Secretary of State Pompeo has argued, or is it just that China, according to Pompeo, is not a member of this club—despite its recent assertions? And if China is not a member, then who might qualify? A downward look at the globe from the North Pole reveals there are, indeed, near-Arctic polities adjacent to the Arctic region all around the world, such as the island state of Iceland in the high North Atlantic. Iceland abuts the Arctic Circle, but its territory otherwise lies to its south, though its exclusive economic zone does, at the northwestern end of the island, extend beyond the Arctic Circle to as far north as the 69th parallel. As such, it nicely illustrates the concept of near-Arctic statehood. By virtue of its northern geography and significant glaciation, its millennium-long sovereign experience and centuries of subsistence fishing culture, and its possession of a sliver of land adjacent to as well as a sector of sea to the north of the Arctic Circle, it is increasingly accepted by the world community as a bona fide Arctic state—so much so that it is a founding member of the Arctic Council and current holder of its rotating chairship. But to its disappointment, Iceland was excluded from the 2009 Ilulissat conference of the Arctic 5 states held in Greenland, reflecting a view held then (but less so today) that not all Arctic states were equal in their Arcticness. Iceland has worked diligently to correct this impression among its peers and since 2013 has annually hosted its must-attend Arctic Circle assemblies, where even China, as part of its campaign for recognition as a near-Arctic state, hosts a swanky, deliciously catered reception. While China's near-Arctic aspiration is little more than wishful thinking, geography, climate, and culture align to suggest Iceland would be an archetypical near-Arctic state if ever there were one.

There's another logical member of the near-Arctic club, I believe, and that is Japan. It was briefly an Arctic power, when at the zenith of its global empire in 1942 it held the outer Aleutians for over a year. While it is true that the Aleutians are well south of the Arctic Circle, they are included by virtue of their climate, geography, and Indigenous cultures in many definitions of the Arctic region (including the United States' own official definition), making their possession a qualifying precondition for being a bona fide Arctic state. Their subsequent dispossession restored Japan to its previous status of near-Arctic statehood by virtue of Hokkaido's adjacency to the Russian Far East and the eastern terminus of the Northern Sea Route. Its many other northern climatic, geographic, and cultural features, including winter sea ice, record snow falls, and long cold winters, instill in Japan a northernness familiar to residents of the circumpolar Arctic and sub-Arctic, as well as its own Ainu Indigenous culture, whose traditional homeland extends all the way up to Sakhalin, Kamchatka, and the northern Kurils. That Russian president Vladimir Putin recently described the Ainu as an Indigenous people to Russia further reinforces Hokkaido's near-Arctic status (being part of the same northern, Indigenous homeland that Putin now recognizes as part of Mother Russia), much the way the long presence of Indigenous Unangan (Aleut) culture reinforces the Aleutians' (and by extension, the United States') Arcticness and the long Inuit cultural presence reinforces Canada's Arcticness.

So there is indeed much merit to the concept of near-Arctic statehood, if properly qualified by climate, geography, and culture. But for China, as Secretary Pompeo observed, the claim of being a near-Arctic state is, at least since the Qing dynasty's retreat from outer Manchuria, out of date. However, with Russia weakened and isolated by its war in Ukraine, its attempt to change borders in Europe by force widely perceived as a strategic miscalculation, it depends more than ever on China. This could in time serve China's Arctic ambitions, and the restoration of its earlier control over sub-Arctic Northeast Asia is now a possibility to consider. China, as a rising power, should not only participate in the Arctic region's economic development—so long as it remains transparent and in the interest of northerners—but perhaps by virtue of its investments in the region also strengthen its historic claims as Moscow's own sovereign hold over the region weakens.

Japan is an archipelagic state comparable to the island state of Iceland for its geostrategic importance adjacent to emergent Arctic sea lanes and increasingly active Russian naval bastions. Thanks to its own northward expansion to Hokkaido around the same time as the Manchu retreat from outer Manchuria, it presents us with an important, allied gateway to the eastern entrance to the Northern Sea Route. These island states will become especially important in wartime as fortified bastions from which to take the fight to the enemy (just as we experienced in World War II in the case of Iceland), for which we prepared diligently during the Cold War. Fortunately for the United States, Japan and Iceland remain dedicated allies and longtime strategic partners that help bring stability to both the North Pacific and the North Atlantic, and to the increasingly important Arctic basin. And, just as fortunate, Japan and Iceland stand out for their uniquely important role in world politics as bona fide near-Arctic states, a status that China lost more than a century and a half ago but that it could, if it plays its cards right in the years ahead, regain.

China hawks concerned with Beijing's continued rise as a great power and its growing ambition to be universally recognized as an unthreatened superpower worry that Beijing is not content to accept the boundaries of the world shaped by the first wave of globalization in 1867, when China was in retreat and Japan was rapidly ascending as a world power. As CNN reported in 2020,

> In a speech on the 150th anniversary of the birth of Sun Yat-sen—seen by many as modern China's founding father—President Xi Jinping pledged to "resolutely oppose" any attempt to divide the country. "We will never allow anyone, any organization or political party to rip out any part of our territory at any time or in any form," he said, standing under a giant portrait of Sun. A key tenet of Xi's rule has been his pledge to restore the country to greatness, undoing the "century of humiliation" during which the Qing Empire and later the Republic of China were laid low by foreign powers, with territories including Hong Kong, Manchuria and much of Shanghai shaved off into colonies and concessions.[125]

What if Beijing were not only unprepared to tolerate future fragmentation but also intent on restoring its undisputed sovereignty over all of its onetime domain at the height of its (or more properly, the preceding) Qing empire? Beijing's hard line against dissent at home, its mass incarceration and reeducation of Uighurs in Xinjiang, its unilateral de facto abrogation of Hong Kong's long-running "One Country, Two Systems" constitutional autonomy in place since the British handover and recent imposition of its new national security law, and its increasingly bold rhetoric regarding Taiwan's long, implicitly acknowledged sovereign autonomy could be but a prelude to a forceful Chinese expansion into its old imperial footprint.

Were this to happen in a tectonic transformation of East Asia, China could, through force of arms and/or the threat thereof, reemerge as a truly near-Arctic state, fulfilling Beijing's 2018 claim of such a status. While not considered a probable outcome, it is nonetheless a potential scenario and one that would introduce a new threat to Arctic exceptionalism, particularly in light of Beijing's nonrecognition of Indigenous rights in mainland China, in marked contrast to the Arctic states, which to varying degrees have embarked along a historic path of state-tribe reconciliation. Even Japan, whose hegemonic rule across the Asia Pacific during its years of empire was, as viewed by most, brutal, has shown progress both at home with its gradually increasing recognition of Ainu rights and in world politics through its participation on the Arctic Council, an effort matched at some levels by Russia (albeit with much room for forward progress given the comparably low starting point).

If China were to expand northward without reforming its Indigenous rights recognition at home, it would put Arctic exceptionalism under new strain or at the least limit its applicability to the non-Arctic world. But if Beijing were to meaningfully reform its approach to its own Indigenous minorities after engaging respectfully and proactively with Indigenous peoples of the Arctic, as Beijing continues to win diplomatic points at the Arctic Council for its engagement with the Permanent Participants, we might see Arctic exceptionalism transform world politics through such a transformation of China. (Another related potentiality, albeit highly improbable, is for Taiwan, already a key if unsung player in mainland China's successful economic modernization, to come to power on the mainland through a collapse of one-party communist rule there, comparable to what happened in the Soviet Union at its end. After its failed coup against Mikhail Gorbachev in August 1991, in which co-plotter Gennady Yanayev briefly assumed the role of acting president, the Communist Party lost its monopoly control and unwittingly precipitated the collapse of Communist Party legitimacy in Russia. Taiwan has made great strides in its recognition of Indigenous rights on the island compared to Beijing's track record on the mainland, though recent judicial setbacks on Indigenous hunting rights in Taiwan show there is still room for improvement.)

Whether China internalizes the state-Indigenous reconciliation that is for me the most salient feature of Arctic exceptionalism remains to be seen. Thus far their evolution since asserting great power interest in the Arctic region earlier in the twenty-first century has demonstrated a willingness to respect Arctic principles and not to disrupt the Arctic system as much as critics have feared they might—though the night is still young, as the saying goes. As Lackenbauer and Dean observe,

> For some scholars, China represents an inherently benign actor, either as a country with no pernicious designs for the region (perhaps a naïve case of "Arctic exceptionalism" given its behavior elsewhere in the world) or as one seeking to play a constructive role in circumpolar affairs and Arctic development in accordance with established norms. Other authors have cast strong suspicion at Beijing, arguing that this Asian great power is embarking on a "long-con" or "bait-and-switch" strategy where it will seek to undermine the sovereignty of Arctic states and co-opt regional governance mechanisms to facilitate access to resources and new sea routes to fuel and connect its growing global empire.[126]

Lackenbauer and Dean further explain,

What Western commentators saw as an initial Chinese push to internationalize the Circumpolar North a decade ago was promptly rebuffed by the Arctic States and ran contrary to Chinese efforts to nationalize the East and South China Seas, leading China to recalibrate its approach. Pushing for regional change beyond the tolerances of the Arctic States would risk major trading relationships that already supply cheaper natural resources from elsewhere than can be secured from the Arctic. China has little to gain from upsetting the Arctic—a region of limited consequence to it compared to other parts of the world—and much to lose. Instead, by refraining from overt repudiations of "Arctic exceptionalism" and playing within the regional governance rules set largely by Arctic states with prestige and influence within the international system, China can win trust and accrue "political capital" through good international behaviour. As part of a global strategy, China may choose to forego its preferences to "internationalize" the Arctic, play by the regional rules to showcase how it abides by international law and norms, and then make a decisive revisionist move closer to home.[127]

Notes

1. Declaration on the Establishment of the Arctic Council (the Ottawa Declaration), preamble, 1.
2. Ibid.
3. Ibid.
4. Ibid., section 1, 1–2.
5. Ibid., section 2, 2.
6. Ibid.
7. Ibid.
8. Ibid., section 3, 2.
9. Young, "Governing the Arctic," 9–15.
10. See: White House, Presidential Decision Directive 26 (PDD/NSC-26), June 9, 1994. https://irp.fas.org/offdocs/pdd/pdd-26.pdf; White House, National Security Presidential Directive-26 / Homeland Security Presidential Directive-25 (NSPD-66/HSPD-25), January 9, 2009, www.fas.org/irp/offdocs/nspd/nspd-66.htm; White House, National Strategy for the Arctic Region, May 2013, https://obamawhitehouse.archives.gov/sites/default/files/docs/nat_arctic_strategy.pdf; and White House, National Strategy for the Arctic Region, October 2022, https://www.whitehouse.gov/wp-content/uploads/2022/10/National-Strategy-for-the-Arctic-Region.pdf. For an overview of the new policy, see Andrew C Revkin, "Ice Retreat Prompts Bush Shift in Arctic Policy," *New York Times*, January 13, 2009.
11. Dalee Sambo Dorough, "A New Gold Standard for Arctic Policies," Arctic Circle Assembly, as presented on YouTube, May 7, 2020, www.youtube.com/watch?v=ItbUl_gmS5M.
12. Crown-Indigenous Relations and Northern Affairs Canada, foreword from the minister, Arctic and Northern Policy Framework, www.rcaanc-cirnac.gc.ca/eng/1560523306861/1560523330587.

13. Peter Kikkert and P. Whitney Lackenbauer, "Canada's Arctic and Northern Policy Framework: A Roadmap for the Future?," Briefing Note, *2019 Arctic Yearbook*, https://arcticyearbook.com/arctic-yearbook/2019/2019-briefing-notes/332-canada-s-arctic-and-northern-policy-framework-a-roadmap-for-the-future.

14. Ibid.

15. Ibid.

16. Ibid.

17. Ibid.

18. Ibid.

19. Ibid.

20. Ibid.

21. US Coast Guard, *Arctic Strategic Outlook* (April 2019), 14, 32, www.uscg.mil/Portals/0/Images/arctic/Arctic_Strategy_Book_APR_2019.pdf.

22. Ibid., 5, 7, 18, 21, 35, 38, 40.

23. Ibid., 5, 14, 19, 21, 28, 29, 34, 39–42.

24. US Air Force, *The Department of the Air Force Arctic Strategy: Ensuring a Stable Arctic Through Vigilance, Power Projection, Cooperation and Preparation* (July 21, 2020), 4, 12, www.af.mil/Portals/1/documents/2020SAF/July/ArcticStrategy.pdf.

25. Office of the President of the Russian Foundation, *Foundations of the Russian Federation State Policy in the Arctic for the Period up to 2035*, translated by Anna Davis and Ryan Vest, Russia Maritime Studies Institute, US Naval War College, https://digital-commons.usnwc.edu/cgi/viewcontent.cgi?article=1004&context=rmsi_research: see General Terms, 4; Assessment of National Security in the Arctic, 5; Goals, Primary Trends, and Objectives of the State Policy of the Russian Federation in the Arctic, 6, 7, 9, 10.

26. Ibid., 4, 14.

27. State Council Information Office of the People's Republic of China, *China's Arctic Policy*, 1st ed., January 2018, https://english.www.gov.cn/archive/white_paper/2018/01/26/content_281476026660336.htm: see Part I: The Arctic Situation and Recent Changes, 3; Part III: China's Policy Goals and Basic Principles on the Arctic, 5–6; Part IV: China's Policies and Positions on Participating in Arctic Affairs, 9, 11.

28. Headquarters for Ocean Policy, Prime Minister's Office of Japan, Japan's Arctic Policy (October 16, 2015), 1–3, www8.cao.go.jp/ocean/english/arctic/pdf/japans_ap_e.pdf.

29. Rob Huebert, "The New Arctic Strategic Triangle Environment (NASTE): Ramifications for Canada," presented to Special Senate Committee on the Arctic, Ottawa, April 3, 2019, https://sencanada.ca/content/sen/committee/421/ARCT/Briefs/RobertHuebert_e.pdf.

30. On Singapore's courtship of Arctic Indigenous peoples through its soft power approach to diplomacy, see the University of the Arctic's Singapore–Arctic Council Permanent Participants (AC PP) Cooperation Package, "Postgraduate Scholarships in Singapore for Arctic Indigenous Students," May 22, 2015, www.uarctic.org/news/2015/5/postgraduate-scholarships-in-singapore-for-arctic-Indigenous-students. Also see Bennett, "Singapore," 289–310; Hema Nadara, "How Singapore Legitimises Its Presence in the Arctic Council," *Today Online*, July 24, 2018; Storey, "Arctic Novice," 66–72. For a critical if now somewhat dated discussion of Japan's Arctic Council observership and its Ainu rights policy, see Hossain and Maruyama, "Japan's Admission to the Arctic Council," 169–187.

31. Snook, Cunsolo, and Dale, "Conceptual Representation of the 'Shared Space' of Co-management," 52–56.

32. "Blinken to Head North on Arctic Trip," NPR, *Weekend Edition Saturday*, May 15, 2021.

33. Zellen, "Arctic Strategies and Indigenous Engagement by the Numbers." Unpublished Paper, March 2021.

34. *Foundations of the Russian Federation State Policy in the Arctic for the Period up to 2035*, 4.

35. Ibid., 10.

36. *China's Arctic Policy*, 5.

37. Ibid., 5–6.

38. Ibid., 6.

39. Jan Husdal, "The Final Frontier: The Northern Sea Route," Husdal.com, May 22, 2011; Schøyena and Bråthenb, "Northern Sea Route." Also see "More Oil and Ore Along Northern Sea Route," *Barents Observer*, February 11, 2011.

40. See "Russian Sub Plants Flag at North Pole," Reuters, August 2, 2007; Mike Eckel, "Russia Defends North Pole Flag-Planting," Associated Press, August 8, 2007; Dodds, "Flag Planting and Finger Pointing," 1–11; CTV News Staff, "Canada, Russia Locked in Dispute over Arctic Border," CTV.ca, September 16, 2010.

41. As cited in Graham Chandler, "Pole Position," *Up Here*, December 18, 2007. Also see Commission on the Limits of the Continental Shelf (CLCS), "Outer Limits of the Continental Shelf Beyond 200 Nautical Miles from the Baselines: Submission by the Russian Federation," updated June 30, 2009, www.un.org/depts/los/clcs_new/submissions_files/submission_rus.htm.

42. Chandler, "Pole Position"; CLCS, "Outer Limits."

43. Yereth Rosen, "US-Canadian Mission Set to Map Arctic Seafloor," Reuters, August 1, 2010; Arctic Governance Project, "The Canada-US Arctic Cooperation Agreement," http://arcticgovernance.custompublish.com/the-canada-us-arctic-cooperation-agreement.4668242-142904.html; "Canada and Denmark to Share Hans Island? 37-Year Dispute over One-Kilometre Square Rock May Be Coming to an End," *Macleans*, November 9, 2010. Also Anido, "Greenland Is an Important Neighbour"; Peter Juel Thiis Knudsen, "Strong Arm Not Wanted in Arctic," *Ottawa Citizen*, May 18, 2011; John Ivison, "How to Keep a Cool Head in the Arctic," *National Post*, March 30, 2010; Elin Hofverberg, "The Hans Island 'Peace' Agreement Between Canada, Denmark, and Greenland," *Library of Congress Blog*, June 22, 2022.

44. Luke Harding, "Russia and Norway Resolve Arctic Border Dispute: Treaty Allows for New Oil and Gas Exploration and Settles 40-Year Row over Barents Sea," *The Guardian*, September 15, 2010; "Russia, Norway Border Agreement Opens Arctic Up to Exploration," *New Europe*, September 19, 2010.

45. On the renewal of Cold War–like tensions between Russia and the West, see M. K. Bhadrakumar, "US Breathes Life into a New Cold War," *Asia Times Online*, June 7, 2011.

46. As described on a Northern Sea Route website (http://northern-sea-route.co.tv), and by Brigham, "Northern Sea Route, 1997," 219–224. In the years since, interest in and use of the route have been steadily rising.

47. See Mackinder, "Round World," 595–605. Also see Zellen, *Arctic Doom, Arctic Boom*; Antrim, "Next Geographical Pivot."

48. See Brigham, "Russia Opens Its Maritime Arctic"; Antrim, "Next Geographical Pivot."

49. "Russia Outlines Arctic Force Plan," BBC News, March 27, 2009; "Moscow to Deploy More Forces in Arctic," *Russia Today*, March 30, 2011.

50. See Dan Shapley, "Global Warming, the Arctic Thaw and the New Cold War: Why Russia's Incursion into Georgia Bodes Ill for the Climate," *The Daily*

Green, August 18, 2008; Robert Coalson, "A Year After Russia-Georgia War—a New Reality, but Old Relations," *Radio Free Europe/Radio Liberty*, August 5, 2009.

51. For details on the final years of Russian America, see Mazour, "Prelude to Russia's Departure from America," 311–319; Golder, "Purchase of Alaska," 411–25; Farrar, *Annexation of Russian America*. For more on the lingering rawness, see Matthew Holroyd and Sophia Khatsenkova, "'Alaska Is Ours': Local Russian Advertisement Sparks Outrage on Social Media," *Euronews*, July 13, 2022.

52. Vladimir Solovyov and Elena Klepikova, *Zhirinovsky: Russian Fascism and the Making of a Dictator* (New York: Perseus Books, 1995); Timothy Egan, "The World; Alaskans Don't Want to Be Anyone's Siberia," *New York Times*, December 19, 1993; David McHugh, "Zhirinovsky Snubs Court Date Again," *Moscow Times,* December 18, 1997; "Zhirinovsky Beat: At Home with Russia's Top Ultranationalist," *Time,* February 28, 1994. See Holroyd and Khatsenkova, "'Alaska Is Ours,'" for more on the billboards in today's Siberia.

53. P. Whitney Lackenbauer and Ryan Dean, "Arctic Exceptionalisms," in *The Arctic and World Order*, edited by Kristina Spohr and Daniel S. Hamilton (Washington, DC: Foreign Policy Institute/Henry A. Kissinger Center for Global Affairs, Johns Hopkins University, 2020), 337, citing Olga Khrushcheva and Marianna Poberezhskaya, "The Arctic in the Political Discourse of Russian Leaders: The National Pride and Economic Ambitions," *East European Politics* 32, no. 4 (2016): 547–566.

54. Ibid., 338, citing the *Barents Observer*, "Ice on Russia's Northern Sea Route Has Disappeared, Opening Up Arctic Shipping Lanes," as it appeared in the *Moscow Times*, August 29, 2019.

55. Ibid., 338, citing both Ekaterina Klimenko, Annika Nilsson, and Miyase Christensen, *Narratives in the Russian Media of Conflict and Cooperation in the Arctic* (Stockholm: Stockholm International Peace Research Institute, August 2019); and Alexander Sergunin, "Russia and Arctic Security: Inward-Looking Realities," in *Understanding Arctic Sovereignty and Security*, edited by Wilfrid Greaves and P. Whitney Lackenbauer (Toronto: University of Toronto Press, 2021).

56. Ibid., 338–339.

57. Ibid., 339, n42, citing BBC, "Putin: Russian President Says Liberalism 'Obsolete,'" June 28, 2019.

58.. See the Ilulissat Declaration, Arctic Ocean Conference, Ilulissat, Greenland, May 27–29, 2008.

59. Mikhail Gorbachev, "Speech in Murmansk at the Ceremonial Meeting on the Occasion of the Presentation of the Order of Lenin and the Gold Star to the City of Murmansk," Murmansk, Russia, October 1, 1987, www.barentsinfo.fi/docs/Gorbachev_speech.pdf. Also see Kristian Åtland, "Mikhail Gorbachev, the Murmansk Initiative, and the Desecuritization of Interstate Relations in the Arctic," 289–311.

60. Lackenbauer and Dean, "Arctic Exceptionalisms," 341.

61. Ibid.

62. "AP Interview: NORAD Commanders Say Melting Ice and Energy Development Will Make Arctic Busier," Associated Press, May 7, 2011.

63. John Vandiver, "Arctic Pact May Herald Cooperation in Region," *Stars and Stripes*, May 13, 2011.

64. Ibid.

65. As noted on the web page of the National Museum of the US Air Force, in 1924, "Mitchell made an inspection tour of the Pacific and the Far East, after which he made his famous prediction that the Japanese would attack Pearl Harbor on a Sunday morning." See "Brig. Gen. William 'Billy' Mitchell," NationalMuseum.af.mil, February 11, 2010.

66. Reuters, "Russian Sub Plants Flag at North Pole"; Eckel, "Russia Defends North Pole Flag-Planting"; Dodds, "Flag Planting and Finger Pointing"; CTV News Staff, "Canada, Russia Locked in Dispute."

67. CLCS, "Outer Limits."

68. Hattendorf, *Evolution of the US Navy's Maritime Strategy, 1977–1986*; Gray, *Maritime Strategy*; Mearsheimer, "Strategic Misstep," 3–57.

69. Astri Edvardsen, "The US Department of Defense Announces New Arctic Strategy in Early 2024," *High North News*, October 25, 2023.

70. Ibid.

71. Ibid.

72. "The Battle of the Atlantic," *New York Times Magazine*, July 10, 2020; Williams, *Battle of the Atlantic*; Nikolay Gerasimovich Kuznetsov, *Memoirs of Wartime Minister of the Navy,* trans. Vladimir Krivoshchekov (Moscow: Progress Publishers, 1990); Pope, *73 North*; Woodman, *Arctic Convoys*.

73. "Statement by the Department of State on the U.S.-Danish Agreement on Greenland, April 10, 1941," US Department of State, Publication 1983, *Peace and War: United States Foreign Policy, 1931–1941* (Washington, DC: US, Government Printing Office, 1943), 641–647.

74. "Statement by the Department of State."

75. Ibid.

76. Ibid.

77. Ibid.

78. See Defense of Greenland: Agreement Between the United States and the Kingdom of Denmark, April 27, 1951, https://avalon.law.yale.edu/20th_century/den001.asp.

79. Ibid., Article I.

80. Kingdom of Denmark, "History Made in Southern Greenland," Archives 2003–2004, Denmark.dk.

81. Ibid.

82. Ibid.

83. See Zellen, *Arctic Doom, Arctic Boom*, 167; Cloe and Monaghan, *Top Cover for America*, 21; Naske and Slotnick, *Alaska*, 122.

84. Marine Corps Historical Reference Pamphlet, *The United States Marines in Iceland, 1941–1942* (Washington, DC: US Marine Corps Historical Division, 1970), 1, citing J. R. M. Butler, *Grand Strategy Volume II: History of the Second World War* (London: HMSO, 1957).

85. Richard Stengel, Laurence I. Barrett, and Barrett Seaman, "How Reagan Stays out of Touch," *Time*, December 8, 1986.

86. Jonsson, *Friends in Conflict*; "The High Seas: Now, the Cod Peace," *Time*, June 14, 1976; Hugh Clayton, "Dispute with Iceland Means Dearer Fish This Weekend," *The Times,* November 28, 1975, 6.

87. See Kieran Mulvaney, "A New Fisheries Frontier in the Arctic," *Discovery News,* May 25, 2011; PEW Environment Group, "Oceans North International," OceansNorth.org.

88. Ironically, as coastal areas become more accessible, interior regions will become less so—potentially offsetting the benefits of a thaw in terms of net economic effects. Timothy Gardner, "Ice Melt to Close Off Arctic's Interior Riches: Study," Reuters, May 30, 2011.

89. "Terra Daily: Arctic Summer Sea Lanes Open by 2015 Forecasts ONR," SpaceDaily.com, February 14, 2002; Jill Burke, "Brainstorming the Bering Strait Bottleneck," *AlaskaDispatch*, May 12, 2011.

90. Joe Friesen, "Russian Ship Crosses 'Arctic Bridge' to Manitoba," *Globe and Mail*, October 18, 2007.

91. Rebecca Woodgate, Knut Aagaard, Tom Weingartner, Terry Whitledge, and Igor Lavrenov, "Bering Strait: Pacific Gateway to the Arctic," http://psc.apl.washington.edu/HLD/Bstrait/bstrait.html#Basics.

92. Heather Exner-Pirot, "Eye on the Arctic: How Fast Will Arctic Shipping Grow?," *AlaskaDispatch*, April 11, 2011. Also see "Alaska Beat: Korean Gas Execs Explore Arctic LNG Shipping Plan, in Canada," *AlaskaDispatch*, April 20, 2011.

93. Rowena Mason, "Financial Crisis: Iceland Gets €4bn Russian Loan as Banks Collapse," *The Telegraph*, October 7, 2008; also see Konstantin Rozhnov, "Russia's Role in Rescuing Iceland," BBC News, November 13, 2008.

94. Rozhnov, "Russia's Role in Rescuing Iceland."

95. "Greenland Takes Step Toward Independence from Denmark," *The Telegraph*, June 21, 2009; Marianne Stigset, "Greenland Steps Up Its Independence Calls as Oil Ambitions Grow," *Businessweek*, January 11, 2011; Shaun Walker, "Revealed: The Secret Battle for the Riches of the Arctic: Leaked Cables Show How Nations Are Carving Up Pristine Wilderness," *The Independent*, May 13, 2011; Bloomberg, "Business Report: Greenland Bets on Oil, Metals, Cows as Ticket to Independence," SFGATE, May 11, 2011.

96. "Shaping Greenland's Future," Cable 07COPENHAGEN1010, WikiLeaks, November 7, 2007, http://wikileaks.org/cable/2007/11/07COPENHAGEN1010.html.

97. Ibid.

98. Bockstoce, *Furs and Frontiers*; Davis, ed., *Rupert's Land*. Also see Zellen, "Cold Front."

99. Zellen, *On Thin Ice*.

100. Zellen, "Cold Front."

101. I first discussed President Trump's overture to purchase Greenland outright from Denmark and the various constitutional and sovereign forms Greenland might take in the future in 2019. See Barry Scott Zellen, "Greenland as an American Territory: A Bad Idea, a Bold Idea . . . or a Beyond-the-Horizon Idea Worth Studying?," Center for Arctic Study and Policy White Paper, US Coast Guard Academy, August 18, 2019, and Barry Scott Zellen, "Donald Trump Is Thinking of Buying Greenland. That's Not Necessarily a Bad Idea," *Globe and Mail*, August 18, 2019. I revisited the topic two years later while a visiting Fulbright Scholar at the University of Akureyri in 2020. See Zellen, "High Stakes in the High North."

102. Wintermute, "War in the Fog."

103. See US Coast Guard Historical Section, Public Information Division, *The Coast Guard at War, Vol III: Alaska*, February 15, 1946, 65–69.

104. See Flynn, *First Strike*.

105. See Cloe, *Attu*, 113, for a comprehensive discussion of these losses; also see Garfield, *Thousand-Mile War*.

106. See Cloe, *Attu*, 21. For additional discussion of the Aleutians campaign as a feint coinciding with Japan's attack on Midway, see Garfield, *Thousand-Mile War*.

107. Wartime censorship, inclusive of the Aleutians war, is described in Alvin William McDaniel Jr., "The War on Words: The Office of Censorship in World War II" (master's thesis, Department of History, University of the Pacific, May 1972), 95; and Cloe, *Attu*, 25.

108. Kathleen A. Cooper, "'North to Alaska': The Geostrategic Importance of the Last Frontier" (thesis, School of Advanced Air and Space Studies, Air University, Maxwell Air Force Base, Alabama, June 2012).

109. Perras, *Stepping Stones to Nowhere*; Cloe, *Attu*.

110. See Vidal, *Williwaw*; Goldstein and Dillon, *Williwaw War*; Feinberg, *Where the Williwaw Blows*.

111. Garfield, *Thousand-Mile War.*

112. Wintermute, "War in the Fog."

113. See Cohen, *Forgotten War*; Chandonnet, ed., *Alaska at War*; Cloe, *Attu*; David Huntoon, *The Aleutians—Lessons from a Forgotten Campaign* (Fort Leavenworth, KS: US Army Command and General Staff College's School of Advanced Military Studies, 1988).

114. Tyler Parsons, "When the U.S. Fought a Deadly Battle with No One," KnowledgeNuts.com, May 18, 2014. Also see Wintermute, "War in the Fog."

115. Rearden, *Cracking the Zero Mystery*. Also see Elizabeth Hanes, "The Akutan Zero: How a Captured Japanese Fighter Plane Helped Win World War II," History.com, June 4, 2012.

116. Rob Szasz, "Thoughts on the Sacrifice [Gyokusai] of the Attu Garrison," October 25, 2014, https://shatteredjewels.wordpress.com/2014/10/25/thoughts-on-the-sacrifice-gyokusai-of-the-attu-garrison; "The Battle of Attu (アッツ島の玉砕)," October 26, 2013, https://shatteredjewels.wordpress.com/2013/10/26/the-battle-of-attu.

117. Morison, *History of United States Naval Operations*. For an excellent summary of the battle, see Michael Morgan, "Battle of the Komandorski Islands: Close Call in the Aleutians During World War II," History.net.

118. See "The Forgotten War: KOREA," US Army Quartermaster Corps, https://web.archive.org/web/20040531114711/http://www.quartermaster.army.mil/oqmg/Professional_Bulletin/2000/spring2000/The_Forgotte_War_KOREA.htm.

119. Ricks, *Fiasco*.

120. Indeed, as Cloe describes, "The Battle of Attu drove home to the Japanese leadership the fact that they were no longer on the offensive, with the real possibility of losing the war." See Cloe, *Attu*, 114.

121. Sarah Ferguson, "The Unknown War," Sunday Book Review, *New York Times*, January 31, 2014.

122. Mason, "You Can't Go Home Again," 17–29.

123. I first discussed this in Zellen, "China Lost Chance to Be 'Near-Arctic' 150 Years Ago"; Zellen, "China and the 'Near-Arctic': An Opportunity Lost Over 150 Years Ago"; and Zellen, "A Missed Opportunity."

124. Mike Pompeo, "Looking North: Sharpening America's Arctic Focus," US Department of State (2017-21 Archives), May 6, 2019, https://2017-2021.state.gov/looking-north-sharpening-americas-arctic-focus/.

125. James Griffiths, "From the Qing Empire to the People's Republic, China's Worries About Separatism Run Deep," CNN, June 29, 2020.

126. Lackenbauer and Dean, "Arctic Exceptionalisms," 339–340, citing, among others, Roger W. Robinson Jr., "China's 'Long Con' in the Arctic," *Commentaries* (September 2013).

127. Ibid.; and Lackenbauer, Lajeunesse, Manicom, and Lasserre, *China's Arctic Ambitions* ; Ryan Dean and P. Whitney Lackenbauer, "China's Arctic Gambit? Contemplating Possible Strategies," North American and Arctic Defence and Security Network Policy Brief (April 2020), www.naadsn.ca/wp-content/uploads/2020/04/20-apr-23-China-Arctic-Gam-bit-RD-PWL.pdf.

6

The Realist Foundations of a Collaborative Arctic Order

With the largest Arctic state—with the most developed Arctic economy and largest Arctic Indigenous population—presently at war in Ukraine, and with the consensus-driven, inherently cooperative Arctic Council's operations involving Russia suspended by the remaining Arctic states (with a mix of both support and concern expressed by most of the Permanent Participants), many are rightly concerned with the future of Arctic exceptionalism, and whether there will in fact be a future for this hitherto seemingly enduring feature of Arctic international relations. My belief is that, after more than 100,000 years of Arctic globalization—from the prehistoric ice age through to the historic medieval, colonial, and modern worlds—the Arctic will not only remain at the heart of the globalized world but will (eventually) continue to drive Arctic exceptionalism's cooperative dynamics, the present war and its resulting diplomatic tensions notwithstanding.

Geopolitics of a Warming Arctic

This is not the first time war has come to, through, or near the Arctic, and some wars were protracted global cataclysms far graver in geographic reach and in their levels of destruction. And yet Arctic exceptionalism endured throughout. Were it not for the alignment of the entire Western world against fellow council member Russia, and the pause on Arctic Council operations implemented near the start of the war (clarified a few months later to apply only to those operations in which Russia was involved), one could expect Arctic cooperation to have otherwise continued in spite of this regional war to Russia's southeast. Indeed, the foundation for Arctic exceptionalism since the Arctic Council's formation in 1996 was laid during the Cold War, a conflict that managed to help unite the Arctic despite its global bifurcation of international relations.

It was almost four decades ago, at the height of the Cold War, that esteemed Arctic scholar Oran Young described the world as "entering the age of the Arctic, an era in which those concerned with international peace and security will urgently need to know much more about the region and in which policy makers in the Arctic rim states will become increasingly concerned."[1] Any Cold War–era map of the polar region showed the two superpowers standing face-to-face across their common polar frontier, as the logic of geopolitics placed the North Pole at the very center of this bifurcated world, making the Arctic Ocean the modern-day analogue of the classical world's Mediterranean.

But the region's harsh climate and permanent polar ice pack kept development of the region and its integration with the world system limited to just a few large-scale energy, natural resource, and military projects. Apart from nuclear submarines passing silently beneath and exceedingly expensive heavy icebreakers crunching their way across, the polar sea ice remained largely impenetrable during the very Cold War that positioned the Arctic at the seeming center of global geopolitics. Two decades after Young introduced the concept of the "age of the Arctic" to the lexicon of northern studies, something transformative finally did begin to happen as the long frozen polar sea began its rapid thaw, unexpectedly fast to nearly all observers, opening up larger and larger portions of the Arctic Ocean to seasonally ice-free conditions for longer and longer periods of time.[2]

Geopolitical theorist Halford J. Mackinder long ago recognized the strategic interconnections between the Eurasian "heartland" (history's primary "pivot area"), the "inner" or "marginal crescent" (called by many, most famously Nicholas J. Spykman, the "rimland"), and the more distant "outer crescent" of the world island, but he dubbed the frozen Eurasian Arctic "Lenaland" for its strategic isolation.[3] The Northeast Passage (of which the Northern Sea Route is part) and the offshore islands to its north were, before the current era of profound and accelerating Arctic climate change, every bit as remote as Lenaland—of much future anticipated value but doomed by climate and geography to remain of minimal strategic importance while Ice Age conditions persisted. But with our era's polar thaw, Lenaland has the potential to become a new extension of the heartland and to gain equivalency with the rest of this all-important pivot area. Similarly, even more remote archipelagic regions of the Arctic long impenetrable to state power and thus of limited importance to world politics, particularly prior to the militarization of the global air, space, and subsurface domains during the nineteenth and twentieth centuries, may one day emerge as strategically salient rimlands.

That the Arctic was largely insular and archipelagic mattered less in the prior era of near-permanent polar pack ice, which was perceived as a strategic buffer akin to the vast wall of sand of the Sahara. But this is decreasingly the case and will in time profoundly transform the geopolitics of the Arctic and its strategic significance to the world. In particular, sovereign claims to both the Northeast Passage (by Russia) and the Northwest Passage (by Canada) rooted in Section 8, Article 234 of the Law of the Sea Convention (LOSC) will be weakened once they're no longer perceived to be ice-covered areas. Indeed, with the disappearance of much formerly permanent ice, the Arctic's defining geographical feature is its archipelagic nature comprising multitudinous islets and islands that encircle the entirety of the circumpolar region—as if stepping stones for future northern state expansion and development, and even potential conquest. These tightly clustered archipelagos form increasingly salient island chains that must be held in order to defend against external threats or to deter them from rising altogether. In their aggregate, they form super island chains, each component cluster with its own inherent tactical or regionally strategic value.

Key islands can be found all the way from the Aleutian Islands in the west to Wrangel Island in the east—between which are thousands of others, from Herschel Island (44 square miles) just off the Yukon Territory's North Slope, with its storied whaling history and worrisome (to Ottawa) year-round US commercial presence, to the even smaller Hans Island (0.5 square mile), a rocky outcropping midway between Canada's Baffin Island and Denmark's island colony of Greenland, which has been amicably contested by Ottawa and Denmark through what some have dubbed their "whiskey wars"[4] as the armed forces of each nation leave their liquor of choice—Canadian whiskey or Danish schnapps, along with their respective national flags—for their rival to enjoy, to the slightly larger Grimsey (2 square miles), just off Iceland's northern shore, straddling the Arctic Circle and making Iceland a bona fide Arctic state. There is Greenland, of course, the world's largest island, greater in area than the next three largest islands combined, New Guinea, Borneo, and Madagascar. To the immediate west are Canada's

Queen Elizabeth Islands, where a controversial Cold War resettlement of Inuit families from northern Quebec took place, which critics have described as a callous deployment of "human flagpoles" to strengthen Canada's sovereign claims to this lightly settled region.[5] Baffin Island lies toward the south, and farthest west in Arctic North America is the relatively newly settled Banks Island at the western entrance to the Northwest Passage, which has known modern Inuit habitation only since the early twentieth century, when schooners brought Inuit trappers over in search of white fox.[6] To Greenland's immediate east is the independent and sovereign island state of Iceland, and north of Iceland but still east of Greenland is Jan Mayen island, a remote outpost of Norwegian sovereignty. Even farther north lies the multinational treaty-governed island of Svalbard (formerly Spitzbergen), formally under Norwegian rule but bound by treaty to remain unmilitarized and to provide full and equal economic access to all treaty signatories, resulting in a permanent though declining Russian economic and consular presence there. With its proximity to Russia's northwestern frontier and gateway to both the Northern Sea Route and Russia's submarine bastion at Murmansk, it has long been recognized as a strategic island. Farther eastward are the numerous islands and archipelagos off Russia's mainland, including Franz Josef Land, Novaya Zemlya (so isolated that it served as one of the Soviet Union's aboveground atomic testing grounds during the early Cold War years), the Zapovednik Islands, Severnaya Zemlya, the New Siberian Islands, and Wrangel Island, among others, where Russia has been busy fortifying its defense installations after they were mothballed at the Cold War's end.

Maintaining control over these many islands of the Arctic and adjacent gateway regions is of increasing importance not only to the security of the Arctic region but to global stability and world order itself. They present fixed geographical nodes—much like the many "unsinkable aircraft carriers"—that are essential to homeland defense and to the dissuasion and deterrence of aggression and other coercive behaviors by external rivals. Here, the Arctic states are especially well positioned by geographical endowment, owing to their firm and globally recognized sovereignty over their respective lands, waters, and offshore islands, which ring the Arctic basin like a necklace and provide a protective barrier to each Arctic state's territorial mainland. It is true that much of North America north of Canada's mainland is either lightly settled or unsettled, with minimal state presence and with complex histories of resettlement whose soreness still lingers generations later—and this could be a weakness to otherwise recognized claims of

Arctic sovereignty for potential exploitation, such as, hypothetically, by China much the way it did in the South China Sea. But in the latter, China fortified unoccupied islands adjacent to much weaker states that lacked effective means of asserting sovereignty against a rival claim by China, while in the former, the islands of Canada's high Arctic, like those off Russia's mainland or the sovereign and semi-sovereign island polities of the high North Atlantic, are internationally recognized. Canada's Arctic neighbors recognize its claims, just as Canada reciprocally recognizes the claims of its Arctic neighbors, with few, and largely insignificant, exceptions, mostly border disputes with neighbors who are allies first and foremost. This unity of purpose and commonality of values, reflected in governing institutions and international fora across the Arctic (like the Arctic Council and the many organizations associated with and/or parallel to the council, such as the Arctic Economic Council and the Arctic Coast Guard Forum), strengthen the Arctic as a whole and the bonds between the eight sovereign states of the Arctic region.

This circumpolar collaboration has been enduring and will likely withstand new pressures arising from the triangular competition for power between the United States on one hand and Russia and China on the other—described, as noted above, by Rob Huebert as the "New Arctic Strategic Triangle Environment" (NASTE) in 2019.[7] It would be immeasurably harder for China to replicate its tactics from the South China Sea, and doing so would almost certainly draw universal rebuke from the entire membership of the Arctic Council, state and Indigenous alike. It would likely lead to China's isolation from not only the democratic Arctic states but its partner-of-the-moment, Russia—a consequence Beijing would find humiliating and that would show the fragility of Beijing's current entente with Moscow. And while China has sought to influence, through its checkbook, the loyalties of remote Indigenous and settler communities across the Arctic, its early efforts only succeeded in catalyzing a renewed effort by the democratic Arctic states to invest in the development of their northern frontier communities, as experienced in Greenland,[8] which ironically precipitated the 2019 White House overture to buy Greenland from Denmark. President Trump's most ambitious real estate deal was quickly rebuffed by Greenland's foreign minister, who famously quipped, "We are open for business, but we're not for sale."[9] But this heightened strategic interest by the United States in Greenland did lead to a longer-term and more mutual diplomatic reengagement between the two, as I called for in August 2019,[10] including numerous high-level visits by US dignitaries and culminating with a $12.1 million aid package in April 2020, followed by the July reopening of a US consulate in Nuuk for the first time since the 1950s, and another $10 million aid package in September 2021.

This suggests that Beijing will ultimately have to accept its place in the Arctic order as an outsider, a third-party observer state with maritime and commercial interests and many potential economic opportunities, but limited strategic, military, or diplomatic space for expansion. A more concerted effort by the democratic Arctic states to court Moscow, through existing international institutions like the Arctic Council, could greatly help toward this end. By strengthening ties to their Indigenous communities, and their relationship with fellow sovereign Russia—currently under Western sanctions for its Ukraine aggression but still a vital member of the circumpolar family of nations—the seven democratic members of the Arctic Council could greatly reduce the likelihood of a new polar Cold War. The Arctic Council members, who already have established a solid foundation for enduring pan-Arctic collaboration, are well positioned to take the lead on these initiatives once tensions over Ukraine cool, with deep and broad traditions of Indigenous engagement to build upon. While the previous Cold War divided not only the Arctic but much of the planet into competing military-diplomatic-economic blocs, today's world is far more integrated and thus seems much less likely to fully bifurcate once again. The added unity fostered by the long and continuing processes of Arctic globalization and economic integration should ultimately trump whatever regional advantages China may seek—so as much as Beijing may persist in its pursuit of such advantage, it may well emerge both humbled and disappointed by the results of its efforts, much the way Russia has taken a bruising in the wake of its Ukraine invasion, becoming isolated from the rest of its Arctic neighbors for its aggression.

Unsinkable Aircraft Carriers: From Cold War to Collapsing Cryosphere

The distinctive geopolitics of islands and archipelagos—perceived as "unsinkable aircraft carriers" since the early Cold War and as foundation stones of global security today—undergird and reinforce much strategic thinking in both the East and the West with regard to emerging zones of maritime and naval competition around the world. This is particularly evident in recent years in the Arctic Ocean amid the continuing and intensifying polar thaw, with its chain of archipelagos forming natural bridges between the continents, where they have played an outsized role in human history from prehistoric times up to the present. These maritime geopolitical structures are reemerging from the ice in their primordial insular form, transforming the Arctic region and fostering its reconnection to the world ocean.[11] By understanding the geopolitical significance of these islands and their enduring importance to a stable world order, we can better contextualize the emerging strategic importance of the Arctic region, in addition to other remote and peripheral regions in the world.[12]

Island Chains in Naval Thought

There has been much attention paid to island chains in discussions of Chinese naval strategy in recent years, and to Beijing's ongoing fleet modernization and naval expansion as the People's Liberation Army (PLA) Navy continues its evolution full throttle from regional brown water fleet to an impressively robust blue water naval power.[13]

Increasingly, PLA Navy strategists look to the island chains along China's maritime perimeter, with the first connecting the island of Borneo, Taiwan, the Ryukyus (Okinawa), and Kyushu (Japan's southernmost main island) and the second extending from eastern Indonesia to Japan's main island of Honshu via Palau, Guam, the Northern Marianas, and Iwo Jima. More recently, a third island chain has entered the lexicon, running north from Wake Island to Attu in the western Aleutians,[14] and a possible fourth connecting the Gilberts, Midway, and Adak.[15]

Island chains are an old feature of maritime geopolitics, dating back to well before the colonial era, and have been as central to US strategic expansion across the Pacific as they were to the imperial Japanese, and before them, the British. US appreciation of island chains intensified during World War II and achieved its zenith during the Cold War, integrating the continued security of the United States' island and archipelagic partners from Britain to Taiwan at the doorstep to Asia's mainland.[16] While preferred Allied partner Britain withdrew as a global hegemon, London nonetheless augmented US efforts through its own interventions from oceanic Southeast Asia, where it defended an independent Malaysia and Singapore from expansionist Indonesian nationalism, to its liberation of the British-ruled Falkland Islands from Argentinian conquest (or, as Argentina viewed it, unification of mainland Argentina with its lost offshore possession—a struggle that parallels in many ways US efforts to defend an independent Taiwan from what Beijing perceives as its rightful reunification).

Securing strategic island chains has been a fixture of US containment strategy since the dawn of the Cold War, and China has embraced our own doctrine of island chain security with gusto. Beijing now views island chains as strategically vital outposts for asserting its own military control and containing US influence. The term "unsinkable aircraft carrier,"[17] as Taiwan was famously described during the early Cold War, has been applied to a diverse constellation of strategic islands that include Britain, Malta, Iceland, the Aleutians, Japan, and Singapore in addition to myriad South Pacific islands and atolls, as well as the many islets of the South China Sea fortified by Beijing in recent years. Island chains punctuate the world's oceans much the way frontier forts punctuated the seemingly endless expanse of forest and grassland of the American West. Islands provide essential forward offshore supply depots; safe harbors for repairs, recovery, and maintenance; and air strips for power projection and over-the-horizon air defense. These strategically advantageous and well-fortified zones of persistent presence, force resilience, and control of surrounding air and sea space are as central to recent Expeditionary Advanced Base Operations[18] strategies as they were to our island-hopping efforts in World War II.

Well-defended islands and archipelagos can be costly to neutralize during war, and in time of peace become de facto zones of unrivaled economic, diplomatic, and political influence—a stepping stone toward further strategic expansion. This importance of island and archipelagic control to the ability of larger states to project military power, defend trade routes, assert diplomatic influence, and contain regional rivals explains why Beijing has fortified so many islands and archipelagic clusters in what has been dubbed its "Great Wall of Sand"[19] in the South China Sea, and why it has shown a comparably strong interest in the strategic-economic integration of its "String of Pearls" that arcs across the Indian Ocean. It is also why Moscow has done much the same to its own chain of Arctic islands to the immediate north of Russia's mainland.[20] That both major powers and leading rivals to Western influence sense this same vulnerability and opportunity suggests they share a view of geopolitical theory and its intersection with naval strategy, one that the West is cognizant of and moving to counterbalance, as it did during the Cold War. This also explains why the White House, amid the many paralyzing challenges of the global public health and economic crisis, mustered the renewed energy, foresight, and policy attention to reassert and clarify its Arctic (and Antarctic) interests, as expressed in its June 9, 2020, memo on Arctic security[21]—which preceded by just one day the reopening of the US consulate in Nuuk, Greenland. Around the same time the United States anted up over a billion dollars in funding for its long-anticipated icebreaker modernization program, renamed, appropriate to the contemporary challenge of projecting US sovereignty across an increasingly contested Arctic region, the Polar Security Program, providing a platform for sovereign assertion with the mobility to reach into the deepest of pack ice, from the shores along the Arctic basin all the way to the North Pole.[22]

Stepping Stones to Everywhere

Beyond the Arctic, in places where tensions remain between US armed forces and the communities that host their foreign bases, China rightly perceives an opportunity for strategic expansion without conquest. Its overtures to elites in a long list of remote island, archipelagic, and coastal nations as part of its global Belt and Road Initiative (BRI), from its "String of Pearls" to its "Great Wall of Sand," will invariably yield new economic partnerships from which closer strategic integration will follow. The onus thus falls to the United States and its allies to ensure the many Indigenous polities that host its global military presence feel appreciated for their contribution to the current international order. It is their hearts and minds that are essential to continued US ability to project military power globally and secure Western values along the way.

This will require a continued and determined diplomatic effort, and a generosity of investment that can match China's BRI. It is an arena of strategic competition with which the United States is intimately familiar, from its early global expansion to the long and successfully managed Cold War. More recent challenges during the global war on terror presented a series of setbacks to US power on the global stage and precipitated a humbling crash course in counterinsurgency warfare and human terrain mapping. Considered amid China's rise and Moscow's resurgence, the coming global struggle for the hearts and minds of remote, Indigenous islanders around the world, whose homelands are critical nodes for effective global naval operations, will require continued attention and dedication of resources. With a concerted effort and clarity of strategic vision, fueled by a proper understanding of maritime geopolitics and the continued (indeed, increasing) strategic saliency of island chains to international order, another success story for Western values remains every bit as possible as it was during the Cold War.

Beyond Strategic Triangularity: Indigenous Peoples, Transnational Polities, and the Human Terrain

As noted above, there has been much recent discussion of triangular strategic competition in the Arctic between the United States, whose interests have aligned historically with its democratic allies in NATO, and Russia, in partial alignment with China. The latter pair of rivals to Western influence are widely perceived to have the advantage of momentum, with the United States playing catchup. In this sense, as the Arctic region rejoins the world ocean with the continuing polar thaw, it mirrors the strategic alignment between Beijing and Moscow evident across much of maritime and continental Eurasia. The strategic triangle of the United States, Russia, and China dates back to before the Cold War and has realigned several times since then. The rise of China as a great power with ambitions in the Arctic creates a new wrinkle in the already complex state-tribe interface and interconnected diplomatic environment of the polar region.[23]

While Huebert's NASTE is an elegant and influential concept[24] (helped by its distinctive and witty acronym, as perfect as one could imagine for the increasingly Westphalian Arctic), and it was quickly embraced by the national security community and increasingly reflected in numerous strategy and policy documents that followed, the reality of Arctic geopolitical competition is much more complex and asymmetrical than the parsimony of trinitarianism can explain. Indeed, instead of a tidy triangle of competing sovereign powers, the Arctic region's strategic competition can be more accurately visualized as an irregular strategic polygon with a dynamic mix of bilateral, trilateral, and multilateral interstate relations, with the added complexity of an overlapping but largely invisible (to outsiders) set of internal and transnational fault lines of conflict. Many of these underlying dynamics are indeed triangular, but at the regional level or interlinking a different trinity of actors.

By focusing systemically on triangular great power dynamics, one can miss the many important local and regional triangular dynamics that can pit state and subnational interests against one another. Identifying regional and local power competitions across the Arctic can be helpful to clarify what at a higher level looks like complexity—for instance, the Washington-Tokyo-Ryukyus (Okinawa) triangle, the Beijing-Tokyo-Moscow or Beijing-Tokyo-Washington triangle (each a subset of a larger strategic quadrangle), or the Washington-Moscow-Aleut triangle. Each of these regional strategic triangles complicates the overarching great power strategic triangle that Huebert colorfully dubbed NASTE.

These yield a diverse but collaborative group of stakeholders comprising Arctic and non-Arctic states (including their national, regional, and local governments and major economic actors), Indigenous peoples' organizations (some holding regional and local governing powers), and numerous issue-specific NGOs. They operate within an environment marked by shifting alignments of interests and a patchwork of governing systems with extreme variance and volatility over time. This is especially evident when examined village by village, region by region, and country by country, yielding a complexity that eludes easy explanation or simple strategic dictum.[25]

While triangularity may elegantly describe one of the many salient levels of analysis in Arctic geopolitics, this trinity of major states is anything but equal when it comes to their power and influence in the Arctic. China is, in the most important ways, not even a significant player, with no Arctic territory of its own and no seat at the Arctic Council's table owing to its status as an observer state. This is in marked contrast to Russia, whose Arctic territories are the world's largest, and to the United States, which in alignment with its Arctic NATO partners (Canada, Denmark/Greenland, Iceland, and Norway) presents a formidable and historically united bloc. It is along these sovereign shores that all proposed marine shipping routes in a warming Arctic will pass.

If there is going to be a rumble for the Arctic and for control over its emergent sea lanes and vast, untapped repositories of natural resources, it will be a more complex, nuanced, and asymmetrical struggle for the few un-agreed-upon sectors of maritime borders and extended continental shelf claims that remain under contention. Given the universal embrace of the UN Convention on the Law of the Sea by all five Arctic states littoral to the Arctic basin, it is likely to be defined by a concerted multilateral effort (if not fully aligned on every maritime boundary or oceanic subsurface detail) by all eight sovereign Arctic states (Russia's current isolation notwithstanding) in response to rising external interest and pressures from an expanding group of non-Arctic states (inclusive of China) and nonstate actors (many of which have now gained observer status at the Arctic Council).[26]

All eight states will align not only in their mutual interest in asserting sovereign control over their respective portions of the Arctic but also to persuasively demonstrate their individual and collective capacity to defend their Arctic claims from future external threats. And at the same time, the seven Arctic states (all but Iceland) with an Indigenous population will endeavor to successfully manage Indigenous claims and aspirations for greater domestic engagement and inclusion that, if left unresolved, could be exploited by these very same external states looking north to their future economic and strategic needs. It is this potential for exploitation of Indigenous claims and aspirations by non-Arctic states that presents the greatest wildcard relating to future Arctic stability. While the situation with Russia is now complicated by its isolation, past history and future logic suggest it is only a matter of time before Russia returns to its traditional position in Arctic international relations—as a dedicated pillar of Arctic exceptionalism rather than a threat to it.

With all eight of the Arctic states accepting the international legal framework established by the Law of the Sea Convention (LOSC), even without, as in the US case, formally signing on to the treaty, the region as a whole remains remarkably stable and largely uncontested with the exception of several persistent but relatively low-intensity border disputes over the finer details of baselines and maritime boundaries. And though of some economic significance (such as the effect of the boundary demarcation on ownership of offshore petroleum reserves in the Beaufort Sea along the Alaska-Yukon northern boundary), with the LOSC, these lingering border disagreements present virtually no escalation risk, in marked contrast to the transformation of the maritime geopolitics of the South China Sea due to Beijing's unilateral fortification of numerous contested islands.

Consequently, the United States and Canada can continue to disagree over their Beaufort Sea boundary and the legal status of the Northwest Passage while at the same time jointly participating in undersea mapping (in preparation for their respective LOSC claims), search and rescue, and joint military exercises across the region. And Canada and Denmark (Greenland) can continue to disagree over who governs Hans Island, the tiny outcropping of rock midway between the two, while maintaining a strong, friendly, and growing bilateral diplomatic relationship and partnership within NATO. Additionally, Russia and Norway can amicably settle their historic border dispute amid intensifying tensions between Russia and NATO,[27] and Russia and the United States, principal adversaries during the Cold War era, can jointly manage their long, and thus far stable, maritime boundary between Alaska and Siberia.[28] (In 2017, Russia and the United States established a joint management regime for safe shipping,[29] through what might otherwise present a geopolitical vulnerability with potential to become every bit as significant and worrisome as that found in the Malacca Strait, presenting a narrow strategic chokepoint right at the Pacific gateway to both the Northern Sea Route and the Northwest Passage.[30])

Local tensions may rise and fall, but the broader tenor of amicable international Arctic relations remains notably stable when compared to other regions of the world, and China is at best a peripheral player in the Arctic basin. The multilateral cooperative tradition between Arctic states is nicely illustrated by the 2011 Arctic Search and Rescue Agreement agreed to by the eight member states of the Arctic Council, which coordinates search and rescue coverage and response in the Arctic and establishes areas of responsibility for each state party—again without any role to play for non-Arctic states. Parallel to the Arctic Council in structure is the Arctic Coast Guard Forum, which provides for collaboration among the coast guards of the eight Arctic states, which despite rising diplomatic and military tensions, have sustained strong bilateral cooperation in their maritime relations.

As described by Andreas Østhagen on the website of the Arctic Institute,

> For both Norway and the US, coast guard cooperation with Russia is placed in the context of long-standing bilateral relationships as maritime neighbors in the Arctic. Having formalized cooperation through the exchange of information and joint exercises, the cost of tearing down decades of relationship-building in the Arctic was considered too high by both Norway and the United States. Neighboring states are dependent on dialogue across borders which, in turn, must be kept somewhat separate from the domain of overarching international diplomacy. That this cooperation survived the tensions over the ongoing conflict in Ukraine is indicative of as much.[31]

And yet, even amid this strong tradition of intra-Arctic cooperation between once and future rivals, there has been a marked shift in strategy by the Arctic states in recent years, and the return of a more pragmatic and defensive realism, with national interests again driving decisionmaking. This is illustrated by Washington's recent policy and budgetary recommitment to an icebreaker modernization program to enhance its capacity to assert sovereignty farther offshore in the rapidly warming Arctic, and in its flurry of new bilateral diplomatic initiatives with the world's largest island, Greenland, standing guard over North America's northeast flank.

This is equally illustrated by Moscow's recent fortification of its long-ignored offshore Arctic islands. A key driver of these efforts is, of course, the Arctic climate and in particular the rapidity of the polar thaw. One need not make any assumptions about either the cause or the end state of the dynamic climate instability observed in recent years across the Arctic to recognize the military implications of a warming Arctic, and the risks, uncertainties, and insecurities revealed with the continuing thaw.

If there is an Arctic-wide vulnerability that could be exploited by a diplomatically skilled and economically powerful external actor, it would most likely be found in the region's least populous and most remote areas such as Canada's vast Arctic archipelago north of the mainland. There the Inuit have steadily gained more powers and in 1999 achieved the formation of autonomous Nunavut, a vast territory of increasingly valued strategic lands and waters with a tiny population of just over 40,000. Also vulnerable to external machinations is neighboring Greenland, which has been undergoing its own incremental and amicable process of decolonization between its majority Inuit populace and its colonial sovereign, Denmark—and whose small population of just over 50,000 lacks the domestic capacity to effectively defend its vast exclusive economic zone (a challenge even for Denmark, itself a small state with a population under 6 million).

One can envision a comparable vulnerability on the mainlands of Arctic North America and Eurasia, where many struggling interior and coastal villages dot these vast, lightly populated, and infrastructure-light regions, forming islands of humanity separated by great distances of often unoccupied space, as isolated as the offshore Arctic islands. In these vast archipelagos, a struggle to win over the hearts and minds, and thus the loyalty, of Indigenous peoples could emerge as the next salient fault line of conflict, internally dividing each of the hitherto stable Arctic states.

A similar fault line is now retrospectively apparent in the outer Aleutians, three-quarters of a century ago when Japan invaded and occupied the predominantly Unangan (Aleut) islands of Attu and Kiska and bombed Dutch Harbor on Amaknak Island, home to the Dutch Harbor Naval Operating Base and the US Army's Fort Mears.[32] As an example of what was perceived as a triangular strategic contest between the armed forces of imperial Japan on the one hand and the United States and Canada on the other, the Indigenous Unangan people were caught in the middle and at the time perceived by combatants on both sides as peripheral to the conflict, the opposite of the "human flagpoles" resettled to Canada's unoccupied high Arctic Queen Elizabeth Islands. Instead they were displaced by both warring parties to detention centers, with some forcibly relocated to Japan for the remainder of the war, where around half perished, while most were forcibly relocated, ostensibly for their own protection, by American forces—their very own sovereign protectors—to distant, cold, and excessively damp detention camps in southeast Alaska. There, some 10 percent of the population died, and many of the survivors were later prevented from ever returning to their home villages, deemed to be economically unviable in the postwar era.[33]

Such tragic consequences to an Indigenous populace caught in a war zone between the armed forces of imperial Japan and the United States were, unfortunately, all too common in World War II, from the liberation of the Aleutians, the first island chain to be retaken from the Japanese in 1943, all the way to the final battle of Okinawa, where once more the determined forces of the United States wrestled back from Japanese control the Ryukyu Islands at high cost to the local Indigenous population that found itself trapped between these two warring parties much the way the Unangan were. The Ryukyuans endured staggering losses estimated at half their prewar population of 300,000, far more than the 110,000 Japanese and 12,500 US soldiers lost. In the postwar years, the US military presence in Okinawa, essential to both Cold War and post–Cold War security in the Pacific, has continued to face determined local opposition, as the rift created during the Battle of Okinawa has never completely healed, even three-quarters of a century later.

Similar tales of dispossession and exile took place across the Canadian Arctic and in Greenland, creating rifts between the Inuit and Ottawa as well as Copenhagen (and Washington, too, which was responsible for the defense of Greenland during World War II and the Cold War, resulting in some displacement of the Inuit of Thule).[34] The consequences were universally tragic for the Inuit and Unangan. As the Qikiqtani Truth Commission has described, "For Inuit, the loss of home is more than the loss of a dwelling—it is a disruption of a critical relationship of people with the land and animals. It represents the loss of independence and replacement of a way of life."[35]

Understanding the historical context of this internal fault line across the Arctic thus presents a potentially exploitable vector for external manipulation and destabilization that could disrupt the internal balance of power between faraway national and regional governments of the "center" and the isolated Indigenous villages of the "periphery," forging a new axis of conflict that could come into play during a period of intensified strategic competition in the Arctic. This is true whether it is a campaign of hybrid warfare of the sort mastered by Moscow in recent years, the "pocketbook" diplomacy (or the more malicious "debt-trap" diplomacy) favored by China, or the less likely scenario of a formal state of war between rival states for control of the Arctic.

Colonial States and Sovereign Dualities: Indigenous Polities and Political Order

Whether in the Arctic or farther south throughout the world ocean, such tensions and vulnerabilities are commonplace. They are felt by US forces in the Pacific, from Guam to Okinawa; by their Allied partners who maintain their own offshore bases on remote islands across Oceania; and by their rivals—China, as it expands its influence, faces similar tensions and vulnerabilities along Indigenous fault lines. The result is one vast, asymmetrical, and increasingly global theater of strategic competition where local Indigenous polities are fundamental to the international order. While their populations and territories may seem relatively small, they are locally overwhelmingly predominant and thus essential partners to the security of not just the Arctic basin but the adjoining Pacific and Atlantic basins as well.

The widely perceived (and I believe generally overstated) triangular strategic rivalry pitting Washington's interests against those of Moscow and Beijing at the heart of several recent US and allied Arctic strategy documents presumes an inherently Westphalian unity of the Arctic states, but this is far from the case in much of the Arctic, where most of the states are not unitary nation-states at all. They are for the most part former colonial states cobbled together in earlier centuries by the maritime expansions of unitary (and predominantly European) states in the Westphalian core as they reached across the seas, leaving—with the exception of the colonies that became the Lower 48—Indigenous peoples and their local governing structures in the more remote northern locations largely intact, thus enabling colonial rule via local and corporate trading proxies.

Limited in manpower and dependent upon native hunters and trappers to exploit the region's bounty of furs, the colonial-era chartered companies preserved precolonial power relationships and networks that were successfully leveraged in the interest of ascendant colonial powers. Because this remains a defining feature of most of today's Arctic states, a lingering fault line remains between center and periphery: between the settler elites in command of the state apparatus to the south and the Indigenous communities in the remote hinterland that have been gradually restoring their self-governing powers. (The one notable exception is Iceland, which was settled prior to the arrival of the eastward-migrating Inuit, leaving this Arctic state as a truly unitary Westphalian polity.) Understanding this internal dynamic and achieving a stable balance of interests through inclusive and respectful policies of native enrichment and empowerment may be of momentous consequence in the event of external agitation by a non-Arctic state.

This historic struggle for the human terrain of the Arctic is thus of great importance to the future stability of the region and requires forward-thinking investment, respectful relationship-building and sustainment, and a continuous process of confidence-building measures to ensure that the legitimacy of the rule of the sovereign states of the Arctic remains intact and uncontested, should a foreign interloper such as China seek to destabilize the status quo. Because of the many socioeconomic challenges facing northern villages from one end of the Arctic to the other, some argue they have already become a target for exploitation by Beijing. Because these northern Indigenous homelands have been imperfectly integrated with the political economies of the Arctic states, despite much progress and ongoing efforts in recent years, this remains a challenge faced by the seven Arctic states that have Indigenous populations engaged in long-term processes of cultural renewal, economic development, and restoration of land rights.

Progress on this front varies greatly by region and by state, offering an uneven opportunity for external exploitation. While Russia had, prior to its onslaught of Ukraine, mastered the art of hybrid warfare below the threshold of formally declared war,[36] with particularly effective results in Crimea in 2014, and Beijing similarly deployed checkbook diplomacy[37] to co-opt elites along the global network envisioned by its BRI, including its northern component, the Polar Silk Road, it was China that faced initially strong blowback against what the United States and its allies reframed as "debt-trap diplomacy."[38] At the same time Russia generated near universal distrust throughout the states of the former Soviet Union, particularly among small states bordering Russia that feared they could become the next Crimea (and now even more gravely fear becoming the next Ukraine). In short, tactical blunders by both Moscow and Beijing, through clumsy and overconfident efforts to coerce small polities and peoples, have blunted their capacity to project power into the Arctic—with the exception, of course, of Moscow's own Arctic territories and waters, where its sovereignty remains uncontested, but where it remains far behind its democratic Arctic counterparts in reconciling state and tribal interests.

Indigenous Engagement, Autocratic Containment, and Arctic Exceptionalism

Strengthening the alignment of interests between Indigenous peoples and their sovereigns from Alaska to Finland can provide the democratic Arctic with an advantage over Russia, whose own native peoples remain marginalized, their lands and resources encroached upon or expropriated, and their leaders either exiled or effectively marginalized.[39] One can even imagine the democratic Arctic states mastering the art of hybrid warfare, just as many of them by necessity relearned the art of counterinsurgency warfare during the long global war on terror.[40] By confidently and proactively turning the tables on Moscow, the process of winning the battle for the hearts and minds of Russia's own oppressed native peoples is already underway to a limited degree with the warm diplomatic reception enjoyed by Russian Indigenous leaders in governing institutions like the Arctic Council, where Indigenous organizations enjoy their distinct membership status as Permanent Participants, second only to the founding member states and superior in organizational status to the many observer organizations and states, among which China is included.

But normatively, one can reasonably hope Russia will come to realize that its security will be best strengthened by achieving parity with its democratic counterparts on the Arctic Council in the area of native rights and empowerment; indeed, this may be underway, and could explain the notable prominence of Indigenous issues in Russia's latest Arctic strategy, where they are mentioned at least seventeen times.[41] It could also explain some interesting but little reported pronouncements by Vladimir Putin several years ago recognizing the Ainu people, whose homeland has been partitioned by Russia and Japan since World War II's end, as indigenous to Russia. This caught the attention of Tokyo, where Ainu recognition has been advancing incrementally, leaving Japan vulnerable to a more proactive approach by Moscow on Ainu rights issues and potentially planting a seed for an eventual sovereign claim by Russia of Hokkaido as part of the Ainu homeland, newly perceived as a Russian

Indigenous people. Amid the present circumstances in Ukraine, and with Russia's continued occupation of most of the Kuril Islands since the final hours of World War II when it grabbed Japan's former colonial island territories, this could be a harbinger of future worry.

But it may also signal a convergence in recognition of the Indigenous foundations of northern political stability, and the starting point for a conversation on how Russia can join its Arctic neighbors in supporting and defending the rights of Indigenous peoples. While a double-edged sword, it is no less a potential tool to solidify Arctic international solidarity. Engaging in the battle for influence of the Indigenous Arctic human terrain could thus be to Russia's advantage, helping to neutralize a diplomatic and political vulnerability vis-à-vis the West. And in so doing, the inter-Arctic collaborative dynamic can also be strengthened—further neutralizing any risk associated with the increasingly competitive strategic triangularity described above.

With its deep pockets, China may take the opportunity to retool its approach as well, shifting away from the perceived "power grab" of debt-trap diplomacy to foster a more mutually beneficial model of Arctic economic development. This shift could position Beijing to more adeptly exploit any failures by the Arctic states to sufficiently support and re-empower their own Indigenous peoples, who are intimately aware of any unevenness in Arctic social, cultural, and economic development. The shift is already articulated in China's 2018 Arctic white paper, which references Indigenous issues seven times—particularly noteworthy for a country that does not even recognize its own domestic Indigenous peoples and continues to impose its will on its many diverse minorities, particularly those aspiring to greater autonomy.

A triumph by the democratic Arctic states is by no means guaranteed with regard to the battle for Indigenous hearts and minds, but they still have many advantages over Russia and China that could make it impossible for either of these rivals to meaningfully undermine Western influence in the region or to dilute the sovereignty they have over their respective Arctic territories. Thus, if there is indeed a new cold war in the Arctic region—and many believe there already is one, more so since the war in Ukraine began—continued gains in native development on the home front in each Arctic state will be crucial for their political legitimacy and territorial integrity. It is an important arena of engagement, indeed perhaps the most salient human terrain, where the United States and its allies can consolidate victory and ensure unrivaled regional supremacy through more inclusive and effective governance in partnership with the Indigenous peoples of the Arctic. The United States can help to sustain Arctic exceptionalism through its continued leadership in aligning its interests and values, and those of its fellow Arctic states, with the Indigenous peoples of the North, and awaiting the time it becomes possible to welcome Russia back to the table at which it long cooperated with its seven counterparts and the six Indigenous Permanent Participants, an anchor to a stable and peaceful Arctic.

Beyond the Arctic: Polar Exceptionalism, North and South

While much of this book has had a distinctly North American focus, such a continental perspective reveals enduring historical influences of the colonial era—particularly the territorial fur empires and the maritime whale trade—on subsequent state formation and expansion to and in the Arctic, collaborative state-tribe relations, and the reemergence of Indigenous self-governance. From this cauldron emerged a deep institutional commitment to co-management and Indigenous governance in Arctic North America, with particularly strong commitment in Ottawa for Indigenous empowerment owing to the Canadian Constitution's protection of inherent aboriginal rights. Indeed, since 2017, Canadian Arctic policy has been reframed explicitly through the lens of state-tribe co-management, evolving beyond its origins in environmental governance to become a pillar of the Canadian Arctic and Northern Policy Framework.

While the Arctic Council has in many ways internationalized co-management principles and structures developed for local and regional environmental governance, bringing them to the entire circumpolar family of nations and embedding them in the international agreements that emerged under the auspices of the Arctic Council, polar governance beyond Arctic North America differs greatly in form and structure, embracing multilateralism much more than multilevel governance. This is most evident in parts of the polar world without Indigenous peoples, enabling states to engage directly with one another without the intermediation of the sub- and transnational dynamics introduced by Indigenous peoples in their efforts to establish a multilateral governance framework for remote regions whether in the Far North or Global South.

Lackenbauer and Dean, in their 2020 discussion of "Polar Exceptionalism: The Arctic-Antarctic Analogy,"[42] recount how

> early twenty-first century discussions on climate change, the protection of the marine environment, and the "opening" of the region pre-

> cipitated various calls for a new comprehensive international legal regime to govern the Arctic Ocean, often predicated on another form of Arctic or polar exceptionalism. [But after] some academics began to assert that the soft-law approach to regional governance could not effectively manage challenges related to climate change, resource development, and increased shipping . . . advocates across the ideological spectrum promoted stronger regional institutions with legal powers or an ambitious new Arctic treaty architecture modeled on the Antarctic Treaty.[43]

They further observe,

> The Antarctic Treaty had been designed to deal with the exceptional circumstances around the south pole. By linking the Arctic to its southern counterpart, the implication was that a stable and unique regime designed for Antarctica could be applied to the other polar region.
>
> The "polar exceptionalism" argument fell apart when commentators emphasized the simple geographical reality that Antarctica is a continent with no permanent human residents, while the Arctic Ocean is a maritime space already covered by the UNCLOS where coastal states enjoy well-established and internationally-recognized sovereign rights. It was unreasonable to think that the Arctic states could see the Antarctic Treaty as an appropriate model, given that it was deliberately designed to hold sovereignty claims in abeyance.[44]

But there is still merit in "polar exceptionalism" as a concept extending Arctic exceptionalism to other remote regions such as the Antarctic, so long as the concept's users remain cognizant of these fundamental asymmetries. Also uniting the north and south polar extremes of our planet is their mutual predisposition to resolve disputes without war and force, and a receptiveness to treaty structures (and negotiations in pursuit thereof) as a principal method for resolving disputes between stakeholders.

This is evident from two distinctive treaty systems negotiated at opposite ends of the world: the 1920 Svalbard Treaty (originally known as the Spitsbergen Treaty) and the 1959 Antarctic Treaty System. There are no Indigenous peoples residing in either territory, and signatory states have mutually constrained numerous pillars of traditional Westphalian sovereignty to keep these remote polar regions demilitarized and free of great power conflict and war. Each achieved a diplomatic workaround to the question of unitary state sovereignty, and both treaties greatly constrain the limits of state sovereignty in the polar regions—the latter designed to keep the Cold War from engulfing the southern continent, and the former designed during the heyday of the League of

Nations to prevent a recurrence of a World War I–like escalation. And yet both of these treaty systems also leave sovereign aspirations (and prior claims of sovereign reach) intact for a later day. The many changes associated with climate change, particularly increases in marine access and the potential for accelerated living and nonliving resource extraction, could supercharge these original, unresolved sovereignty claims left hovering (and, some might fear, festering) in the background.

For the Antarctic Treaty System, signed in 1959 (the same year Alaska gained statehood) and coming into effect in 1961 (only a year before the Cuban Missile Crisis brought the world dangerously close to the nuclear abyss), it's Antarctica's seven pretreaty territorial claims as well as potential future claims that could emerge if the treaty collapses upon arrival of its renewal date in 2048 (not too far in our futures to consider now). For Svalbard, it's Norway's growing sovereign aspirations that parallel advances achieved in the Law of the Sea Convention beyond where things stood when the treaty was signed just over a century ago in 1920—pitting its own national interests (and allied interests) against those of its rivals. During the Cold War, it was nearly the entirety of the old Warsaw Pact but is now mainly Russia that objects to Norway's partial assertion of an exclusive economic zone around Svalbard. Norway is not the only state to experience friction with a neighbor as a result of expanded territorial claims under the Law of the Sea Convention. Iceland, as it expanded its offshore claims from 4 to 12, and then to 50, and finally to 200 nautical miles, collided forcefully with Britain's fishing fleet, coming to blows in a series of low-intensity skirmishes known collectively as the Cod Wars, pitting Iceland's diminutive coast guard against the might of the Royal Navy. Some worry that Norway and Russia may well face a comparable future showdown while others worry that Russia may attempt a Crimea-inspired lightning expansion and annexation, or endeavor to grab the entirety of the archipelago modeled on its 2022 Ukrainian gambit (but with better contingency planning), in order to protect a "vulnerable" Russian population through Russian state reexpansion.

Can either the Svalbard Treaty or the Antarctic Treaty System survive the many challenges of increased economic activities in both remote regions as they become more integrated with the rest of the world? Each treaty is built upon mutual agreement to keep these remote islands demilitarized and free of war, in addition to restraining territorial ambitions and accepting limits on sovereign assertions, demonstrating a willingness to compromise in the interest of world peace. The Svalbard Treaty has endured for over a century of global change, and the Antarctic

Treaty System over a half century. With a diverse and growing assortment of signatories, these two polar treaties may indeed continue to endure the test of time, helping to strengthen the credibility of extending "Arctic exceptionalism" to a more universal "polar exceptionalism."

Looking to these two treaty systems and their shared embrace of multilateral polar governance, can we find commonalities with Arctic North America? One can recognize a collaborative dynamic reminiscent of co-management in Arctic North America, even in the absence of Indigenous peoples. In both treaties we find a comparable, and vibrant, spirit of multilateral interstate collaboration that has endured periods of tension and war, the division of world politics into blocs and alliances, and ideological duels between competing states. And we have found states willing to constrain their ambitions in the interest of international peace and stability, even if they haven't renounced their ambitions outright. Could these treaties endure, forever, despite what may over time seem to be an antiquated structure and foundation? Is there room for ascendant powers to assert themselves within the treaty structure? Yes, and yes. But there is still much reason for hope, and to extrapolate from these successful treaties on opposite ends of our planet that a transcendent polar exceptionalism, and not just Arctic exceptionalism, has reasonable prospects.

Since taking effect in 1925, the Svalbard Treaty formally grants sovereignty to Norway while ensuring nondiscriminatory access and resource exploitation rights to other treaty signatories, the most prominent and active of which has been Russia. Coal mining, the original reason for the treaty's access rights clause, has been in steady decline, limiting the economic benefit to Russia, but the diplomatic and strategic importance to Russia of a continued presence has thus far outweighed economic considerations. Because the archipelago is distinct from Norway, it's not included in the EU's Schengen Area, so anyone arriving in Norway from the archipelago must go through longer passport lines. So, while Norway is the formal sovereign over Svalbard, it's not 100 percent integrated into the Norwegian polity. Norway would like it to be and over time has sought to expand its sovereign reach, declaring a fisheries protection zone in 1977. Rivals, particularly Russia, believe the Svalbard Treaty stops at or near the waterline, extending no farther out than the twelve nautical miles of its territorial waters since the treaty predates exclusive economic zones and extended continental shelf claims. As with many Arctic disagreements, it has remained possible to agree to disagree, since until recently, Svalbard, like Antarctica, was remote and of only limited interest to the world community. Curiously, China has been a signatory since 1925, one of the first states to sign on. China remains a strong defender of the treaty's "non-discriminatory rights" allowing scientific research, resource exploitation, fishing, hunting, and commercial activity.

Half a world away, there are seven claims to sovereignty over much of the southern continent from either former colonial powers or post-colonial independent states adjacent to the southern continent, and one unclaimed sector, but all of these claims remain in abeyance for the duration of the Antarctic Treaty (at least until 2048). Three of the claims overlap, though many world powers (including both the United States and Russia) do not have any formal claims to the southern continent—so a nonrenewal of the Antarctic Treaty could cause much consternation and contention among enough of the world community to make its endurance beyond 2048 a reasonable expectation. Its distance from Moscow and Beijing certainly helps, as does the persistence of the Antarctic ice cap, though increasing signs of instability, mostly in West Antarctica, could portend future risks of accelerated climate change, and this could in turn catalyze misbehavior by treaty dissidents hoping to benefit from a collapse of the Antarctic Treaty System.

As in Svalbard (and, incidentally, Iceland too), there are no Indigenous peoples in Antarctica. Indeed, its only human inhabitants are the small groups of people at research stations, creating a powerful international scientific diaspora, which is itself a particularly collaborative subpopulation of the human family that favors both Arctic exceptionalism and its polar extension. There's no active hydrocarbon exploitation (as it's still too cold, and the ice is too thick), though scientific research does include seismic surveys to map the southern continent's hydrocarbon potential, and in the event of a collapse of its ice pack, things could change quickly on this front. There is fishing in the Southern Ocean, which is regulated under the Convention on the Conservation of Antarctic Marine Living Resources (signed in 1980, coming into effect in 1982). Mining was banned under the Antarctic treaty's environmental protocol (signed in 1991, coming into effect in 1998), before which it was impractical enough not to be an issue.

Antarctica, like Svalbard, can be collaboratively governed because of its remoteness from population centers and major markets, its extremely small and sparsely distributed human population, and the unlikelihood of profitable resource exploitation. With climate change, one wonders if these conditions will hold. And if so, for how long? Until 2048 seems all but certain, despite recent data suggesting an acceleration in the extent of Antarctic thaw. Viewing through a new lens of international treaty governance rather than through the old lens of sovereign state expansion and domestic treaty-making with Indigenous peoples introduces a new paradigm to consider as we imagine the Arctic's future. Some advocate for an Antarctic-like approach for the Arctic while others, primarily the Arctic 8, prefer to maintain their own undisputed and mutually reinforced sovereign supremacy, and to co-manage the region in partnership with the Indigenous peoples of the Arctic, minimizing the opportunity for external states to participate in the region's governance. The Arctic Council allows external states to serve as observers and participate as technical experts on working groups but doesn't allow them to share in governance. Expanding the opportunity for Indigenous peoples to participate in governance in areas where states enjoy monopoly control, such as in the use of force and the right to enter into international treaties, is a tough slog with much state resistance, though some progress has been made with Greenland's evolution from home rule to self-rule in 2009. This means that an Antarctic system for the Arctic is less likely, owing to the degree of sovereign control exercised under international law by the Arctic 8.

But more localized treaties like in Svalbard, to manage equal access for scientific research, fishing, hunting, mining, and other means of economic participation, could, in theory, be crafted to align with existing land claims treaties for their respective regions of the Arctic. As Norway, in 1920, accepted a diluted form of sovereignty over the Svalbard archipelago, it is conceivable that other Arctic states may find their way to similar compromises. In some ways, the land claims experience in Arctic North America mirrors this, with sovereignty evolving from absolute and unitary to collaborative and multilevel, within a domestic context. Greenland, and its constitutional evolution from colony toward independence, also shares some parallels as Danish sovereignty becomes increasingly diluted there. One connective thread, from Arctic North America all the way to Antarctica, is the globalizing expansion of colonial-era resource trade, from the fur empires to marine mammal harvesting (whaling and sealing, in particular).

These resources compelled states to expand farther and farther afield, whether north or south. Both polar regions experienced state expansion around the same time, and both became integrated through complex, multisignatory international treaties to world politics and the global economy. They differ in many details, but a key commonality is in how two remote, resource-rich frontier regions on opposite sides of the planet have been successfully integrated into world politics via treaty-making between diverse and asymmetrical stakeholders, and how these treaties have endured substantial tests of time. How they withstand the effects of continuing and intensifying climate change in the years ahead remains to be seen. That is, in part, the reason why so many states in the years preceding Russia's invasion of Ukraine updated their Arctic strategies and policies to reflect their common struggle against climate change, and to align their national policies with the collaborative underpinnings of global Arctic governance by the time the Arctic Council celebrated its twenty-fifth anniversary in 2021. Only a year later Russian tanks rolled through neighboring Belarus in their ill-fated effort to encircle Kyiv, bringing major war to Europe for the first time in decades and throwing the Arctic's long collaborative, multilateral, and multilevel order into crisis.

Notes

1. Young, "Age of the Arctic," 160–179.

2. I first discussed the maritime geopolitics of the thawing Arctic in *Arctic Doom, Arctic Boom* (2009), and revisited the topic over a decade later in abbreviated form in "Stepping Stones to a Secure Arctic," 26–20; and in greater depth a year later in "Geopolitics, Indigenous Peoples and the Polar Thaw."

3. For a current discussion of Arctic geopolitics, see Nebojša Vuković, "Do We Need Revision of the Key Geopolitical Paradigms?," *Medjunarodni Problemi* (International Problems, published by the Institute of International Politics and Economics, Belgrade, Serbia) 72, no. 1 (2020): 15–36.

4. Dan Levin, "Canada and Denmark Fight over Island with Whisky and Schnapps," *New York Times,* November 7, 2016.

5. For detailed histories of Canada's Inuit "exiles," see McGrath, *Long Exile*, or for a more critical view of the conventional narrative, see Kenney, *Arctic Smoke and Mirrors.*

6. For a detailed history of the island, see Bockstoce, *White Fox and Icy Seas*; and Peter J. Usher, *The Bankslanders: Economy and Ecology of a Frontier Trapping Community* (Ottawa: Northern Science Research Group, Department of Indian Affairs and Northern Development, 1970).

7. Huebert, "New Arctic Strategic Triangle Environment (NASTE)."

8. For more on China's rebuffed efforts to co-opt Greenland, see Drew Hinshaw and Jeremy Page, "How the Pentagon Countered China's Designs on Greenland," *Wall Street Journal,* February 10, 2019.

9. Jacob Gronholt-Pedersen, "Greenland Tells Trump It Is Open for Business but Not for Sale," Reuters, August 16, 2019.

10. Zellen, "Donald Trump Is Thinking of Buying Greenland."

11. For an in-depth discussion of these macrolevel systemic changes to the Arctic, see Zellen, *Arctic Doom, Arctic Boom*. For an understanding of the interconnected geopolitics of islands and archipelagos and its significance to world politics, consider the pioneering contributions and analytical framework of Sir Halford J. Mackinder, in addition to more recent critical geographical approaches. For a discussion of the origins of the field of nissology, see Oscar Campomanes, "The

Islandic in the Postcolonial Critique of American Empire," University of California Humanities Research Institute, September 2018.

12. This is true not only of the Arctic Ocean but also as far south as the Southern Ocean, where the security implications of the alliance integration of so many isolated islands and archipelagos bear a striking similarity to the challenges facing the top of the world and remote maritime regions around the world.

13. See the prolific work of James R. Holmes and Toshi Yoshihara, including *Red Star over the Pacific: China's Rise and the Challenge to U.S. Maritime Strategy* (Annapolis: Naval Institute Press, 2010), in addition to their many articles, such as "Responding to China's Rising Sea Power," *Orbis* 61, no. 1 (2017): 91–100; "Asymmetric Warfare, American Style," *Proceedings* 138 (April 2012); and Yoshihara, "China's Vision of Its Seascape," 293–314, among others.

14. Department of Defense, *Annual Report to Congress: Military Power of the People's Republic of China*, May 23, 2006.

15. Martin D. Mitchell, "The South China Sea: A Geopolitical Analysis," *Journal of Geography and Geology* 8, no. 3 (2016).

16. In fact, it was General Douglas MacArthur's public articulation in 1950, amid hostilities in South Korea, of an island-chain strategy for the United States in the postwar Pacific and asserting the strategic importance of Formosa (Taiwan) as part of an allied island chain linking the Philippines to Okinawa, that precipitated his dismissal by President Truman. See "Texts of Controversial MacArthur Message and Truman's Formosa Statement," *New York Times*, August 29, 1950, 16.

17. See the following for selected details: Michael O'Hanlon, "China's Unsinkable Aircraft Carriers: While America Builds Carriers, China Builds Islands," *Wall Street Journal*, August 6, 2015; Brad Lendon and Michelle Lim, "Japanese Island Could Become an Unsinkable US Aircraft Carrier," CNN, December 6, 2019; "Here Are the Unsinkable Aircraft Carriers of the World," *Fighter Jets World*, June 28, 2020.

18. For an introduction, see Marine Corps Association and Foundation, *Expeditionary Advanced Base Operations (EABO) Handbook* 1.1, June 1, 2018, https://mca-marines.org/wp-content/uploads/Expeditionary-Advanced-Base-Operations-EABO-handbook-1.1.pdf. For a discussion, see Robbin Laird, "Reshaping Perimeter Defense: A New Pacific Island Strategy," *Defense.info*, August 12, 2019, or Laird's coauthored book with Edward Timperlake and Richard Weitz, *Rebuilding American Military Power in the Pacific*.

19. This phrase generated worldwide headlines after being utilized in a speech by Admiral Harry B. Harris, Commander of the US Pacific Fleet, at the Australian Strategic Policy Institute in Canberra on March 31, 2015. For an example of media coverage of this, see "China Building 'Great Wall of Sand' in South China Sea," BBC News, April 1, 2021. For a recent discussion of the "String of Pearls," see Pankaj Jha, "Countering Chinese String of Pearls, India's 'Double Fish Hook' Strategy," *Modern Diplomacy*, August 8, 2020.

20. Gary K. Busch, "Russia's New Arctic Military Bases," *Lima Charlie News*, April 29, 2017.

21. The White House, "Memorandum on Safeguarding U.S. National Interests in the Arctic and Antarctic Regions," June 9, 2020.

22. For more details, see the US Coast Guard Polar Security Cutter program information page, www.dcms.uscg.mil/Our-Organization/Assistant-Commandant-for-Acquisitions-CG-9/Programs/Surface-Programs/Polar-Icebreaker.

23. Liu Rui, "China-US-Russia Triangle," *Global Times*, October 29, 2018.

24. Huebert's influential concept is introduced in the second sentence of Pincus, "Three-Way Power Dynamics," 40n1.

25. A good introduction to the complexity of Arctic governance can be found in Landriault, Chater, Rowe, and Lackenbauer, *Governing Complexity in the Arctic Region*.

26. Drishti IAS, "Arctic Council," September 17, 2019, www.drishtiias.com/important-institutions/drishti-specials-important-institutions-international-institution/arctic-council.

27. For more details on the joint US-Canada seabed mapping effort, see Sian Griffiths, "US-Canada Arctic Border Dispute Key to Maritime Riches," BBC News, August 2, 2010; for a broader discussion of the multiple overlap disputes across the Arctic region, see Jessica Brown, "Thaw in Accord: As Arctic Ice Melts, Territorial Disputes Are Hotting Up, Too," *The Independent*, February 28, 2018. For a discussion of the history of US Arctic policy since the Cold War, see Zellen, "U.S. Defense Policy and the North"; Zellen, "Where East and West Converge."

28. Known as the Baker-Shevardnadze line since its formal demarcation by negotiation in 1990, it remains unratified on the Russian side due to the rapidity of the subsequent Soviet collapse but has remained nonetheless adhered to throughout the post–Cold War era. See the UN, Agreement Between the United States of America and the Union of Soviet Socialist Republics on the Maritime Boundary, June 1, 1990, www.un.org/depts/los/legislationandtreaties/pdffiles/treaties/usa-rus1990mb.pdf.

29. For an excellent summary of the 2017 agreement, see Emily Russell, "US, Russia Agree on Shipping Standards for Bering Strait," *Alaska Public Media*, June 6, 2018; for the full details of the agreement, see IMO Sub-Committee on Navigation, Communications and Search and Rescue, "Routeing Measures and Mandatory Ship Reporting Systems: Establishment of Two-Way Routes and Precautionary Areas in the Bering Sea and Bering Strait," submitted by the Russian Federation and the United States, 5th session, agenda item 3, NCSR 5/3/7, November 17, 2017—building upon the even longer tradition of amicable maritime border relations that has been sustained between the Cold War's principal adversaries, both during and after the Cold War.

30. For more details on the joint US-Canada seabed mapping effort, see Griffiths, "US-Canada Arctic Border Dispute"; Brown, "Thaw in Accord."

31. Andreas Østhagen, "Coast Guard Cooperation with Russia in the Arctic," Arctic Institute, June 6, 2016.

32. Perras, *Stepping Stones to Nowhere*.

33. Carlene J. Arnold, "The Legacy of Unjust and Illegal Treatment of Unangan" (master's thesis, University of Kansas, 2011).

34. Zacharias Kunuk, *Exile* (Igloolik Isuma Productions, 2009), www.isuma.tv/isuma-productions/exile-0; and *Kivitoo: What They Thought of Us* (Igloolik Isuma Productions, 2018), www.isuma.tv/hwma/kivitoo/ep4kivitoo1.

35. Qikiqtani Truth Commission, "Nuutauniq: Moves in Inuit Life. Thematic Reports and Special Studies Qikiqtani Inuit Association," http://docplayer.net/55079820-Qikiqtani-truth-commission-nuutauniq-moves-in-inuit-life-thematic-reports-and-special-studies-qikiqtani-inuit-association.html.

36. Jim Garamone, "Military Must Be Ready to Confront Hybrid Threats, Intel Official Says," *Defense.gov*, September 4, 2019; Alex Deep, "Hybrid War: Old Concept, New Techniques," *Small Wars Journal*, March 2, 2015; William Safire, "On Language; The Near Abroad," *New York Times Magazine*, May 22, 1994.

37. Daniel S. Hamilton, "Checkbook Diplomacy," *The Diplomatist*, 2016, as reposted by the Center for Transatlantic Relations, https://archive.transatlanticrelations.org/publication/checkbook-diplomacy-daniel-s-hamilton.

38. Hadeeka Taj, "China's New Silk Road or Debt-Trap Diplomacy?," *Global Risk Insights*, May 5, 2019; Marc Lanteigne, "The US Throws Down the Gauntlet at the Arctic Council's Finland Meeting," *Over the Circle,* May 7, 2019.

39. See Thomas Nilsen, "Russia Removes Critical Voices Ahead of Arctic Council Chairmanship, Claims Indigenous Peoples Expert," *Barents Observer,* November 27, 2019.

40. Zellen, *Art of War in an Asymmetric World.*

41. Russian Maritime Studies Institute at the Naval War College, *Foundations of the Russian Federation State Policy in the Arctic for the Period up to 2035*, March 5, 2020.

42. Lackenbauer and Dean, "Arctic Exceptionalisms," 334.

43. Ibid., citing Ed Struzik, "As the Far North Melts, Calls Grow for Arctic Treaty," *Yale Environment 360*, June 14, 2010; Rob Huebert, "The Need for an Arctic Treaty: Growing from the United Nations Convention on the Law of the Sea," *Ocean Yearbook* 23 (2009); Hertell, "Arctic Melt," 565–591; Timo Koivurova and Erik J. Molenaar, *International Governance and Regulation of the Marine Arctic: Overview and Gap Analysis* (Oslo: World Wildlife Foundation, 2009).

44. Lackenbauer and Dean, "Arctic Exceptionalisms," 335.

7

A Cooperative Arctic in a Contested World?

As chronicled above, for many centuries, interest in the Arctic and the commercial and strategic potential of its sea lanes and natural resources has been enduring, from the fur-trading empires of Rupert's Land and Russian America to our own times. However, climatic conditions had prevented the region's full potential from being achieved, holding back its development and limiting its contribution to the world economy. It has been neither a "Rimland" nor a "Heartland" but something more closely resembling what geopolitical theorist Mackinder called "Lenaland"[1]—after the isolated Lena River valley in Russia, which captured the unique geostrategic insularity of the Far North. The Cold War's two armed and often hostile superpowers came face-to-face along this ice curtain with very little risk of war, in great contrast to the Central Front in the once divided Germany, where a million men stood armed and ready for war for a generation, not far from today's conflagration engulfing most of eastern and southern Ukraine, imposing a new firewall between East and West.

Geopolitical Turbulence Returns to the Arctic

The long isolation of the Arctic dates back before the dawn of humanity and accounts for the region's unique fauna, like the polar bear and beluga whale, blending into an environment defined by ice and snow for millennia. What defined the region's biological evolution also shaped its geopolitical stability and limited an otherwise heavy human footprint. Now all this seems to be changing, or at least the prospect of such a change has tipped from the implausible to the possible, as a result of the rapid warming of the Arctic climate and the measurably accelerated summer ice melts. Even the most alarmist of ice scientists were caught off guard back in 2007, when summer sea ice set a record low more than a month before the end of the melt season, with such a steep and rapid decline in sea ice coming years if not decades earlier than most experts had imagined possible.[2] This put the region in play strategically for the first time since the end of the Cold War after a period of strategic disinterest, which had nurtured an Arctic calm. The promise of unlocking the Arctic's full potential and the simultaneous global natural resource rush stimulated interest in the region among numerous stakeholders, many of whom had been content to ignore the polar region throughout the 1990s.[3]

The Arctic Ocean's dominant characteristics are changing rapidly, particularly the massive, permanent, continent-sized barrier of multiyear ice that sits atop the pole, which could in time disappear. It has certainly shown a capacity to retreat farther and faster than anticipated, presenting us with a strategic surprise that suggests further surprises could arise. As the ice pack retreats, the polar barrier that marked the very ends of the Earth (or "Ultima Thule") has the potential to become a transpolar crossroads, or what mapmakers long ago imagined to be the "Midnight Sea." Shipping companies are increasingly utilizing new routes across the top of the world, linking Northeast Asian ports with their counterparts in Europe. What Rob Huebert and Brooks Yeager called a "New Sea" in their January 2008 World Wildlife Fund report, and was dubbed the "Blue Arctic" in the US Navy's 2021 Arctic strategy, will one day emerge, with huge geopolitical consequences.[4] What was once the ends of the Earth has shown increasing potential to become the world's new geostrategic center, a literal *medi-terranean*. But as we have seen in recent months, there can—and will—be geopolitical bumps along the way.

The Russian Bear Reawakens

Before Russia's intensifying military activities in eastern Ukraine and Crimea threw cold water on the earnest efforts of the Obama administration to "reset" relations with Russia and introduce a more cooperative and mutually trusting relationship, Russia's activities in its sector of the

Arctic had focused on its vast, resource-rich, and uniquely shallow continental shelf—which it hoped the world would recognize as Russia's own extended continental shelf and under the UN Convention on the Law of the Sea (UNCLOS) would then be considered Russian and not high seas.[5] And while Russia's aggression in Ukraine has led to its international isolation, its pragmatic and notably cooperative approach to the Arctic may not be in full retreat even now[6] (despite threats made early in the war to its northern neighbors Finland, Sweden, and Japan on the matter of their support for sanctions throttling Russia's economy, and in the case of the former two, their unexpected abandonment of neutrality and policy pivot toward joining NATO, the very end Putin had hoped to forestall through his coercive display of military might in Ukraine).

Indeed, a primary Arctic tension—in addition to that between Indigenous peoples and the predominantly non-Indigenous states that assert sovereign claim to the high North—has been over differences in the Arctic states' maritime boundaries and extended continental shelf claims, the former mostly quite modest and none existential in risk. And while this is often portrayed as an emergent fault line between Russia and the West, thus far Russia has not only played by the rules set forth under UNCLOS (which remains unratified by the United States, alone among the Arctic states), but has amicably resolved its maritime border disputes with neighboring states, NATO members Norway and the United States both included.

If state rivalry continues to overshadow cooperation as it now does in so many regions of the world, and East and West again face off in an epic and protracted showdown, the Arctic could become one of the next regions in which a newly assertive Russia (and, to a lesser extent, China) challenges the West (as much recent policy rhetoric suggests is on the rise), forcing the United States and its democratic Arctic partners to rethink or regionally truncate their long-standing collaborative approach to Arctic security—and to prepare, once again, for a less universally cooperative and more regionally aligned circumpolar world. This would not forever doom Arctic exceptionalism but would instead constrain it within blocs, much the way it was during the Cold War, where we saw recurring efforts to break down the various "Ice Curtains" that had built up between East and West in the North.

One encouraging reason to hope for Arctic exceptionalism's endurance amid the reawakening of the Russian bear was the 2021 transition of the rotating chairship of the Arctic Council to Russia from Iceland. It started out auspiciously and with much hope (since 2022 largely demolished) that this collaborative spirit would continue.

As Dr. Lawson W. Brigham observed in the Wilson Center's Polar Institute blog, *Polar Points*,

> The Arctic Council Ministerial meeting in Reykjavik, Iceland on 20 May 2021 had a strong message for the globe: the Arctic is a peaceful and stable region of broad cooperation, and the eight Arctic states intend to keep it that way. The Arctic states also recognized they have the primary responsibility to address the many environmental, economic, social and political challenges ongoing in the region, most importantly those issues faced by its inhabitants.[7]

Brigham noted that

> Russia's Arctic Council Chairmanship (2021–2023) program aligns very closely with the Council's new Strategic Plan. Priority areas include: Arctic People, including Indigenous Peoples of the North; Environmental Protection, including Climate Change; Socio-Economic Development in the Region; and, Strengthening of the Arctic Council. Eleven thematic clusters or themes are spread among the four priority areas with a focus on responsible governance for the sustainable development of the Arctic. Notably the Russian Chairmanship calls for greater connectivity of the Council with the Arctic Economic Council, the Arctic Coast Guard Forum, the University of the Arctic, and other regional bodies addressing Arctic issues.
>
> The narrative most of the world hears today is that tension, potential military conflict, and resource races dominate the Arctic. The message in Iceland led by the members of the Arctic Council stands in stark contrast by reminding everyone that the Arctic remains one of the most peaceful places on Earth. Certainly, the Arctic is an integral part of the global security arena as it has been for the past seven decades. Yet continued stability and close international cooperation are the cornerstones of a shared vision of the Arctic states.[8]

Brigham further noted that "Russia's Foreign Minister Sergey Lavrov vowed to keep up the Council's spirit of cooperation and he reiterated the Council's shared vision that the 'Arctic is the territory of peace, stability and cooperation.'"[9] But Putin's gambit to conquer Ukraine a year later upended the unity between all eight Arctic states and undermined the preceding quarter century of Arctic cooperation predicated on Russia's responsible rejoining of the Arctic community of nations at the Cold War's end. While popular sentiment in the West quickly turned against Russia in the wake of its armed assault of its neighbor, bringing an end to most of the goodwill restored between

Russia and the other Arctic states, one should not overlook all the successes of the last quarter century, or Russia's important contribution to Arctic exceptionalism. Indeed, there is much to be proud of, and this is illustrated in the binding treaties to which Russia is a signatory, which, despite tragic events in the Ukraine, it has not violated. Indeed, were it not for Russia's diplomatic isolation resulting from the Ukraine war, it seems likely it would otherwise have chaired the Arctic Council as it had intended at the time of Iceland's handoff in 2021, cooperatively with its neighbors. It is up to the rest of the Arctic world to remember this, and to ensure a space remains for Russia to continue to demonstrate its good faith commitment to Arctic exceptionalism, a concept that was very much shaped and defined and even elevated to the highest levels of global statecraft at the end of Gorbachev's tenure.

The Arctic Council at War: Invasion of Ukraine and the Arctic Council Boycott

On March 3, 2022, seven of the eight Arctic Council member states—known by some as the Arctic 7 (A7)—announced a historic, unanimous boycott of council participation in protest of Russia's invasion of Ukraine, after just over twenty-five years of continuous operations since its inaugural meeting in Ottawa on September 19, 1996.[10] While this is the first time all seven of the democratic Arctic states agreed to suspend participation in council activities, in April 2014, after Russia's first assault upon Ukraine, the United States and Canada jointly boycotted a Moscow meeting of the council but soon rejoined fellow council members, who had not suspended their participation. The 2022 boycott decision by the A7 was one of many similar decisions by countless organizations around the world, part of a quickly achieved diplomatic effort to isolate Russia in protest of its aggression against its neighbor.

However, the Arctic 7's decision caught several of the Arctic Council's Permanent Participant organizations, representing the Indigenous peoples of the region, by surprise, as they were not consulted. This broke with the spirit and long tradition of the Arctic Council, which stands first among the world's many intergovernmental forums for its efforts to unite state and Indigenous interests and for elevating state-tribe consultation to the highest of normative values. While the eight founding member states hold all of the formal institutional power, the Permanent Participants are essential partners in building the consensus that defines Arctic Council governance, and they have played a vital role in both the formation of the Arctic Council in 1996 and its operations in the quarter century since. Indeed, the stability of the Arctic region owes much to the spirit of collaborative governance that aligns Indigenous and state interests, as reflected in the Arctic Council's structure.

While surprised, most of the Permanent Participants endorsed the decision made by the democratic Arctic states but not all with the same level of enthusiasm, and most expressed concern for the future of Arctic cooperation. One of the Permanent Participants, the Russian Association of Indigenous Peoples of the North (RAIPON), viewed by many as a mouthpiece for Putin's government in Moscow, came out in full and enthusiastic support of Moscow's "peacemaking" effort in Ukraine. This outraged a network of Indigenous leaders in involuntary exile from Russia who were formerly associated with RAIPON, who in turn issued their own counterstatement ten days later while also announcing the formation of their own parallel organization to fill the vacuum created by RAIPON's lost legitimacy and what they believe is its outrageous support for Russia's unjust assault on Ukraine.

A Temporary Pause or Permanent Collapse of Arctic Exceptionalism?

According to the March 3, 2022, Joint Statement on Arctic Council Cooperation Following Russia's Invasion of Ukraine, released by officials in the seven Arctic states, "Canada, the Kingdom of Denmark, Finland, Iceland, Norway, Sweden, and the United States condemn Russia's unprovoked invasion of Ukraine and note the grave impediments to international cooperation, including in the Arctic, that Russia's actions have caused." They reasserted their conviction "of the enduring value of the Arctic Council for circumpolar cooperation and reiterate our support for this institution and its work. We hold a responsibility to the people of the Arctic, including the indigenous peoples, who contribute to and benefit from the important work undertaken in the Council."

Their brief statement concluded by explaining,

> The core principles of sovereignty and territorial integrity, based on international law, have long underpinned the work of the Arctic Council, a forum which Russia currently chairs. In light of Russia's flagrant violation of these principles, our representatives will not travel to Russia for meetings of the Arctic Council. Additionally, our states are temporarily pausing participation in all meetings of the Council and

its subsidiary bodies, pending consideration of the necessary modalities that can allow us to continue the Council's important work in view of the current circumstances."

Just one day before, both the Finnish and Swedish ambassadors to Canada announced they would be attending an upcoming session at the Arctic360 Arctic Infrastructure Investment Conference in Toronto, after Russia's planned participation was reversed. The Finnish and Swedish ambassadors had initially withdrawn from the event in protest on February 25, soon after the war began.

After Russia invaded Ukraine in late February 2022 and unleashed chaos in the heart of Europe, eventually exceeding the kinetic destruction experienced during Yugoslavia's collapse and at least matching its scale of human atrocities, on the question of how or even whether to engage with Russia, most of the free world responded with an instinctive, passionate, and near unanimous *nyet*. In the case of the boycott of the Arctic Council by the Arctic 7, this *nyet* would optimally mean *not yet* rather than *never*, and the quest for new modalities in the wording of the boycott indicates a future reset remained conceivable. But even a *not yet* could be (and thus far seems to be) for an indeterminate period, putting into question for the first time since 1996 the very future of the Arctic Council.

This was not the case in 2014, when the boycott was for a single meeting in Moscow. To ensure there is another twenty-five years for the Arctic Council, it's imperative that the Arctic 7's *not yet* be as brief a period as diplomatically possible. But two years on, the higher-level stakeholders are still unwilling to work with Russia. At the working group level, virtual meetings have just been authorized, opening the door to a modest resumption of cooperation. Given the protracted nature of the war in Ukraine and its ever-present potential to draw in NATO, a resumption of face-to-face meetings of principles representing the whole of the Arctic 8 remains as unlikely now under Norway's turn as chair as it was during Russia's two-year rotating chair, which concluded in 2023. All things considered, this ongoing (albeit narrowing) boycott need not mean an end to the Arctic Council or its mutual vision of a cooperative and peaceful Arctic with Russia's full and active participation, which only became possible at the end of the Cold War. But any cessation of the Arctic Council's exemplary consensual alignment of interests and values among a diverse range of states and Indigenous organizations, which includes an even wider array of observers, both state and nonstate entities, from around the world, is worrisome. If Russia remains a pariah beyond such a limited time frame, this could spell the end of a truly exceptional vision of Arctic collaboration and an innovative experiment in inclusive, multilevel, multilateral diplomacy. It was among the best of what the post–Cold War achieved and worth preserving.

Preserving a Peaceful Arctic in a Contested World

This is not to diminish the gravity of events in Ukraine, the clear and present danger to NATO itself, or the specter of total war that could result from a collapse of the post–Cold War order, let alone the human tragedy unfolding across Ukraine since Russia launched its invasion. But the Arctic Council is a unique organization, with legitimacy that extends across the entirety of the circumpolar world, from the western tip of the Aleutian Islands all the way to the eastern tip of Siberia spanning a diverse mosaic of states, tribal peoples, remote environments, and fragile ecosystems undergoing a historic climatic transition.

Moreover, the issues facing the Arctic, of which climate change is perhaps the most pressing for all stakeholders, small and large, cannot be paused. Nor should they. And excluding Russia from any discussions, with the Russian Arctic representing some half the Arctic's geographical extent, would render the council's efforts substantially reduced. There are no halfway solutions to the future of the Arctic, whether it's peacetime or wartime. After a quarter century of Arctic cooperation, this is the first time this unique intergovernmental forum, which I believe is an exemplary model for the world to emulate, has been suspended by all seven democratic member states. Even during the peak of the Covid-19 pandemic, the council managed to meet virtually, finding like so many others that vast distances could be overcome through digital connections. Somehow the Arctic felt less vast, less isolated, at a moment when it was more cut off from the rest of the world than it had been for generations. There's simply no good reason why this wasn't the case after Russia invaded Ukraine, given the high stakes to the Arctic and to the world.

And while Russia's actions in Ukraine are reprehensible, putting at risk the post–Cold War order of which the formation of the Arctic Council was an exemplary and illustrative moment, suspending the council's operations just because Russia was at the time holding its rotating chair, seemed as illogical as boycotting the UN General Assembly—or putting a pause on meetings of the Security Council, shuttering what was originally conceived as an incubator for thawing the East-

West Cold War and providing an off-ramp for the bipolar nuclear order. In hindsight, it is reminiscent of the decision during the Carter administration to boycott the 1980 Moscow Olympics in protest of the 1979 Soviet invasion of Afghanistan—a boycott that only harmed the young American athletes who had devoted years of their lives preparing to defeat their Soviet-bloc rivals on the playing field, and further inflamed East-West tensions. Intergovernmental bodies are one space in world politics where rivals and opponents can meet face to face, even in times of war. Their business does not stop when hostilities commence; oftentimes, their responsibilities multiply manifold at such times.

This is particularly true for the Arctic Council, as it has provided a forum for not only East-West cooperation but state-tribe cooperation as well. Boycotting the council at such a time only puts these two important gains at risk. We thus need the Arctic Council no less today than we did before the Ukraine invasion and may indeed need it more than ever. Arctic Council members find unity in their diversity and approach their Arctic borders with a collaborative spirit seldom found along borders farther south. The challenge, of course, is in minimizing tensions across Arctic borders, at a time when one of the Arctic states is at war with a neighbor that is on increasingly friendly terms with the other seven Arctic states. This is no easy feat. But it's not the first time there's been strategic dissension at the top of the world; indeed, with five of the eight Arctic Council members part of the NATO alliance, there can hardly be a day without strategic dissension in the Arctic even at the best of times. And yet, the Arctic has been one of the most stable regions of the world, despite this preexisting condition of strategic competition.

Indeed, there was a time not long ago when the Arctic Council confronted a deep division in its ranks that threatened the very consensus that serves as the bedrock of its successful first quarter century. That member challenged the consensus accepted by all the council's other stakeholders (member states, Permanent Participants, and observers alike). After two impressive decades of sustained consensus in good times and bad, that member state broke ranks with that unanimity—and in so doing, made it impossible for a joint declaration to emerge from a ministerial meeting for the very first time. That time was just five years ago, in 2019. The founding member state was not Russia, but the United States. And the issue that drove a wedge between the United States and its fellow Arctic Council members and other stakeholders was a change in policy on climate change, long a unifying issue for all Arctic Council stakeholders and the most pressing and salient issue before the council and, many believe, confronting humanity. Despite this collapse in consensus, the council survived, and only a few short years later consensus was fully restored. The organization proved as resilient as the diverse collective of Arctic peoples, states, cultures, and organizations it represents.

Over a decade earlier, that same Arctic Council outlier, the United States, went to war half a world away from the Arctic, bringing along a coalition of partners in its quest to rid Iraq of a fictitious weapons program and to depose its autocratic leader. The Middle East region descended into chaos that lasted a generation. Iraq collapsed into a failed state and became a breeding ground of terror, ultimately providing al-Qaeda with new opportunities and yielding the emergence of the Islamic State caliphate, which required a brutal air war to dismantle. Across Iraq and Syria, a generation has endured unprecedented bloodshed and destruction, all a direct result of a US war of choice built on a foundation of lies. At the top of the world, however, the Arctic remained united and cooperative, and the United States and Russia each contributed their part to the enduring Arctic peace, even as their proxies battled violently across the Middle East. US wars may have been framed as wars of self-defense, much the way Moscow now frames its current war, and wars to preempt weapons programs, much as Moscow now echoes, seizing control of nuclear plants while cautioning against seemingly fictitious Ukrainian chemical weapons programs. It's as if we've seen this story before. But because it's Moscow on the offensive and not the United States, the world is reacting differently to a similarly tragic and disingenuous masquerade.

With Russia on the warpath, having launched what the world sees as an unprovoked war of aggression that once again challenges the consensus that has so long united the Arctic world, we are back to this very same precipice of collapsing Arctic consensus we saw just three years earlier when the United States unilaterally quit the global coalition against climate change. The brutal conflict presently unfolding is different in nearly every way from that previous disagreement at the Arctic Council's Rovaniemi ministerial in 2019, but the stakes are perhaps comparable, if imperfectly so—with the future of humanity once again at stake, and consensus on unifying values once again eluding a single yet essential member of the circumpolar family. And while this comparison will not be greeted with equal receptiveness amid the current crisis unfolding across Ukraine at the hand of Russian aggression, the moral scale of both global challenges, and their mutual risk to humanity, have an equivalency of scale and risk even if not the same sense of immediacy.

The Arctic Council and the Collapse of Consensus

If the Arctic Council consensus can survive collapse as it did at the Rovaniemi ministerial in 2019 over a split between the United States and the rest of the council regarding climate change, there is no reason it couldn't do the same again as the council's diverse membership unites in opposition to Russia's military aggression. Indeed, it has been my belief since this new collapse in consensus emerged that it must do so. Russia is the largest of the Arctic states, with the largest Arctic population, and most diverse Arctic economy and mosaic of cultures. Its portion of the Arctic represents fully half the circumpolar world.

While it took the collapse of the Soviet Union to allow for the close collaboration that the Arctic Council nurtured, with its new model of inclusive diplomacy bridging the state-Indigenous interface, the foundations of the Arctic Council took root well before the Cold War ended. Its founding vision was articulated eloquently by the last Soviet premier in his famed Murmansk speech in 1987. The boundary line separating the United States from the USSR (and now Russia), which was negotiated by the last Soviet foreign minister with his American counterpart, has been respected by both post–Cold War United States and Russia since 1990, laying a stable foundation on what could otherwise be a volatile border. The current shipping lane through the Bering Strait was jointly negotiated by the United States and Russia, reflecting the ongoing stability on this Arctic frontier, with the US Coast Guard and the Russian Border Service developing a laudable bilateral relationship that contrasts with so many of Russia's other borders. This says something about the fundamental importance of Arctic collaboration to world order, regardless of which party or individual controls either country's government.

As bad as things were, and as bad as they became (and may yet become), this was not the time to shutter the Arctic Council's operations altogether, or to stop meeting with all of the Arctic Council's diverse and important stakeholders, whether states, Indigenous peoples' organizations, or NGOs. Framing the A7's suspension as a pause in quest of new modalities did not succeed in creating a window of hope that the council would find its way toward a resumption of its important business. Indeed, the threats made by President Putin to fellow Arctic Council member states Finland and Sweden on the question of their consideration of NATO membership were shocking[11]; Russia's large-scale nuclear triad exercise in Norway's exclusive economic zone (including the launch of a Tsirkon hypersonic missile on February 19, 2022, near the midpoint between Norway's mainland and Svalbard, "the first time ever such launch has taken place in waters between mainland Norway and Svalbard"[12]) is worrisome; and Russia's decision to position a Russian flotilla just offshore the vast and vulnerable island colony of Greenland is foreboding. But it was then and still remains imperative that the Arctic Council find its way back together, and for meetings between *all* of its diverse stakeholders inclusive of Russia, at *every* level, even in the absence of a unifying consensus as they once enjoyed, to continue.

Among some potential modalities that could have been embraced as early as March 2022 for resuming Arctic Council meetings in wartime under Russia's term as chair were boycotting only in-person meetings until peace is restored in Ukraine, utilizing remote technologies to maintain a virtual connection until then among all council stakeholders (partly embraced by Norway in 2024, one year into its turn as council chair, at the working group level); deploying junior proxies in place of senior officials as a subtler but no less obvious rebuke to Russia under its chairship than a complete cessation of meetings, modeled on White House policy regarding China's hosting of the Olympics, where top officials were notably absent in protest of Beijing's long occupation and mistreatment of the Uighur homeland while athletes were free to compete; and adoption of symbolic yellow and blue attire by attendees of such meetings to echo the world's embrace of Ukraine's flag and its spirit of independence and resilience, values embraced across the Arctic. Additionally, the Arctic 7, united in its opposition to Russia's aggression, could have easily assisted Ukraine with an application to become an Arctic Council observer, a move that Russia would surely have opposed—but the point would have been made. Additionally, if Moscow ultimately succeeds in extinguishing Ukraine's sovereignty by forcibly absorbing it into an expanded Russian state, Ukrainians, in their occupation and subjugation, will share a historic experience with Indigenous peoples, as acutely experienced by the Aleuts under Russian colonization and Japanese occupation, the Sami under state formation and expansion, and the Athabaskans and Inuit under the economic domination of the fur monopolies, which colonized so much of Arctic North America.

The council's six Permanent Participants would thus be in a helpful position side by side with the Arctic states in helping to frame Ukraine's experience and to utilize the council as a venue for reconciling the conflict's parties. The Arctic Council could thereby have leveraged its rich mosaic of perspectives, and perhaps have helped the world find a way toward a multilateral solution to this grave world crisis. How to continue this important collaboration in a time of war would, of course, require diplomatic agility and ingenuity, but it's not beyond the capabilities of those who have managed the world's response to the present crisis and was worthy of consideration. While meeting in Russia remained a diplomatic nonstarter, the council could have met in the Russian coal-mining community of Barentsburg on Svalbard, formally part of Norway and whose populace, owing to Soviet history, is in near equal parts Russian and Ukrainian, offering additional symbolic resonance; or in an Aleutian Island community in western Alaska, once a part of the Russian Empire and rich in Russian heritage, and occupied by Japan in World War II; or even in a northern, nonmember state, such as in Hokkaido in northern Japan, which long maintained a constructive diplomatic relationship with Russia even as its northernmost islands in the Kuril chain just offshore remained under Russian occupation, as they have since the final days of World War II—until Tokyo joined the West in its global isolation of Russia. Any of these meeting spots would have be a powerful metaphor for the Arctic Council should the Arctic 7 have chosen to end its boycott, in such a historic venue where the history of strategic competition with Russia is still palpable.[13]

Indigenous Responses to the Arctic Council Boycott

The boycott of Arctic Council meetings by the A7 was announced without first consulting with the Permanent Participants, whose valued knowledge and perspective have long differentiated the council from other intergovernmental fora.[14] The Arctic Council, like the Arctic, is a place where Indigenous and state meet and share a table. That unity is its essence and as such is truly essential to maintain. But as reported by Radio Canada International, and very quickly cached on Google before being unpublished from the Internet in a rare and surprising stealth retraction of a thoughtful and important current affairs analysis, "Climate change, resource development and increased global interest in the North is putting unprecedented stress on the culture and lifestyles of Arctic indigenous peoples, and for northerners, work and collaboration on these issues can't necessarily be 'paused,' says Bill Erasmus, Canadian Chair of the Arctic Athabaskan Council."[15] As Erasmus explained, "'Condemnation of Russia's invasion of Ukraine needed to be recognized at the Arctic Council, but pausing the forum's work may be a mistake,' says a Canadian Arctic indigenous leader. 'The severity of the war needs to be acknowledged at the Arctic Council, but I don't know that a work pause is the answer.' He added,

'None of the Permanent Participants were consulted or involved in the discussions and the [Arctic Athabaskan Council, AAC] is worried about that.'"

Radio Canada International further reports the AAC

> increasingly voiced their concerns over the escalating Ukraine crisis last month, most recently writing Nikolay Korchunov, Russia's ambassador at large for Arctic Affairs and the Chair of the Senior Arctic Officials, on February 23 warning of war's catastrophic impacts on human life, the environmental and economic costs, and the potential spillover on Arctic cooperation. [The AAC is] now examining what impact the Arctic Council pause may have on their projects and collaborations. "Normally, the Permanent Participants are an active part of the Arctic Council and have a right to be there," Erasmus said. "I spoke to [Heidi Kutz], Canada's Senior Arctic Official, on Thursday and told her we are quite concerned we weren't included." . . . Stressing that he's only speaking for the AAC and not the other Permanent Participants, Erasmus said his organization will continue to pursue their projects and collaborations as they judge necessary. "We weren't part of the decision, so we can't be held to something we weren't included in," he said. "As northerners we have to take care of ourselves, we can't rely on others to do it for us."

Two weeks before war broke out, Erasmus observed in *News North*, "We're concerned this may have implications for how the Arctic Council functions as an entity." With much of the original boycott still ongoing (with the exception of virtual working groups, now authorized to meet), such concerns remain ever salient. While the ACC has articulated its concerns about its exclusion from member state deliberations on the boycott, and its concern about the undiminished pressing issues facing Arctic Indigenous peoples in wartime as well as peacetime, *Nunatsiaq News* has reported, "Some Arctic organizations are distancing themselves from Russia after the country's sudden invasion of Ukraine late last week. 'We are monitoring the situation on a day-to-day basis,' Inuit Circumpolar Council Canada president Monica Ell-Kanayuk said."[16] As *Nunatsiaq News* further reported, "At the moment, members of ICC Canada are planning not to travel to Arkhangelsk, Russia, at the end of May for Arctic Council meetings, said Ell-Kanayuk. But she added ICC Canada has kept in contact with ICC Chukotka and the two members are committed to continuing to work peacefully and co-operatively."[17]

Chief Erasmus has been one of the few public critics, albeit briefly so, of the Indigenous exclusion from the A7's decision to boycott the

Arctic Council, and while most of the Permanent Participants endorsed the boycott after it was announced, they did so with some qualification, cognizant of how great Indigenous gains have been and how much Arctic Indigenous peoples have to lose if the Arctic Council does not find its way back together. Moreover, the response of the Permanent Participants has been somewhat less unified than the A7's, itself a reflection of their diversity within the council.

The Russian section of the Saami Council issued its own statement on February 27, among the first, commenting they "cannot ignore the current situation in the country or remain silent about it" while "in no case will we touch upon the question of who is right and who is wrong." The section "understands that there is no justification for military action. In any case, all this touches us, so the Section on the Russian side considers it necessary to comment on this topic."[18] They note, "The citizens of the Russian Federation, including the Saami people in Russia, are in a situation where no one knows what awaits us in the future. We cannot plan anything and we find ourselves in a very unstable situation. Sanctions already introduced by different countries, and possible future sanctions, will primarily hit, not businessmen and owners of mega-corporations and banks, but ordinary residents of the country."[19]

Such impacts were immediately felt by the Sami:

> "Already, prices on the electronics market have increased by 30% in one day, and we expect the prices to increase even more, not only for electronics, but also, for food and essential goods. The sanctions and the measures introduced do not separate the citizens of the Russian Federation by area of work or nationality, so the Saami people in Russia find themselves in an extremely unstable, one might say, dangerous, situation. None of us can predict how the aggravated situation will end, but already now we must be prepared for additional difficulties affecting international work.[20]

The Sami discussed the effect of sanctions on Russian banks, including Sberbank serving Russia's north, "which means that transactions to Russia will be difficult. This involves both projects and salaries and makes cooperation more difficult. Sberbank has conducted transactions in Norwegian kroner through a US bank, transactions in euros through a German bank, and both of these countries imposed sanctions on working with Sberbank and many other banks."[21]

Further impacts were felt in the everyday form of a "partial blocking of Facebook" to the consequences of suspensions "from membership in the Council of Europe, the Committee of Foreign Ministers, the Parliamentary Assembly of the Council of Europe, the European Council for Human Rights, many sporting events" and even "from participating in Eurovision." Amid this dizzying cascade of suspensions, "the fact remains that international cooperation for Russian citizens, in any direction, is now as difficult as possible," and "the possibility of sanctions that will annul existing visas for Russian citizens" was identified as a concern of the Sami: "In many documents, the Saami Council states that the Sami are one people who live regardless of state borders. Now, this is high on the agenda, to make sure that the Sami people from the Russian side can continue to participate in international meetings and conferences, including visiting other countries."[22] Indeed, "Now, more than ever, the Sami people in Russia need international support to continue cooperation between the Sami of the four countries. We hope that this difficult situation will soon be resolved in the least painful way."[23]

Gwich'in Council International, representing the Gwich'in communities in the northernmost forests of Alaska, Yukon, and the Northwest Territories, in its "Joint Statement on Arctic Council Cooperation Following Russia's Invasion of Ukraine" on March 3rd announced that it "welcomes the collective pause of activities of the Arctic Council as we explore new modalities for pursuing peace and cooperation in the north" and noted its "grave concern for the people of Ukraine, particularly the indigenous peoples, due to the invasion by Russia. We stand with our partners around the world in calling for peace in Ukraine, a peace which can only be achieved by Russia recalling its armed forces immediately."[24] The council reiterated that it "remains committed to engage in productive dialogues that advance the collective aim and responsibility of stewarding a peaceful Arctic region built on cooperation and our shared value of mutual respect."[25]

Four days later, the Inuit Circumpolar Council (ICC) released its "Statement from the Inuit Circumpolar Council Concerning the Arctic Council," acknowledging the joint statement from the seven democratic Arctic states "calling for a temporary pausing of participation at all meetings of the Arctic Council and its subsidiary bodies" as well as a "message from the Russian Chair of the Arctic Council agreeing to the request of the other countries."[26] ICC noted that four of the six Permanent Participants have Russian constituents: "Inuit in Chukotka, Russia, are members of ICC," and the Aleut International Association and Saami Council also have Russian members, while the Russian Association of Indigenous Peoples of the North is entirely Russian in membership. Noting ICC's proud heritage "as a unifying voice for Inuit across our collective homeland" from the Cold War to the present, it expressed concern for "the future of the Arctic Council which is based on peaceful cooperation and mutual respect."[27] ICC reiterated its commitment "to the Arctic remaining a zone of peace, a phrase coined by former USSR President Mikhail Gorbachev," a message that "ICC has repeatedly echoed."[28] Lending its approval to the boycott, ICC noted it's "monitoring the situation closely and agrees with the [Senior Arctic Officials] that this temporary pause will allow time to consider 'the necessary modalities that can allow us to continue the Council's important work in view of the current circumstances.'"[29]

While AAC's Chief Erasmus was openly critical of the lack of Indigenous consultation in the decision to boycott the council, this criticism was only briefly aired to the public prior to the unusual withdrawal of the article from the Internet without public explanation or even editorial revision or correction. Indeed, it became clear there was no error in the article, and in correspondence with this author, the journalist suggested she had erred and conveyed the opposite of what Chief Erasmus intended to say, even though Chief Erasmus's comments were consistent with the openness, transparency, and moral leadership for which he is renowned. The internal pressures not to break rank with the other Permanent Participants, which all refrained from public criticism of the boycott when it started, likely contributed to the withdrawal of Chief Erasmus's thoughtful and earnest observations. But only a few months later, another Permanent Participant organization, the Inuit Circumpolar Council, did mirror Chief Erasmus's concerns in public comments in *Nunatsiaq News* made by James Stotts, president of ICC Alaska and head of the ICC delegation to the Arctic Council.

On June 22, *Nunatsiaq News* reported that the Arctic Council "is resuming work on a limited scale—but Russia won't be allowed to participate," and "according to James Stotts, head of the Inuit Circumpolar Council delegation to the Arctic Council" (and just as experienced in March), the "seven Arctic Council nations other than Russia made the decision without consulting the council's other permanent participants, the six Indigenous groups in the forum."[30] As Stotts described, "We were notified after they made their decisions," to which he added, "We don't like it that we weren't consulted."[31] Stotts further described that "within his organization, 'not everybody is in agreement, I think, on reopening the Arctic Council right now—or even in a limited way.'"[32] Stotts added that "it's not clear how activities can resume without Russia: 'There are more questions than answers with what's going on now for us.'"

This decision for the Arctic Council to resume meeting, without Russia's participation, suggests that Westphalian security interests were once again ascendant in the Arctic. Will the council remain bifurcated along the old East-West fault line—with the A7 representing a "free Arctic" now united by their common NATO membership, and exclusive of the vast Russian Arctic? Will the whole council, inclusive of Russia, ever find its way back together, or has Arctic security returned to Cold War bipolarity? And in its future form, how will the council reconcile the interests of Indigenous peoples and modern states? The Arctic Council has since its inception in 1996 been about more than Arctic states, and its distinct contribution to statecraft has been rooted in its diverse, multilevel collaboration between Indigenous peoples from across the circumpolar world, working in partnership with a diverse group of states—from microstates to middle powers to superpowers, and from democracies to colonial states to autocracies—to reimagine responsible statecraft as a synthesis of national and Indigenous interests, and Arctic security as a distinctly northern synthesis of hard and soft security interests.

With Finland and Sweden exchanging generations of official neutrality for NATO membership, it appears that national security interests are once again trumping Indigenous reconciliation, putting the long, peaceful era of Arctic exceptionalism under renewed pressure. Finding a path back to the restoration of the council's important, indeed sacred, reconciliation of tribe and state is imperative, ideally with the eight founding member states, inclusive of Russia, and the six Permanent Participants all at the table once again—providing the world with an innovative, inclusive, and, importantly, viable model for regional stability.

In the end, as mentioned above, only one of the Permanent Participants came out officially in support of Russia, and this contrarian reaction was put forth by the Russian Association of Indigenous Peoples of the North, a group that many observers have described as no longer authentically representing the voice of Russia's Indigenous peoples.

RAIPON's March 1 statement, addressed to Russian President Vladimir Putin and cosigned by thirty-three associated regional Indigenous leaders, read,

> Respected Vladimir Vladimirovich! A peaceful sky, land of our ancestors and the safety of children—nothing can be more important for every inhabitant of our planet. For everyone. No exceptions. Regardless of ethnicity or native language. The North, Siberia, and the Far

> East remembers with gratitude those who have dedicated their destinies to the formation of our regions. This group includes people born in Ukraine, including Donetsk and Luhansk. For eight long years, and with pain in their hearts, the indigenous inhabitants of places far from these "hotspots" kept hoping for the restoration of human rights in Ukraine and the equality of all citizens without exception. Peacemaking is never easy. The Russian civil organization "Association of the Indigenous Peoples of the North, Siberia, and the Far East" supports your aspiration and decision to protect the rights and interests of the residents of the Donetsk and Luhansk Peoples Republics and the security of a multiethnic Russia. We, representatives of 40 small indigenous peoples of the North, Siberia, and the Far East express hope for quick mutual understanding to ensure peace and harmony.[33]

Ten days later, on March 11, a new organization called the International Committee of Indigenous Peoples of Russia put out its own statement rebutting RAIPON, signed by seven Indigenous leaders in involuntary exile from Russia and contrasting greatly with RAIPON's endorsement of Putin's peacemaking efforts:

> WE—the undersigned representatives of the Indigenous peoples of the North, Siberia and the Far East living outside of Russia against our will—are outraged by the war President Putin has unleashed against Ukraine. At the moment, the entire population of Ukraine is in grave danger. Old people, women and children are dying. Cities and towns of an independent country are being destroyed because their inhabitants did not want to obey the will of a dictator and a tyrant.[34]

They "express solidarity with the people of Ukraine in their struggle for freedom and are extremely concerned about ensuring the rights of Indigenous peoples during the war on Ukrainian territory, including the Crimean Peninsula that remains illegally occupied by Russia." They "are outraged by statements of the Russian Association of Indigenous Peoples of the North (RAIPON) on March 1, 2022 . . . in support of the decisions of President Putin. Such public statements can only be considered as direct support for the military aggression against the Ukrainian people, and their signatories are accomplices of the murderers of civilians in Ukraine."[35] They affirm their belief that RAIPON "has become an adjunct of the Russian Government domestically and internationally," becoming "an organization whose main purpose is to justify the actions of the Government and extractive companies that feed President Putin's regime and destroy the traditional lands of indigenous peoples."[36]

Moreover, they believe "the leaders of RAIPON . . . who signed the letter should be also treated as war criminals, not merely accomplices of the war, because they voted for military action as members of the Russian Parliament."[37] This new group will henceforth "consider any statements on any international platforms made by RAIPON and its representatives or by representatives of other organizations of Indigenous peoples which supported Vladimir Putin's decisions regarding Ukraine as falsehood and propaganda as their main purpose is maintaining the prestige of the Russian authorities at the international level." They call upon the international community, including the UN, Council of Europe, and Arctic Council, "to ignore the statements of RAIPON representatives and spokespeople of other organizations which supported Vladimir Putin's decisions."[38] In closing, they announce their own withdrawal "from all Russia-based organizations and networks of Indigenous peoples of Russia in which we were previously members. We announce the creation of a new, independent organization—the International Committee of Indigenous Peoples of Russia."[39]

The Russians are also, to no one's surprise, disappointed by the decision of the other Arctic states, and for their obvious exclusion from deliberations (being the aggressor, and the ultimate cause of the breach in consensus with fellow council members.) As Gloria Dickie reported in *SaltWire*, "Russian Arctic officials questioned on Friday the decision of their peers on the Arctic Council to boycott future talks held in Russia, calling their actions 'regrettable.' . . . Nikolay Korchunov, Russia's senior Arctic officials chair and an ambassador-at-large of the Russian Ministry of Foreign Affairs, warned that a temporary freeze on council activity would 'inevitably lead to the accumulation of the risks and challenges to soft security in the region.'" He further "stressed the council's strong history of depoliticized dialogue in high latitudes. 'The Arctic should remain as a territory of peace . . . and thus, this unique format should not be subject to the spill-over effect of any extraregional events.'"[40] With Russia representing "roughly a third of the entire Arctic region" and "nearly 70% of economic activity in high latitudes," Korchunov explained that, "For us there is no alternative to uninterrupted sustainable development of our Arctic territories," and that Russia would "refocus" its "Chairmanship toward addressing our domestic needs in the region." Korchunov added, "It is of utmost importance to safeguard the project activities of the Arctic Council in order to be able to pick up where we paused and step up cooperation."[41]

Restoring Circumpolar Cooperation: Ending the Arctic Council Pause Before Russia Ends Its War

This is not the first time world politics has intruded into the otherwise calm spirit of Arctic cooperation. Aidan Chamandy recounts that the council "faced a similar problem after Russia invaded Crimea in 2014," when both "Canada and the U.S. boycotted an April 2014 council meeting in Moscow, but Canada was keen 'to support the important work of the Arctic Council' in future, according to a statement that year by former Environment minister Leona Aglukkaq. The 2014 boycott was the only one, however," until now.[42] Chamandy cites "Nicole Covey, a fellow with the North American Arctic Defence and Security Network," who explained that the limited 2014 boycott by Washington and Ottawa contributed to the widely held "belief that the Arctic Council could withstand a lot of international tensions. So what happened with the pause is very substantial" with its "unified response. . . . The fact that they're only pausing, and that they haven't ended the Arctic Council, is important, because that shows they're hoping things might resolve in some way."[43] Indeed, as Covey further explains, "If Russia is no longer involved in the Arctic Council, you no longer get that circumpolar, holistic approach."[44] Such a view was shared by then ICC chair Dalee Sambo Dorough, who is also cited by Chamandy as explaining, "Everything (in the Arctic) is interconnected. . . . It (could) be difficult for seven other Arctic states . . . to be effective and move forward in a constructive fashion," and the Arctic Council "wouldn't be the same if one of our clear and genuine members is absent."[45]

Russia held the rotating Arctic Council chairship until 2023, when it passed to Norway, and thus far there's been much continuity with the rotation, a testament to the endurance of consensus among its diverse stakeholders. To completely boycott the council under Russia's chairship only undermined the spirit of collaboration that gave form to the Arctic Council during more optimistic times. The agenda for the Arctic Council, even under the early days of Russia's chairship (before the pause forced a shift from transnational to national-level projects), showed much continuity with the previous chair (Iceland), and this continuity alone could have served as an important, symbolic bridge to the restoration of that founding cooperative spirit.

And because consensus is the lifeblood and governing paradigm for the council, there was nothing Russia could achieve as chair without the full consensus of the other council members. Each biannual ministerial meeting, each semiannual Senior Arctic Officials meeting, each working group session, would have provided an opportunity to rebuke Russia and deny it consensus on any issue that deviated from the collective will of the Arctic Council as a whole; and, on issues where consensus was preserved, it could have likewise demonstrated that Russia, even at the worst of times, remained committed to the values and principles of the Arctic Council. This channel of ongoing diplomatic interaction could have served as a bridge to the future, and the restoration of a world where Russia was a responsible member of the world community.

Ironically, even as Russia embarked on the path of expansionist war to its south, it remained committed, for the most part, to cooperation to its north. Consider the words, and life experience, of Russia's Senior Arctic Official, Nikolay Korchunov, who held the rotating chair of the Arctic Council (Russia's second tenure at chair since the Arctic Council was formed). The Arctic Council website described Korchunov as "a career diplomat who has served as the Ambassador at Large for the Arctic Cooperation at the Ministry of Foreign Affairs (MFA) of Russia and the Senior Arctic Official of the Russian Federation to the Arctic Council since December 2018," with prior service at Russia's Ministry of Foreign Affairs headquarters, and in overseas postings to "embassies of the Russian Federation to Sweden and Finland," and as "Head of Russia's delegation to the Council's Task Force on Arctic Marine Cooperation and the Task Force on Improved Connectivity in the Arctic from 2015 until 2018."[46] In these past positions we see a depth of knowledge, experience, and commitment to the Arctic, and to its cooperative legacy. As he assumed the chairmanship of the Arctic Council, he was "especially interested in matters related to sustainable development, in finding the right balance between environmental protection and socio-economic development,"[47] positioning him in the very sweet spot at the intersection of Arctic globalization and the preservation of the sublime, undisturbed Arctic free from the manifold impacts of modernization.

In an interview on the Arctic Council website, he observed, "I have visited many places in the Arctic and gotten acquainted with many people and issues. I can see now that most issues, questions and challenges in the Arctic are interconnected. So there is an obvious need for an integrated, crosssectoral, and inclusive approach to development in the Arctic."[48] A year ahead of assuming the rotating chairmanship of the Arctic Council in 2021, he noted,

> We are in the early stages of the preparations for the forthcoming Arctic Council Chairmanship. There will be a number of priorities on the

> agenda of our Chairmanship, among them of course environmental protection and sustainable development, as well as new technologies for safeguarding the Arctic environment. The human dimension, the Arctic inhabitants including Indigenous peoples, will of course be stressed and underlined throughout our forthcoming Chairmanship.[49]

He noted the Russian term for Arctic Council is "Арктический совет—организация, которая должна обеспечить умное управление Арктикой. И это то, чем мы занимаемся," which translates to, "The Arctic Council is the organization that is supposed to provide the conditions for and contribute to responsible governance in the Arctic. That is what we are doing here."[50] Indeed, that continues a legacy that started in 1996 (or with the Arctic Environmental Protection Strategy, five years earlier) and was carried forward by each of the Arctic states during their periods of service as Arctic Council chair—a tradition Russia had pledged to uphold.

That Russia could nurture a collaborative Arctic even as it engaged in regional conflicts as far afield as Syria, Libya, Crimea, and eastern Ukraine (with only one brief interruption when in 2014 the US and Canadian Arctic Council representatives boycotted a Moscow meeting) was the accepted view from its 1996 formation to its twenty-fifth anniversary in 2021, when Russia assumed the rotating chair for its second time. While it is universally acknowledged that Russia's full invasion of Ukraine is a game changer, and a systemic risk to global security, and while threats to Finland and Sweden on Russia's far northwestern frontier and to Japan on its far northeastern frontier indicate a notable shift in diplomatic tone, one cannot readily forget the "If you're not with us, you are against us" philosophy undergirding President George W. Bush's "Bush Doctrine." That philosophy shaped and guided the global war on terror that the United States waged after the 9/11 attacks, bringing wars of regime change not only to Taliban-ruled Afghanistan where planning of the attacks took place but to countries like Iraq that had nothing to do with the initial conflict. A fluctuating "coalition of the willing" aligned with the United States at war, often to their eventual regret (as seen in the United Kingdom).

While the United States did not threaten the existence of those who did not join it, it did position opponents to the US war as "against us," which was coercive diplomatically. Yet the regional wars that embroiled the US armed forces for a generation did not impede the enduring cooperation between all of the Arctic states, regardless of where they stood on those wars. Though Russia has threatened supporters of Ukraine's defense more forcefully, including Finland and Sweden for considering joining NATO, it must still be noted that Arctic cooperation has endured many disagreements among the Arctic states on various matters of policy, whether related to foreign wars or not (as seen in the Trump administration's break with fellow Arctic Council members on climate change, which under Obama was conceptualized as an all-of-government war against nature out of balance). The A7 decision to boycott the Arctic Council entirely, as part of the global isolation of Russia, does risk the opportunity presented by Korchunov under his watch as Arctic Council chair, and creates a gap in multilateral cooperation that could endure for many years. Can the Arctic afford such a gap? Was the exclusion of the Permanent Participants from the discussion and decision to impose an Arctic Council boycott, breaching the spirit of state-tribe collaboration for which the Arctic Council rightfully takes much pride, likewise a gap within the gap, putting the Indigenous organizations in the uncomfortable position that President Bush put US allies and friends in during the run-up to war after 9/11? Has the A7 response to Russia's aggression in Ukraine created its own risk to Arctic unity that continued operation of the Arctic Council during wartime might have avoided with various alternative ways of expressing profound disagreement with Russia?

According to *Newsweek*, "Russia's Arctic envoy has told *Newsweek* that international tensions over the war in Ukraine should not spill into the northern region that also borders the United States. But recent diplomatic and military moves by Washington and its allies show the usually serene frontier has already become a frontline in the crisis."[51] As *Newsweek* reported,

> Nikolay Korchunov, who serves as Russia's ambassador-at-large to the Arctic Council, told Newsweek that Moscow found this decision "regrettable," arguing that it ran contrary to the apolitical nature of the intergovernmental forum founded more than 25 years ago. "The Council's mandate explicitly excludes matters related to military security," Korchunov said. "It is enshrined in all its founding and strategic documents that the Arctic should remain as the territory of peace, stability and constructive cooperation. Therefore, this unique format should not be subject to the spillover effect of any extra-regional events."[52]

According to *Newsweek*, Korchunov argued that the pause "'initiated by the Western states, could lead to the accumulation of the risks and challenges to soft security in the region which the Council has been addressing effectively. . . . Under the current circumstances, it is of utmost importance to safeguard the project activities of the Arctic Council in order to be able to pick up where we paused and step up cooperation like we unanimously and timely did in the face of the COVID-19 pandemic.'"[53] With the pause preventing Russia from engaging with its peers on the Arctic Council, Korchunov said Russia would instead "refocus the elaborate mechanism of the Chairmanship towards addressing domestic needs in the region."[54]

Even with the collaborative dynamic of the Arctic Council on pause, Korchunov affirmed that "Russia firmly believes that there is no potential for conflict in the Arctic," and reiterated Russia's commitment to the international legal mechanisms that govern the region: "When it comes to existing feuds, Korchunov argued that 'international law clearly stipulates the rights of coastal and other states and provides a firm foundation for cooperation in addressing various issues, including such sensitive ones as the delimitation of the continental shelf in the Arctic Ocean and the prevention of unregulated high seas fishing in the Central Arctic Ocean.'"[55] He also noted the "strong commitment of all Arctic coastal states to the principle of 'orderly settlement' of any possible disputes and overlapping claims led to adopting the Ilulissat Declaration in 2008 and this commitment was reiterated on its 10th anniversary, and consequently led to signing a respective fishery agreement in 2018 by all five Arctic coastal states, China, Japan, the Republic of Korea, and the EU.'"[56]

Korchunov highlighted the importance of the Arctic to Russia, noting its own Arctic region "does not only produce more than 11% of our national GDP and over 20% of exports," but is also "home to more than 2.5 million people, including Indigenous Peoples of the North. . . . For Russia, there is no alternative other than to maintain sustainable development of its Arctic territories. . . . It accounts for approximately a third of the Arctic region, more than half of the Arctic population, almost half of the whole Arctic coastline and nearly 70% of all economic activities in high latitudes."[57] Korchunov reiterated Russia's original goals for its chairship of the Arctic Council, which he said still "aimed at improving the well-being and quality of life for people in the Arctic, including Indigenous Peoples, adapting the region to global climate change, preserving Arctic biodiversity, ensuring sustainable and safe maritime activities and socio-economic development, exploring solutions to ensure global energy security, promoting scientific cooperation in the Arctic, [and] strengthening the Arctic Council."[58]

Korchunov explained that "'the strategic importance of the Arctic has markedly increased' as a result of emerging risks and opportunities," and that the "attention of many stakeholders is now focused on the Arctic as a global 'climate kitchen,' a unique, yet fragile ecosystem, a treasure trove for Indigenous cultural heritage and, of course, a region with tremendous economic opportunities. . . . While the challenges the Arctic faces are numerous, it is important not only to focus on the Arctic's vulnerability but also to be mindful of the region's resilience and opportunities for sustainable development."[59] *Newsweek* noted with "cooperation at a standstill due to the deadly war raging in Ukraine, the future of multilateral efforts remains deeply uncertain," and that "Korchunov said his country 'reiterates its commitment to close and constructive engagement with all Arctic Council member-states, permanent participants, observers and other interested non-regional partners. We are open for long-term partnerships in the region with any nation . . . be it the Arctic Council member state or any other country, in the interest of its sustainable development and well-being of its inhabitants, including Indigenous Peoples.'"[60] Even in this time of war, Korchunov reiterated that "'Russia is convinced that "the spirit of cooperation" inherent in the Arctic Council will help to strengthen trust and mutual understanding . . . and the Council should remain a solid framework for peaceful mutually beneficial collaboration despite geopolitical tensions elsewhere in the interest of a sustainable and prosperous future of the entire Arctic region.'"[61]

There was thus much opportunity to participate, and in so doing, thwart Russia's ambitions with the power of consensus that undergirds the Arctic Council. Even with Ukraine aflame, the Arctic continues to melt. The challenge of climate change does not go away, nor do the many pressing challenges across the remote and isolated communities of the Arctic region. And, with tensions high in Europe, Moscow may choose to ship more of its petroleum products through the Northern Sea Route to Asian markets rather than to European ports—and if it does, its economy will come to increasingly rely on the stable border with the United States it worked so hard to create and to sustain since the final days of the Soviet Union. When Russia sold Alaska to the United States in 1867, it did so for a good reason: we were the best neighbor to have, in good times and bad. This remained true during World War II when that border provided a lend-lease lifeline to the eastern front, and it remained true during the Cold War even when tensions were high. And despite the uncertainty and chaos unleashed with Russia's invasion of Ukraine, it remains true even today.

There was every reason to isolate Russia as the war in Ukraine continued, and to remind Putin that he has become a pariah that threatens global stability. But no matter how things unfold in Europe, Russia remains more than half of the circumpolar world. Governing the Arctic effectively, and peacefully, will require Russia's participation and consent, regardless of events in Ukraine. The Arctic climate will continue its inexorable thaw; its still pristine ecosystems remain as fragile and sublime now as they did before Russia invaded Ukraine. Arctic Indigenous peoples continue to offer the world their wisdom and traditional knowledge, and require our continued, good-faith efforts to overcome past economic marginalization and political exclusion.

Thus by keeping the pause on Arctic Council participation finite and resuming meetings among its stakeholders virtually and in symbolic protest while trying to restore consensus with Russia on the many important issues that still unite the whole of the circumpolar world, the Arctic Council can instead become part of the answer to, and resolution of, the current conflict. It can offer the very same bridge to a collaborative future that it has promised since 1996 and show that its second quarter century can be as successful as its first. The Arctic Council survived the collapse of consensus once before, on an issue of great import, that of the climate change challenge facing humanity and requiring our collective unity to overcome. The council can—and must—survive the current collapse in consensus that has accompanied the conflict over Ukraine and keep this important bridge to a more peaceful and united future open for the time when Moscow is ready to reset its policies and rejoin the consensus it once embraced.

The Ukraine War and the 2022 US Arctic Strategy Update

With war in Ukraine ongoing, many observers on both sides expect the conflict to persist. The specter of escalation worried many early on as Russia confronted preliminary battlefield setbacks and signaled potential nuclear use. Such worries were ominously articulated by President Biden at a New York fundraiser on October 5, 2022, when he compared the present strategic risk to that experienced during the 1962 Cuban Missile Crisis, the last time the United States "faced the prospect of Armageddon." New fears of Armageddon have been fueled by a cascade of interconnected events, from the collapse of Russian lines in eastern and southern Ukraine—bookmarked by Kyiv's successful and audacious assaults on such prized targets as the warship *Moskva*, sunk April 14, and the Kerch Strait Bridge, damaged on October 8—to the bungled initial implementation of Russia's 300,000-man military mobilization—which succeeded only in motivating the sudden exodus of that many young men from Russia, who voted with their feet in opposition to conscription in what may be Russia's largest mass wave of refugees since the Bolshevik Revolution—to the Wagner rebellion, whose ultimate repercussions remain to be seen in the wake of its founder's assassination.

Far to the north, the war in Ukraine has had a profound effect on Arctic security and diplomacy, causing a deepening rift within the Arctic Council. And while the council represents an innovative, post–Cold War collaboration between East and West across the North, as envisioned by Gorbachev in 1987 in Murmansk, with the pioneering inclusion of Arctic Indigenous peoples, as sought by the ICC since its formation in the 1970s, it took the war in Ukraine to put both of these noteworthy achievements at risk.

Just as Russia, with over 24,000 kilometers of Arctic coast (over 50 percent of the Arctic Ocean's coastline) and an Arctic population of over 2 million (around 50 percent of the entire population of the circumpolar Arctic), finds itself isolated from the other seven Arctic states, the six Indigenous Permanent Participant organizations at the Arctic Council all noted a decline in consultation from their state partners as these important changes unfolded. This suggests a Westphalian reconceptualization of Arctic security as a state prerogative was underway, in notable contrast to the multilevel, multilateral cooperation that hitherto united all eight of the council's state members with the Permanent Participant organizations in a triumph of inclusive diplomacy.

This Westphalian resurgence was evident in the sudden change of heart of both Finland and Sweden, which both turned to NATO for protection after Russia invaded Ukraine—another dramatic setback to President Putin's ambition to keep NATO from further encroachment toward Russia's borders. Both hitherto nonaligned Nordic states completed accession talks at NATO headquarters on July 4, a day with much charged symbolism in the United States, formally confirming their willingness and capacity to meet all obligations associated with NATO membership and putting into rapid motion the historic extension of NATO from the Norwegian border all the way to its Baltic member states (all former constituent components of the Soviet state).[62]

Though Sweden and Finland maintained nonalignment through the Cold War era, in the decades since they have both grown closer to the Western alliance, having both joined the Partnership for Peace in 1997. Because of their strengthening collaboration with NATO in the years since, their abandonment of neutrality in favor of alliance membership, while catalyzed by the war in Ukraine, is less tectonic a realignment of

European security architecture than it may first seem. Sweden initially indicated its unwillingness to host NATO assets, in much the same way that Norway reassured Moscow throughout the Cold War that its territory would never be used as a staging ground for a northern invasion of Russia. In contrast, Finland initially placed no such restrictions and even stated its willingness to accept the transit of nuclear weapons across its territory in addition to the construction of NATO bases. But the very fact that the Nordic and Baltic states (with the exception of Russia and its western outpost in Kaliningrad) are now united as NATO allies strengthens Western unity and more tightly aligns the strategic capabilities of the A7, whose divisions in the past often brought Russia and the Arctic coastal states together under UNCLOS, with noncoastal states Sweden and Finland excluded from the maritime Arctic club as they were in May 2008 at the Ilulissat conference.

And just as NATO expands to include these last two nonaligned Arctic states, the United States has refocused more of its strategic and diplomatic attention, investment, and energy on the Arctic as well, culminating in a series of organizational and policy updates. The White House announced in August 2022 its plan to establish a new ambassador for the Arctic region, following in September 2022 with the formation of the new Arctic Strategy and Global Resilience Office at the Pentagon. Iris A. Ferguson, formerly a senior advisor on Arctic and climate security issues for US Air Force Futures, was appointed as the first deputy assistant secretary of defense for Arctic and Global Resilience—a "new position that signifies the importance U.S. leaders place on the region," as described by Jim Garamone at *DOD News*, who added that the United States "is an Arctic power, and the Defense Department has established an office to ensure U.S. strategy and policy protects U.S. interests in that crucial region."[63]

Then, in October 2022, the United States unveiled its new National Strategy for the Arctic Region, updating its 2013 strategy and adapting it to today's more complex, divided, and evolving strategic landscape. As noted in its executive summary, the new strategy "addresses the climate crisis with greater urgency and directs new investments in sustainable development to improve livelihoods for Arctic residents, while conserving the environment. It also acknowledges increasing strategic competition in the Arctic since 2013, exacerbated by Russia's unprovoked war in Ukraine, and seeks to position the United States to both effectively compete and manage tensions."

The strategy update includes four pillars: security, climate change and environmental protection, sustainable economic development, and international cooperation and governance. With regard to security, Washington aims to "deter threats to the U.S. homeland and our allies by enhancing the capabilities required to defend our interests in the Arctic, while coordinating shared approaches with allies and partners and mitigating risks of unintended escalation." Regarding climate change and environmental protection, Washington "will partner with Alaskan communities and the State of Alaska to build resilience to the impacts of climate change, while working to reduce emissions from the Arctic as part of broader global mitigation efforts, to improve scientific understanding, and to conserve Arctic ecosystems." Regarding sustainable economic development, Washington will "pursue sustainable development and improve livelihoods in Alaska, including for Alaska Native communities, by investing in infrastructure, improving access to services, and supporting growing economic sectors," while "work[ing] with allies and partners to expand high standard investment and sustainable development across the Arctic region." And with regard to international cooperation and governance, "Despite the challenges to Arctic cooperation resulting from Russia's aggression in Ukraine," the United States will continue to "work to sustain institutions for Arctic cooperation, including the Arctic Council, and position these institutions to manage the impacts of increasing activity in the region" while "also seek[ing] to uphold international law, rules, norms, and standards in the Arctic." As a headline in the October 10 issue of *High North News* described the updated document, "New US Arctic Strategy Foreshadows Increasing Hurdles for Cooperation in a More Complex Region," with an equally apt subtitle: "The new US Arctic strategy suggests that addressing the adverse impacts from climate change and maintaining international cooperation despite increasing strategic competition vis-à-vis Russia and China will be key challenges in the region over the next decade."

While Arctic Indigenous leaders had expressed their concerns about the recent decline in consultation after the A7 boycott of Russian participation in Arctic Council activities, the updated Arctic strategy offers much reassurance that such exclusion is not expected to continue. Indeed, the executive summary notes, "This strategy is intended to serve as a framework to guide the U.S. government's approach to tackling emerging challenges and opportunities in the Arctic" and "will be guided by five principles that will be applied across all four pillars," with the first of these five being "Consult, Coordinate, and Co-Manage with Alaska Native Tribes and Communities," in which the United States remains "committed to regular, meaningful, and robust consultation, coordination, and co-management with Alaska Native Tribes, communities, corporations, and other organizations and to ensuring equitable inclusion of Indigenous Peoples and their knowledge."

The second of the five principles is "Deepen Relationships with Allies and Partners" including "Canada, the Kingdom of Denmark (including Greenland), Finland, Iceland, Norway, and Sweden" and to "also expand Arctic cooperation with other countries that uphold international law, rules, norms, and standards in the region." The third principle is "Plan for Long Lead-Time Investments," while the fourth is to "Cultivate Cross-Sectoral Coalitions and Innovative Ideas." The fifth and final principle, "Commit to a Whole of Government, Evidence-Based Approach," complements the first principle by observing that the "Arctic extends beyond the responsibility of any single region or government agency," and that "US Federal departments and agencies will work together to implement this strategy [and] will deploy evidence-based decision-making and carry out our work in close partnership with the State of Alaska; Alaska Native Tribes, communities, corporations, and other organizations; and local communities, as well as with the US Congress."

As stated in the new strategy's introduction, "Despite current tensions stemming from Russia's unprovoked, full-scale invasion of Ukraine," Washington still "seeks an Arctic region that is peaceful, stable, prosperous, and cooperative," adding that a

> peaceful Arctic will have guardrails to manage competition and resolve disputes without force or coercion. Stability results from countries acting responsibly and in accordance with international law, rules, norms, and standards, including freedom of navigation. A prosperous Arctic features healthy and vibrant Arctic communities and sustainable economic growth. We aim to secure these objectives in the U.S. Arctic and throughout the region, working primarily with our allies and partners to solve shared challenges. Russia's unprovoked war of aggression against Ukraine has rendered government-to-government cooperation with Russia in the Arctic virtually impossible. Over the coming decade, it may be possible to resume cooperation under certain conditions. Russia's continued aggression makes most cooperation unlikely for the foreseeable future.

But despite the new challenges to Arctic security of a resurgent and increasingly muscular Russia, there is much in the updated US strategy that feels familiar, with an echo of the collaborative dynamics embraced in past US Arctic strategies and policies during more cooperative times. Indeed, as discussed in the pages above, the 2022 update is not the first to note new challenges to the cooperative Arctic posed by Russia's military ambitions and/or China's economic and diplomatic ambitions in the region, with such concerns finding more prominence in policy statements after Russia's annexation of Crimea in 2014 and China's self-declaration as a "Near-Arctic" state in 2018. As such, the updated Arctic strategy offers a reaffirmation of hope that the Arctic will continue to be a region defined more by cooperation than conflict, even if in the near term such cooperation is confined to the expanded footprint of NATO's Arctic members, and that the region's Indigenous peoples will not be forgotten, even if they have not been as frequently consulted as in the recent past.

US National Security Advisor Jake Sullivan tweeted on October 7, 2022, that the updated strategy "demonstrates domestic and foreign policy are connected. We will invest at home, including in Alaska, and deepen Arctic cooperation with allies and partners in ways that make us stronger."[64] Like all tweets, his was short, but its brevity in no way correlates to its historic significance or impressive eloquence (and implicit theoretical nuance). Indeed, Sullivan's recognition of the new Arctic strategy's interconnection of foreign and domestic policy captures the inherent nature of Arctic multilateralism that undergirds and fosters the endurance through good times and bad of Arctic exceptionalism, which as I argue throughout this book has continued forth from the colonial era into the contemporary world, and will continue to do so.

This is not a commonly heard view nowadays and indeed is about as popular as statements in defense of neorealism when the Cold War ended, upending the predominant IR theory of the late Cold War era. A more popular narrative is that Arctic exceptionalism is now dead, or at the very least, on life support. Indeed, Australian polar expert Elizabeth Buchanan argues that the "age of Arctic exceptionalism is gone: the high north region is no longer a protected sphere of Russia-West engagement and dialogue."[65] Buchanan describes the Arctic as a hitherto "protected sphere," but the enduring continuity of Russia-West cooperation in the Arctic is just one by-product of the underlying multilevel linkage of domestic and foreign policy that Jake Sullivan describes.

Indeed, this multilevelism unites the levels of analysis used to parse and comprehend the complexities of international relations: not just "image two" (national-state) and "image three" (international system) in the Waltzian vernacular (known prior to Waltz as J. David Singer's first and second levels of analysis), but also the "first image" (individual) that Waltz introduced to Singer's taxonomy, as well as what I have called for more than two decades now the "fourth image," a tribal level of analysis. Russia-West cooperation is not protected by Arctic exceptionalism, but rather by the universality of multilevel multilateralism across all four levels of analysis (inclusive of tribal polities), with hints of a fifth (transnational structures) and sixth (supranational structures) as generations of my students have, in classroom discussions, added to an ever more complex taxonomy of world politics.

During the colonial era, when states competed for overseas territorial domination aggressively inclusive of the polar world; again in World War II, when great power armed conflict crept into the Arctic; and yet again during the Cold War, when superpower competition divided the world into bipolar blocs, this inherent multilevel multilateralism nurtured an enduring tradition of Arctic exceptionalism, even in times of state competition. And now, with Russia hemmed in by an expanded NATO after Putin's Ukraine gambit, it will, in time, do so once again, albeit under renewed constraints of a newly redivided world.

Chair Rotates but Arctic Council Future Still Uncertain

In contrast to the headlines in 2021 celebrating the Arctic Council and its legacy as it reached its first quarter century, there has been much recent discussion in the media and academic literature of the potential demise of the Arctic Council.[66] As John Last described on CBC News in May 2023,

> Russia's two-year term as chair of the Arctic Council came to an end Thursday, leaving the future of the council's role as a forum for international collaboration in doubt. But in a low-key virtual ceremony attended by senior bureaucrats, Russia ceded the chair to Norway, which will lead the council for the next two years. The style of the ceremony marked a shift for the council. The transfer of the chair is an opportunity for grand declarations and photo ops, as the foreign ministers of the eight Arctic countries mingle with Indigenous representatives and powerhouse observers like China and Japan. It's obvious what has changed—one year into its chairship, Russia went from problematic partner to international pariah with its 2022 invasion of Ukraine.[67]

As Evan Bloom, senior fellow at the Wilson Center, explained to Last,

> "The idea of sitting with the Russians and talking about economic development while they're bombing Kyiv—it just doesn't work." . . . Though some projects have since resumed without Russia, the future of the council's work is in doubt. How can the Arctic Council continue

> when, in the words of the incoming chair, co-operation with Moscow is "politically impossible"? "Right now, it leaves the concept somewhat in tatters," said Bloom. "Russia makes up about half the Arctic. You can't really have an Arctic Council without Russia."[68]

> The first two decades of the council proved so successful that it helped give rise to an idea of "Arctic exceptionalism"—that the region was an arena for benevolent co-operation, immune to the geopolitical struggles occurring in the rest of the world. But as soon as Russia invaded Ukraine in February 2022, "any dream of full-on Arctic exceptionalism . . . was over," according to Whitney Lackenbauer, a Canada Research chair in the study of the Canadian North at Trent University. . . . "In essence, what we're now seeing is that the Arctic, as a circumpolar region, is not isolated from geopolitics."[69]

Indeed, the trust so central to successful operations of the Arctic Council has yet to fully recover. Last cites Timo Koivurova of the University of Lapland, who observes, "It's such a stark change of reality. . . . Many of these projects are now seen as . . . too friendly to Russia."[70] And he cites Russian Arctic ambassador Nikolay Korchunov, who told the Russian news agency TASS, "How can we . . . even if the Arctic Council resumes its work . . . build co-operation on a long-term basis? . . . The most important element is lost—the trust is lost."[71] Adds Last, "That lack of trust has led to fears that the Arctic Council could devolve into an 'Arctic 7,' an alliance of like-minded Western nations that formerly [*sic*] breaks with Russia—and divides the Arctic territory, literally, in half."[72] And without a restoration of trust, there remains much uncertainty how or even if the Arctic Council can return to its collaborative, multilevel, and multilateral foundations. Some fear the Arctic 7 will emerge as a truncated alternate organization, excluding Russia from ongoing Arctic cooperation, while others fear the council will endure with all eight member states, but be stuck in a holding pattern, potentially for years to come—turning the Arctic Council into a "Zombie council."[73]

However, just a few months later in August 2023, *Voice of America*'s White House bureau chief Patsy Widakuswara published a more optimistic assessment in which she reports that the "Biden administration, which has sought to isolate Moscow diplomatically on the world stage, is supporting efforts to re-establish technical cooperation with Russia in one of the world's most challenging geographical regions—the Arctic."[74] While the United States "successfully lobbied to boot Moscow out of various international forums," she points out that "the polar region is the one place where Washington is not icing Russia out completely. Specifically in the Arctic Council." She cites a "senior administration official, who spoke to *VOA* on condition of anonymity to discuss security issues," who explains that the "administration believes the Arctic Council should continue to serve as the premier forum for cooperation among Arctic states."[75] As Widakuswara further describes, "With no end in sight for the war in Ukraine, the administration is now working with other members of the council . . . to re-establish some of the ties to Moscow that were fully suspended shortly after Russia's invasion," and "Morten Høglund, the council's chair of the Senior Arctic Officials, said they are aiming to begin work on a technical level. 'The Norwegian Chairship of the Arctic Council is in the process of consulting with all Arctic States and Indigenous Permanent Participants to develop guidelines for the resumption of Working Group–level work with all Arctic States, including the Russian Federation.'"[76]

Future Unknown, for the Arctic Council and for Arctic Exceptionalism

While *VOA* sounded upbeat on the council's future, there still remains widespread gloom among many scholars and journalists covering the Arctic regarding the future of the Arctic Council, even after it has successfully and without complication transitioned its rotating chairship from Russia to Norway. Consider this sampling of article titles from 2023 addressing the uncertain future of the council, the fate of Arctic exceptionalism, and the consequences of the A7 boycott of the council under Russia's tenure as chair: "The Disruption of Arctic Exceptionalism: Managing Environmental Change in Light of Russian Aggression" (German Council on Foreign Relations, *DCAP Report,* February 2023); "The Ukraine War Is Affecting Vital Arctic Research" (*Bulletin of the Atomic Scientists*, February 17, 2023); "After Russia's Invasion of Ukraine in 2022: Can We Still Cooperate with Russia in the Arctic?" (*Polar Record*, March 17, 2023); "The Arctic Council in Danger" (*High North News*, March 28, 2023); "Arctic Council Under Pressure as Norway Readies for Russian Handoff" (Reuters, March 28, 2023); "Fears Mount for the Arctic as Cooperation with Russia Stalls" (Al Jazeera, May 9, 2023); "Future of Arctic Council in Doubt After End of Russian Chairship" (CBC News, May 11, 2023); "Norway Takes Control of the Arctic Council at an Especially Troubled Time" (*Bellona*, May 30, 2023); "The Rise and Sudden Fall of the Arctic Council" (*Foreign Policy*, May 31, 2023); "Threatened by Putin's War; Geopolitical Fallout from the Conflict in Ukraine Is Threatening an Environmentally Crucial Region" (*The Guardian*, June 11, 2023); "Arctic Council After Russia's Handover: 'We Are Still Here'" (*High North News*, October 19, 2023).

Most headlines convey a consistency of caution, some trending toward the pessimistic (and even fatalistic). *VOA*'s "US Wants Russia Iced Out Everywhere, Except the Arctic" stands among the few somewhat optimistic media reports, indicative of at least one optimistic if anonymous senior administration official in Washington, though *High North News*'s "Arctic Council After Russia's Handover: 'We Are Still Here'" also conveys some hope, albeit modestly.

Is Arctic Exceptionalism Over? After a Year in "Purgatory," Calls for "Something New"

Observing the transfer of the Arctic Council chair from Russia to Norway in May 2023, Heather Exner-Pirot writes,

> The question of what the Council should look like going forward, in the aftermath of Russia's illegal invasion of Ukraine, remains unsettled. Russia, after all, has half of the Arctic's territory, two-thirds of its people and three-quarters of its GDP. What is the point of a regional organization that excludes it? The better question to ask is what is the point of an Arctic Council that includes Russia. The organization that was established in Ottawa in 1996 no longer exists; Russia broke it. If it cannot adapt it will fail. The Arctic Council has spent a year in purgatory. The time to create something new has come.[77]

Exner-Pirot notes how the

> transition to the Norwegian Chairmanship has provoked optimism of the Arctic Council getting back to normal. As the joint statement issued today by all eight Arctic states—yes, including Russia—and the six Indigenous Permanent Participants proclaimed, there is a "commitment to safeguard and strengthen the Arctic Council." The Norwegians have asserted that their top goal as chair is to ensure that the Arctic Council "survives." On paper, it is business as usual.
>
> Outside the Arctic Council, all hell has broken loose. The west has put tremendous resources into defeating Russia in Ukraine, with the USA alone committing $37 billion in military assistance. Finland is building a wall on its border with Russia, and has officially joined NATO. Sweden is in the process of doing the same.

> Vladmir Putin has a warrant out for his arrest by the International Criminal Court for war crimes against Ukraine, and there are allegations of genocide. Canadian Foreign Minister Mélanie Joly told the UN General Assembly in February that "Through its words and its actions, Russia is showing us very plainly that it does not respect this assembly and the members represented here." It is impossible to square this with calls to maintain Russian participation in the Arctic Council.
>
> But there is a reluctance, even a fear, to talk about what comes next. The default position has been to wait it out—to leave the door ajar and hope things smooth over sooner rather than later. It would be hasty to throw out what has taken decades to build, it is argued. The issues it is tackling, on Arctic science and climate change and Indigenous peoples, are too important.[78]

Exner-Pirot challenges those who—as I do here—embrace precisely these points: that the quarter century of cooperation nurtured in the Arctic is worth saving, just as was the case when the United States invaded other countries, primarily in the Middle East and South Asia, or when it undermined the Arctic consensus on climate change due to a pivot in US climate policy as seen in 2019. Russia and the other Arctic states did not pause their participation in council activities or isolate the United States from Arctic cooperation when it went astray. So why now, when the great power behaviors the United States has often displayed to the chagrin of its middle and small power partners are being emulated by Russia?

As Exner-Pirot points out at the start of her blog article, "Russia, after all, has half of the Arctic's territory, two-thirds of its people and three-quarters of its GDP. What is the point of a regional organization that excludes it?"[79] My answer is thus not to reconceptualize Arctic cooperation through the Westphalian lens of the NATO-aligned Arctic 7, but to keep front and center the importance of Russia to the Arctic, given that it represents (with fully "half its territory, two-thirds of its people, and three-quarters of its GDP") a majority of Arctic stakeholders even if only a singular Arctic state. Excluding it de facto delegitimizes Arctic cooperation by silencing the majority, depriving the circumpolar family of its important voice.

Indeed, any governance body engineered to replace the Arctic Council will fall short of its founding vision of Arctic inclusion by leaving out this important, indeed essential, Arctic stakeholder, perpetuating the mistake made in March 2022 when the A7, without consultation with its six Indigenous Permanent Participants, and without consensus with the largest Arctic state, launched a boycott of the council under Russia's tenure as chair. Otherwise, the world will find itself with a bifurcated

Arctic once again, as it did during the Cold War. There has been much movement in this direction that will take much work to reverse.

Moscow's New Foreign Policy Concept and Eurasian Unity: Pivoting Back to Mackinder's Classic "Geographical Pivot"

A close look at the evolution of Russia's policy, with regard to international relations broadly and the Arctic specifically, reveals a country operating within the norms of statecraft, even in the case of its war in Ukraine, which appears to be less a threat to the world order and more a strategic blunder that has strengthened NATO and catalyzed its rapid Nordic expansion. It is comparable, perhaps, to the United States' decade-long Vietnam quagmire, or even its two-decade long war on terror, with its reckless (and unjustified) Iraq invasion and ill-conceived handover of Afghanistan to the Taliban. Whether it's the haunting images of torture and sexual humiliation at Abu Ghraib or of desperate Afghanis falling to their deaths from departing US military aircraft, the Russian playbook in Ukraine, as offensive as it is to Western observers, does not appear to deviate substantially from great power military adventurism, even as practiced by the United States, and it has left the Global South, itself an arena of much colonial expansion and oppression by the West, on the fence.

Russia has responded to the changing strategic landscape, catalyzed by the Ukraine war and its adverse fortunes in battle thus far, with a logical update to its foreign policy concept. On March 31, 2023, Russian president Vladimir Putin issued a new "Decree on Approval of the Foreign Policy Concept of the Russian Federation," superseding the previous November 30, 2016, Decree of the President of the Russian Federation, "On Approval of the Foreign Policy Concept of the Russian Federation." Section 50 of the new decree discusses the Arctic, reiterating Russia's long-standing goal to foster the "peaceful resolution of international issues relating to the Arctic" in addition to its ongoing commitment to UNCLOS and "ensuring the invariability of the historically established international legal regime of the inland maritime waters of the Russian Federation." It adds more muscular wording on "neutralizing the course of unfriendly states towards the militarization of the region and limiting Russia's ability to exercise its sovereign rights" in the Arctic, and "establishing mutually beneficial cooperation with non-Arctic states pursuing a constructive policy towards Russia and interested in international activities in the Arctic, including the infrastructure development of the Northern Sea Route." None of these objectives is particularly concerning, or a threat to the international order.

And the following section, 51 on Eurasia, discusses the strengthening alignment of strategic interests with India and China, which are among the more active new Arctic players benefiting from the West's isolation of Russia after the Ukraine invasion and subsequent Arctic Council boycott: "Of particular importance . . . is the comprehensive deepening of ties and coordination with friendly sovereign global centers of power and development located on the Eurasian continent and committed to approaches that fundamentally coincide with Russian approaches to the future world order and the solution of key issues in world politics."

Accordingly, Moscow "aims to further strengthen the comprehensive partnership and strategic cooperation with the People's Republic of China" (sec. 52) and "will continue to build up a particularly privileged strategic partnership with the Republic of India" (sec. 54). These are part and parcel of Russia's interest in "striving to transform Eurasia into a single all-continental space of peace, stability, mutual trust, development and prosperity," as it doubles down on the region described by pioneering geopolitical theorist Halford Mackinder as the "heartland" of Eurasia and the strategic pivot of world politics for centuries—suggesting that there is nothing fundamentally new or worrisome here as well.

A "Delicate Transfer" of the Arctic Council Chair

With Russia's much disrupted tenure as chair of the Arctic Council now over, attention turns to Norway. According to Gwladys Fouche and Gloria Dickie of Reuters, "Norway said it will prioritise a smooth transition with Russia as it plans to assume the chair of the Moscow-helmed Arctic Council on May 11, but will not commit to restarting stalled cooperation given the Russian invasion of Ukraine."[80] While "Russian Foreign Minister Sergei Lavrov had earlier extended an invitation to Arctic officials to attend a transition ceremony in Salekhard, Siberia," they report Norway decided instead that the ceremony would "take place digitally and . . . be limited to the level of civil servants, not political leaders."[81] Russia's Arctic ambassador, Nikolay Korchunov, "told Reuters the transition would 'presuppose active and responsible participation of all Arctic Council member states in this preparatory process.' Norway's foreign ministry declined to comment on Korchunov's statement."[82] Korchunov added that "because Norway was one of the countries that had decided to put cooperation on hold, 'this creates uncertainty about the eventual future of the Norwegian chairmanship and what approach it will follow with regard to Arctic Council activities.'"[83]

After the transfer of the council chair in May 2023, Jan Olsen at AP News wrote that Norway

> took over the Arctic Council's rotating chairmanship from Russia amid questions about what role the eight-country intergovernmental body can play in protecting the polar region after the invasion of Ukraine prompted Western countries to suspend cooperation with Moscow . . . one of the few settings where Western countries and Russia worked together closely. As a result, research involving Russia on issues ranging from climate change to polar bears has been put on hold, and scientists have lost access to important facilities in the Russian Arctic.[84]

Olsen added that

> Norway vowed to keep the council's work moving forward as it assumed the rotating two-year chairmanship from Russia. "Norway will continue to focus on the core issues the Council deals with, including the impacts of climate change, sustainable development and efforts to enhance the well-being of people living in the region," Norwegian Foreign Minister Anniken Huitfeldt said. "Together with the other member states, we will now explore how this can be achieved in practice." Experts say that's going to be complicated without cooperation with Russia, the biggest Arctic nation. "It is a huge challenge for Norway. They have to isolate Russia and at the same time they have to make sure not to provoke Russia to dissolve the Council," said Rasmus Gjedssø Bertelsen, of the Arctic University of Norway in Tromsø. He worried that Indigenous peoples might "lose an important forum and a prominent platform," adding that many of the groups are cross-border organizations and don't follow national borders.[85]

As Norway had planned, the ceremony was held online, and—accepting one of the modalities I proposed last year to deploy more junior and less senior leaders to council functions—"with member states' Arctic ambassadors rather than foreign ministers attending. The council issued a statement 'recognizing the historic and unique role of the Arctic Council' and acknowledging the commitment to safeguard and strengthen it."[86] In passing the council chairship to Norway without incident, Moscow again demonstrated its responsibility as an Arctic actor and council member state, showing that there has been, and remains, continuity with the rotation, testament to the endurance of consensus among its diverse stakeholders even amid the tensions of the Ukraine war and subsequent A7 boycott.

Trine Jonassen describes in *High North News* the tensions and uncertainties felt in the days and hours ahead of the ultimately successful transfer: "'I am here to tell you that the Arctic Council is not dead,' said the Arctic Council Chair Morten Høglund to a cheering congregation at the Arctic Circle Assembly in Reykjavik—revealing that the takeover from Russia was not certain until the day before. 'A year ago, I attended this conference during the Russian chairship. All work within the Arctic Council was paused, and some considered it dead,' said the Chair of the Arctic Senior Officials."[87] Jonassen recounts the "historic 13th Council meeting last August, where the baton was successfully handed over to Norway," at which

> the Arctic Council stated the unique position it holds, not only in the high North but also globally. . . . Yet, behind closed curtains, Høglund now reveals that the situation with the takeover remained uncertain until the final hour. "It was only the day before I felt like it would happen. But today, I stand here to tell you that the Council is very much alive and that all working groups are committed." The working groups' work has gradually resumed, though not at the normal speed or level. "It is not business as usual," says Høglund.[88]

Indeed, in a worryingly titled commentary in *High North News* ("The Arctic Council in Danger"), Svein Vigeland Rottem raised three important issues concerning the Arctic Council's future. The first is the unique role of Arctic Indigenous peoples, and the paramount importance of protecting their hard-won and exemplary influence:

> Firstly, when the Council was established in Ottawa in September 1996, the eight Arctic states did something unprecedented: they invited representatives of the region's indigenous groups as permanent participants. These representatives lack decision-making powers, but sit at the table with the other members, taking part in discussions when decisions are to be taken. Moreover, they are heard—a unique situation in international cooperation otherwise. If the Arctic Council fails to survive, this unique construction will unravel, and that would be a catastrophe for the Arctic indigenous groups. Their possibilities of influencing and being heard in discussions concerning the future of their Arctic homeland will disappear.[89]

His second is the importance of collaborative networks cultivated over thirty years by Arctic researchers and policymakers, which could not so easily be restored if allowed to collapse:

> Secondly, the Arctic Council, with its various working groups, is a unique producer of valuable knowledge and insights. Over the course of three decades, researchers and members of the civil service in the Arctic states have established networks and gathered information that has served to underpin national-level administration as well as international convention work. Thanks to these efforts, the world has come to realize the enormous climatic and environmental consequences facing the Arctic states. . . . If the Arctic Council fails to survive this crisis, the web of carefully built networks will collapse, and cannot be re-built overnight. Those who disagree with this view cannot realize how long it takes for an institution to build up the credibility and power necessary for it be heard.[90]

And his third addressed the necessity of an eventual restoration of circumpolar cooperation, inclusive of Russia:

> Today it may seem naive to hold that cooperation with Russia can again become possible—but what are the alternatives? The climate and environmental challenges in the Arctic will not disappear of their own accord: cooperation is essential. Russia covers half the territory of the Arctic, and there were signs of growing awareness of climate issues among Russian decisionmakers until the events of 24 February. However, the door must be kept ajar, to enable Russian participation when this becomes politically and morally possible. We need to understand what the Arctic Council [sic], and what it can do. Then we will recognize that the Arctic states must do their utmost to avoid destroying something that has taken 30 years to build up. All the Arctic states—including Russia—must set aside their petty ambitions and find a solution acceptable to all. This is more than a question of the future of the Arctic as such: it also concerns how the world is to meet the global environmental and climate challenges of today—and tomorrow.[91]

I first began to argue for continued cooperation with Russia at the very start of the Arctic Council boycott and have continued to make my case during these long and sometimes very lonely months since then. It is heartening to now see a steadily expanding community of scholars, policymakers, and activists making a similar argument, in awareness of what can be lost if we allow a neo-Westphalian worldview to take root in the Arctic, and increasingly, even Arctic political and diplomatic actors are modulating their message to make room for an eventual resumption of cooperation with Russia.

Consider the reporting in *Bellona* on Norway's assumption of the Arctic Council's rotating chairship, citing officials in both Norway and Denmark:

> "Russia has not been expelled from the Arctic Council. It hasn't exited it. There are no plans to exclude Russia from the council. However, doing business as usual at the political level is now out of the question," Norwegian Secretary of State Eivind Vad Petersson said during a debate in Bodø, Norway this April. Thomas Winkler, the Danish ambassador to the Arctic, agrees: "We keep the door ajar because Russia is still a member of the council. Even if we can't cooperate with it. So, everything depends on Russia. When Russia changes its behavior, the first place where we will resume cooperation is the Arctic."[92]

But as *Bellona* observes, Russia had its own perspective on this latter point:

> However, it is not yet clear that Russia is going to "change its behavior." Thus, on May 11, Sergei Lavrov stated that "the full-scale activity of the council was temporarily frozen by Western countries—members [of the council] under the absolutely far-fetched pretext of the situation in Ukraine, which they themselves provoked. . . . We consider such a step politicized," he added. In the absence of cooperation among the members of the council, Russia is strengthening interaction in the Arctic with non-regional states, primarily China and India.[93]

In addition to Russia's tightening alignment with China as part of its Eurasian refocusing of its foreign policy concept since the Ukraine war began, there is the concern about the loss of Russia's important participation in Arctic scientific research, notably with regard to climate and environment. As Ed Struzik has observed in *Bulletin of the Atomic Scientists*,

> The cessation of scientific collaboration comes at a precarious moment for the Arctic. . . . Scientists from around the globe have collaborated in the Arctic at least since the Cold War. . . . Now, much of this international collaboration is on pause, partly because the other seven Arctic Council states have suspended communication with Russia. Other projects have halted completely as government scientists and non-governmental organizations in Russia have fled the country, been silenced by Russian authorities, or denied the international funds, expertise, and

> infrastructure needed to keep their joint work going. . . . Even if the Arctic Council did find a way to reconcile with Russia, or to forge a different path forward, it's hard to imagine the research community returning to pre-war normal, because so many of Russia's best Arctic scientists have fled the country or are looking for ways to emigrate.[94]

Concluding Thoughts

The very survival of the Arctic Council and doubts concerning its prognosis looking forward have become topics of widespread concern and discussion, as has the future of Arctic exceptionalism that paved the way for the council's formation in 1996, which is now feared by many to be in doubt. If this is indeed so, it is a self-inflicted wound caused not by Russia for its military aggression outside of the Arctic, but by the A7, under the strategic leadership of the United States. Its uncompromising linkage of Russia's external aggression to the peaceful Arctic, where Russia had remained committed to cooperating with its neighbors, and whose chairmanship of the council was likewise so committed, only undermined the council. It is my hope that the death of the Arctic Council ultimately proves prematurely proclaimed, reminiscent of Mark Twain's famed aphorism, "The report of my death has been grossly exaggerated!"

As the Ukraine war enters its third year, I continue to believe this is the case, and to advocate for a return to the council's cooperative circumpolar values and mission, even as the situation in Ukraine remains unresolved—and I hope a new, more fully inclusive modality is developed under Norway's term as chair that balances the desire by the A7 to protest Moscow's international aggression with continuing to work with Russia as a fully included member of the A8. Though this would require some diplomatic flexibility at a time when such flexibility has been absent, and likely needs a passionate champion like Mikhail Gorbachev in Murmansk or the ICC in 1992, as articulated in its visionary *Principles and Elements for a Comprehensive Arctic Policy*, to become politically and diplomatically possible, it is no less a necessity now than at the outset of the war, when the boycott was announced in a manner that betrayed the very sacred principles of multilevel, multilateral collaboration that are in the Arctic Council's DNA.

On March 30, 2023, the Turkish parliament ratified Finland's application to join NATO, taking down the final roadblock hindering Finland's much delayed accession into the expanded Atlantic alliance, just two days after the Hungarian parliament did the same. Both nations continued to thwart Sweden's accession for another ten months, though with signs of softening along the way, leaving this last Nordic state outside NATO for another year before its eventual accession. However, unlike Norway, a NATO founding member, and Finland, its newest member, Sweden does not share a land border with Russia at all; therefore, Finland's accession to the alliance is all the more consequential, further encircling Russia by integrating its last nonaligned Nordic neighbor sharing a common frontier. That brought Western military power to Finland's 1,340-kilometer-long border, which combined with Norway's 196-kilometer border already fully embedded into NATO's strategic architecture yields a 1,536-kilometer fortified frontier to contain any possible future Russian aggression against the Nordic region. Already, border fortifications and fencing are under construction, physically partitioning the Sami homeland without Sami consultation or participation, reminiscent of the original Arctic Council boycott decision and its exclusion of the council's Indigenous Permanent Participants.

The end result, I believe, will be a more brittle frontier at higher risk of escalation, with less strategic flexibility that had long been associated with the neutrality that Finland and neighboring Sweden maintained through good times and bad for many a decade, putting at risk the many shared values that united the circumpolar family since 1996. Indeed, the increasing military integration of the A7 within the NATO alliance could permanently fracture the circumpolar buffer that has so long existed between the Arctic and the non-Arctic worlds—weakening the very inclusive multilevel and multilateral spirit that has infused the Arctic Council since its formation.

As a result of the Arctic Council boycott in March 2022 and its linkage to events in Ukraine in protest of Moscow's aggression there, the de facto expulsion of Russia from the circumpolar family thus jeopardized the very fabric of Arctic cooperation that has for so long showed the world another model of international relations and a vision that brought unity to a diverse ecosystem of states, tribal peoples, and international organizations. In contrast to the many pessimists who ruminate on the demise of the council, I continue to believe it is neither doomed nor on death's door, but in need of bold decisions (and hard work) by the A7 to return to its founding principles, rejoining all of its partners on the council, Russia included, to resume its important work saving the Arctic from the gathering threat presented by climate change. As Arctic climate change expert Ed Struzik eloquently observes,

"The Arctic has long been a model for optimism and international cooperation. A lot needs to be done to keep it that way."[95]

The Arctic became an important and stable buffer between the dueling blocs of East and West during the Cold War, and it was this role as strategic buffer that made the formalization of Arctic cooperation attractive to state and nonstate actors on both sides of the divide. When the Cold War came to its sudden end in 1990, this cooperation solidified into what many hoped would be a permanent state of Arctic cooperation hailed as Arctic exceptionalism, which made possible the consensus that emerged to jointly battle the environmental hazards of the immediate post–Cold War era and the global challenges of climate change that became the hallmark of Arctic cooperation efforts during the quarter-century calm from the Arctic Council's formation in 1996 right up to the collapse of consensus in 2022, when the war in Ukraine metastasized from a low-level "frozen conflict" to a full-throttle conventional war.

The Arctic Council has epitomized Arctic exceptionalism, giving it institutional form as a distinctive multilevel intergovernmental organization with legitimacy that extends across the entirety of the circumpolar world—from the western tip of the Aleutian Islands all the way to the eastern tip of Siberia, spanning a diverse mosaic of states and tribal peoples who share and jointly govern their remote, largely pristine environments with fragile ecosystems undergoing a historic climatic transformation. The result, as the Arctic Council matured during its first quarter century, was an exemplary track record of nurturing Arctic cooperation between a diverse, asymmetrical, and expanding ecosystem of stakeholders, whether Arctic states, Indigenous peoples, or nonstate or state observers. Ideally suited to collectively confront the challenges of climate change, it offered the world a model for inclusive, multilevel multilateralism. Designed to exclude matters of hard-power military and security issues, it was the pressing urgencies of a protracted proxy military conflict between Russia and the West in Ukraine that ultimately froze the Arctic Council and its impressive track record of cooperation in its tracks, under Russia's helm as rotating chair of the council. Arctic exceptionalism, as some pessimists argue in the pages above, is now dead or on life support, its prognosis unknown. The Arctic Council has, they contend, become obsolete. Reminiscent of arguments made at the Cold War's end by scholars like Francis Fukuyama, the long history of Arctic exceptionalism that I trace back several centuries in these pages was suddenly over—seemingly as a result of one state's strategic miscalculation. Or so the argument was made.

But as I argue in the pages above, and still deeply believe—as the Arctic speeds toward a new cold war bifurcation as a result of short-sighted policies preoccupied with extending hard-power contests into this pacific realm rather than leveraging its collaborative tranquility as an off-ramp for runaway strategic competition—Arctic exceptionalism remains the driving force uniting Arctic geography and geopolitics. The ice wall is melting, but it is still formidable to the tools of state power projection. Indeed, it took the intercession of human will to put into reverse the many geopolitical and geographical vectors that aligned for so long to yield Arctic exceptionalism, to undermine the unity of the Arctic Council that emerged from this harmonious alignment, and to begin to dismantle this important, indeed millennial, ice wall separating the Far North from the more conflicted world to its south. The ice wall did not come down under the weight of internal decay but was willfully breached as a matter of state policy. The United States, with its NATO allies in tow, can thus be said to have tried to slay Arctic exceptionalism, in the interest of alliance unity, and to erect a new ice wall along the old East-West frontier.

Had wiser minds with longer time horizons and deeper reservoirs of knowledge been consulted—such as the Indigenous peoples of the Permanent Participant organizations of the Arctic Council who were overlooked when the Arctic Council boycott was rushed into effect, and their wisdom graciously accepted as it had been until recently—perhaps the long-term risks of this hastily conceived "pause" excluding Russia from Arctic cooperation might have been avoided. Instead, we would still have the council's harmonious multilevel multilateralism operating smoothly, as it had for over a quarter century, to guide the world away from its current abyss.

But it's not too late for Russia's isolation from Arctic cooperation to be reconsidered, as I have hoped the Norwegians would do—and which appears to be now under way—during their two-year term as rotating chair of the council. For Russia to be welcomed back fully and equally to the Arctic family of nations, the partisan and increasingly alliance-centric efforts by the A7 to isolate Moscow, even from Arctic governance and diplomacy where its participation is crucial, must be discontinued in the wider interest of our shared humanity, a peaceful Arctic, and posterity itself. With Norway now at the helm of the Arctic Council, it has a unique opportunity as both a founding NATO member and respectful neighbor to Russia, as they share not only a mutually recognized border but a long, cooperative, and constructive relationship as well. Even as the Arctic Council stopped directly working with Russia, annual in-person protocol meetings continued to be held by the chiefs of the Russian FSB Border Guard Directorate for the Western Arctic and the Norwegian Joint Headquarters.[96]

Indeed, less than one year into Norway's term as Chair, and the Arctic Council began taking its first tangible steps toward a restoration of circumpolar unity—initially through a highly symbolic, all-inclusive hybrid (joint virtual/in-person) meeting on January 29, 2024 between the council's Chair and the six Permanent Participant organizations that had been overlooked two years earlier when the ill-conceived boycott was begun; and a month later at the project level through a resumption of all-inclusive (albeit virtual) working group meetings on February 24, 2024. As the Arctic Council Secretariat announced regarding this historic resumption a fully-inclusive meeting of the Permanent Participant organizations in January, "Updates were exchanged on the status of the Norwegian Chairship of the Arctic Council (2023-2025) and Permanent Participants' activities. Other important topics include youth engagement and how the Council can help further Permanent Participants' involvement in Working Group project design and support their active participation in projects."[97] The Secretariat added: "More than 20 participants representing the Council's six Permanent Participants gathered for the hybrid meeting held in Tromso, Norway and online on the margins of the Arctic Frontiers conference."[98] Importantly, the Chair reiterated the Council's commitment to Indigenous engagement and inclusion: "During its Chairship, Norway aims to strengthen cooperation with the Arctic Indigenous Peoples' organizations and stresses the importance of the participation of Indigenous Peoples and their valuable contributions in all areas of cooperation under the Council. To this end, a discussion was held on how to further enable Permanent Participants' involvement in Working Group project design and research question development, and to support their active participation in projects."[99]

And, according to the Arctic Council Secretariat's February announcement on the resumption of Working Group cooperation: "Consensus was reached for the gradual resumption of official Working Group meetings in a virtual format, enabling project-level work to further advance. In February, the eight Arctic States, in consultation with the Indigenous Permanent Participant organizations, reached consensus to gradually resume official Working Group meetings in a virtual format. Prior to this update, Working Groups advanced project work and decision-making only via written procedures after consensus was reached in August 2023. By resuming official Working Group meetings, the Arctic Council aims to take a critical step forward in ensuring that the Council can continue to deliver on its mandate as the preeminent forum for circumpolar collaboration on issues of environmental protection and sustainable development in the Arctic."[100] As Morten Høglund, Chair of the Arctic Council's Senior Arctic Officials, described: "In May 2023 at the 13th Meeting of the Arctic Council, all Arctic States and Permanent Participants reaffirmed their commitment to safeguard and strengthen the Arctic Council. Allowing Working Groups to virtually meet and advance their work is critical in meeting that commitment and the responsibility the Council has to play a leading role in addressing issues of emerging concern as it relates to environmental protection and sustainable development in the Arctic."[101] Høglund added, "With just over one year left of the Norwegian Chairship, I am very pleased that Working Groups can take more steps to advance their projects and initiatives for delivery at the end of our Chairship. This is critical in enabling the Council to effectively respond to rapid climate change and other urgent issues impacting the region and beyond."[102]

As the Arctic Council Secretariat explained, by "resuming virtual Working Group meetings, the Arctic Council aims to improve efficiency and collaboration within project work and enable Working Groups to advance more complex issues. The ability to meet virtually extends to all project-level teams and Expert Groups to ensure project-level work advancement. Observers and external experts are invited to reengage in these meetings as relevant. The resumption of official Working Group meetings will take place gradually over the next three to four months. As Chair of the Arctic Council's Senior Arctic Officials, Morten Høglund will play an active role in the resumption of Working Group meetings, reviewing and providing guidance to Working Groups as they begin to plan schedules for project and Plenary meetings."[103] Patrick Huber, Chair of the Council's Arctic Contaminants Action Program Working Group (AÇAP), echoed Høglund's sentiment: "Engaging with our counterparts in real time is key to the short and long-term success of the Arctic Council. There simply is no replacing the connections and creativity that are generated in real-time interactions with subject matter experts. Engaging with our counterparts in real time is key to the short and long-term success of the Arctic Council. Trust, understanding and respect are the cornerstones on which we build our Arctic cooperation, and we need to strengthen those elements to ensure the viability of our work. As projects approach key milestones and approval of final products are prepared, Arctic Council Working Groups will need to act with certainty and confidence, and they will want to ensure that the communities of Arctic are informed and engaged in a timely manner."[104] Sara Olsvig, International Chair of Inuit Circumpolar agreed: "The Arctic Council is critical for Arctic cooperation and the permanent participation of Indigenous Peoples is what makes this body unique. The ability of the Arctic Council Working Groups to continue to effectively respond to the environmental, climatic and social development is of great importance to the Peoples of the Arctic." She added, "Resuming virtual meetings of the Arctic Council Working Groups is a key step in maintaining and advancing the strong partnerships built over decades, as well as the full and effective participation of Arctic Indigenous Peoples in the work of the Arctic Council."[105]

As the Arctic Council Secretariat noted, this "gradual resumption of virtual Working Group meetings will be the first working-level meetings to take place since March 2022. Diplomatic-level meetings on the Senior Arctic Official-level will remain on pause until such time as consensus is reached by the Arctic States, in consultation with Permanent Participants, on further modalities for their resumption."[106] While consensus has yet to be restored on resuming high level meetings of SAOs and diplomats, one ray of hope in this otherwise modest baby step forward is that in contrast to the decision by the A7 to boycott Arctic Council activities in March 2022, the decision to resume working group cooperation among the entirety of the Council's family of stakeholders (member states, indigenous permanent participants, and observer states and organizations) was made properly, with a consensus among all eight Arctic states (Russia included) in consultation with the six permanent participant organizations, a welcome step forward toward a return of full and inclusive circumpolar cooperation.

As *High North News* noted in its February 29, 2024 coverage of the announcement, "This opens the way for more interaction and activity at the working group level, where the council's main work takes place. Half a year ago, the groups were able to resume their projects, but only via written procedure—a laborious way of working."[107] But now, they working groups will be able to virtually unite via Zoom and comparable technologies that proved their mettle during the pandemic, bringing the vast Arctic together onto a single screen for real-time interaction, a tried and true modality that was readily available to the Council after Russia invaded Ukraine, but not embraced for partisan reasons, despite my urgent and frequent published pleas to the contrary, as reflected in well over a dozen op-eds, articles and conference presentations that I published across the Arctic from April through

December 2022. Most of this time, it felt like no one was listening, so stubborn a united front was presented by the A7 (and even after their initial exclusion, by most of the Permanent Participants as well); but as the war in Ukraine has become a protracted and bloody affair, increasingly described as a stalemate, and the risks to the Arctic posed by continued isolation of Russia have become more widely discussed, my lonely perch as a public critic of the A7's unnecessary blackballing of Russia in the Arctic has become less lonely.

That Norway's first steps toward a restoration of circumpolar unity was through a hybrid meeting of all the Permanent Participants and a resumption of working group meetings via virtual, online conferencing is quite vindicating; I first called for a resumption of virtual Arctic Council meetings like these as early as April 19, 2022—247 years to the day after the American Revolution began its historic fight for full and inclusive representation. I would go on to reiterate this call on 14 separate occasions, in numerous venues including *The Barents Observer* (April 19, 2022), *The Whitehorse Star* and *PolarConnection.org* (April 20, 2022), *The Inuvik Drum* (April 21, 2022); *UConn Daily Campus* (April 26, 2022), *The Globe and Mail* and *Juneau Empire* (April 27, 2022), *Peninsula Clarion* (April 28, 2022), *Nunatsiaq News* (May 7, 2022), *Arctic Circle Journal* (June 14, 2022); the 15th Symposium on Polar Law (October 12, 2022), and *The Northern Review* (December 16, 2022) in addition to *Intersec: The Journal of International Security* (in its May, June and July/August issues).[108]

I find some solace in these modest first steps toward a restoration of fully inclusive, multilevel, multilateral cooperation across the entirety of the circumpolar Arctic. Indeed, the resumption of working group meetings came just two weeks after Moscow announced it had stopped paying its membership dues to the Arctic Council, citing the suspension during the "pause" of normal council business. As reported in *The Moscow Times*: "Moscow has suspended annual payments to the Arctic Council until its work resumes in full, Russia's Foreign Ministry told state media Wednesday. 'Russia's annual contributions to the Arctic Council's budget have been suspended for the time being, pending the resumption of pragmatic work with the participation of all member countries,' the Foreign Ministry said as quoted by the RIA Novosti news agency. The ministry ruled out Russia's full exit from the Arctic Council for now." But now, with Norway's modest, virtual step forward, I feel the initial stirring of renewed optimism, my first in the over two years since the war began. Time will tell if my renewed feeling of optimism is prescient or premature; but in either event, I am no less pleased that all council stakeholders will be interacting once again in real time, even if not yet ready to do so face-to-face.

Every bit as hopeful, I find, is the reassuring commitment to a collaborative Arctic put forth in Greenland's first Arctic strategy, *Greenland in the World: Nothing about us without us* (discussed in chapter four above)—released on February 21, 2024, just three days before the Arctic Council Secretariat announced its resumption of virtual working group meetings. In addition to outlining Greenland's vision of a world bound together by overlapping partnerships that extend from the Arctic around the globe, aligning the interests of Arctic states, non-Arctic states, future Arctic states (as Greenland aspires to become), Indigenous organizations (with which Greenland shares an overlapping connection), and a host of international organizations from the regional to global level. As Greenland articulated in its strategy, "Despite difficult times, there is still a need to look ahead to the long-term development of the Arctic Council. In the spirit of collaboration, the Arctic and its peoples deserve nothing less. We must endeavor to make the Arctic Council's work more relevant and inclusive for the peoples of the Arctic, and new areas of cooperation should be taken into consideration for the Council's mandate."[109] Indeed, Greenland argues that the "role of the Arctic Council must be safeguarded with long-term participation of the entire Arctic region, and its mandate should continue to exclude matters related to hard security."[110] That's because the council "is one of the forums where cohesion in the Arctic can be developed between like-minded states and political entities. The greater the contact and exchanges among the peoples of the region, the more meaningful and constructive the dialogue and cooperation that can be developed within the scope of the Council."[111] Now that the Arctic Council is once again operating at the working group level, albeit only virtually for now, this cohesion can once again be nurtured—with every hope that as communications are increasingly restored across the council's diverse stakeholders, we will witness in turn the restoration of more meaningful and constructive dialogue and cooperation that reaches across the whole of the Arctic.

Notes

1. Mackinder, *Democratic Ideals and Reality*, 199. I first applied Mackinder's Lenaland concept to the Arctic in a 1991 community-based research project funded by the Canadian Institute for International Peace and Security, whose unpublished 1994 report was titled *The Balkanization of the Arctic*. I continued this effort in *Arctic Doom, Arctic Boom*, and again in "Stability and Security in a Post-Arctic World:

Toward a Convergence of Indigenous, State, and Global Interests at the Top of the World," in *Fast-Changing Arctic*, 339–368.

2. Arctic Sea Ice News and Analysis, National Snow and Ice Data Center website, http://nsidc.org/arcticseaicenews.

3. Rob Huebert, "Renaissance in Canadian Arctic Security?," *Canadian Military Journal,* July 14, 2008; Rob Huebert, "Welcome to a New Era of Arctic Security," *Globe and Mail,* August 24, 2010; Brigham, "Think Again," 70–74; Borgerson, "Arctic Meltdown," 63–77; Anderson, *After the Ice*; and Zellen, *Arctic Doom, Arctic Boom.*

4. Huebert and Yeager, *A New Sea.* I first discussed Huebert and Yeager in Zellen, *Arctic Doom, Arctic Boom*, and again in "Stability and Security in a Post-Arctic World."

5. Jan Husdal, "The Final Frontier: The Northern Sea Route," Husdal.com, May 22, 2011, reviewing Halvor Schøyen and Svein Bråthen, "The Northern Sea Route Versus the Suez Canal: Cases from Bulk Shipping," *Journal of Transport Geography* 19, no. 4 (July 2011): 977–983. Also see "More Oil and Ore Along Northern Sea Route," *Barents Observer,* February 11, 2011.

6. Indeed, Russia's historic commitment to Arctic cooperation and its traditionally defensive approach to Arctic security has been noted by top naval commanders in both Canada and the United States as recently as April 2024, as reported in the *Navy Times* by Diana Stancy, who cites Vice Admiral Angus Topshee, commander of the Royal Canadian Navy, who told the Navy League's 2024 Sea-Air-Space conference that "the Arctic represents a significant portion of Russia's gross domestic product, population, and defense infrastructure. As a result, he said it is logical that the Russians would feel the need to defend themselves there." Stancy also cites Vice Admiral Douglas Perry, Commander of USN's 2nd Fleet, who cautioned: "There's no desire to over-militarize or create a theme of a contested battlespace in the Arctic. Rather, the reverse is that we know that 'speak softly and carry a big stick' is an underlying need of free nations that value freedom in an area that will see more and more traffic and more and more competition for resources in the near future and throughout the future." See Diana Stancy, "US Doesn't Want to Make Arctic Contested Battlespace, Admiral Says," *Navy Times*, April 8, 2024, https://www.navytimes.com/news/your-navy/2024/04/08/us-doesnt-want-to-make-arctic-contested-battlespace-admiral-says/.

7. Lawson Brigham, "Reflections on the Arctic Council's Recent Message to the Globe: Peace, Stability and Cooperation," *Polar Points* (blog), July 23, 2021, Wilson Center Polar Institute.

8. Ibid.

9. Ibid.

10. I first articulated this argument in brief news commentaries starting in April 2022. See Barry Scott Zellen, "The World Needs the Arctic Council Now More Than Ever," *Barents Observer*, April 19, 2022, *The Whitehorse Star*, April 20, 2022, PolarConnection.org, April 20, 2022, *The Inuvik Drum*, April 21, 2022; "As World Celebrates Earth Day, Arctic Council 'Pause' Endangers United Stand Against Climate Change," *UConn Daily Campus*, April 26, 2022, *Juneau Empire*, April 27, 2022, *Peninsula Clarion*, April 28, 2022; "Why a Freeze on the Arctic Council Is Short-Sighted," *Globe and Mail*, April 27, 2022; "Cold Shoulder: Barry Scott Zellen Examines Why the Arctic Council 'Pause' Puts Arctic Cooperation into a Deep Freeze," *Intersec*, May 2022, 8–10; "Arctic Council 'Pause' Endangers Humanity's United Stand Against Climate Change," *Nunatsiaq News*, May 7, 2022; "Path to Peace: Barry Scott Zellen Explains Why as the Arctic Thaw Continues, Lifting the Arctic Council Pause Presents Russia a Viable Route to Resolution," *Intersec*, June 2022, 16–18; "The Arctic Council Pause: The Importance of Indigenous Participation

and the Ottawa Declaration," *Arctic Circle Journal*, June 14, 2022; "A Sacred Promise Broken? Barry Scott Zellen Explains Why the Decline in Indigenous Consultation at the Arctic Council Signals a Retreat from the Spirit of the 1996 Ottawa Declaration," *Intersec*, July/August 2022, 22–24; "As War in Ukraine Upends a Quarter Century of Enduring Arctic Cooperation, the World Needs the Whole Arctic Council Now More Than Ever," *Northern Review*, no. 54 (January 2023).

11. Natasha Turak, "Russia Threatens New Nuclear Deployments if Sweden, Finland Join NATO," *CNBC.com*, April 14, 2022, https://www.cnbc.com/2022/04/14/russia-threatens-new-nuclear-deployments-if-sweden-finland-join-nato.html.

12. Thomas Nilsen, "Russia's New Hypersonic Tsirkon Missile Was Fired from Norwegian Sector of Barents Sea," *Barents Observer*, February 23, 2022, https://thebarentsobserver.com/en/security/2022/02/russian-navy-launched-hypersonic-tsirkon-missile-norwegian-sector-barents-sea.

13. When Norway and Russia held their joint 2024 protocol meeting between the chiefs of the Russian Federal Security Service's Border Directorate for the Western Arctic Region and the Norwegian Joint Headquarters, the meeting was held aboard a Norwegian coast guard vessel anchored off the Sør-Varanger municipality in Finnmark, Norway, near the border town of Kirkenes. Reminiscent of the 1989 Bush-Gorbachev summit in Malta, held aboard the Soviet cruise ship *Maksim Gorkiy* (and which came to be known as the "Seasick Summit"), a coast guard vessel could also serve as a symbolic venue for resuming Arctic Council meetings among the A8. See: Hilde-Gunn Bye, "Norwegian-Russian Summit in the Arctic," *High North News*, October 6, 2023, https://www.highnorthnews.com/en/norwegian-russian-summit-arctic.

14. I first explored the Indigenous exclusion from the Arctic Council boycott deliberations in Zellen, "The Arctic Council Pause"; and Zellen, "A Sacred Promise Broken?"

15. Eilís Quinn, "Eye on the Arctic: Condemnation of Ukraine Invasion Needed at Arctic Council but Work Pause May Be Mistake, Says CDN Indigenous Leader," Radio Canada International, March 4, 2022.

16. David Lochead, "Russian Invasion of Ukraine Creates Strain for Arctic Organizations: Russia Is Member of Inuit Circumpolar Council and Arctic Council," *Nunatsiaq News*, March 7, 2002.

17. Ibid.

18. Russian Section of the Saami Council, "The Russian Section of the Saami Council Has Issued a Statement Regarding the Current Situation in Russia (27.02.2022)," February 27, 2022.

19. Ibid.

20. Ibid.

21. Ibid.

22. Ibid.

23. Ibid.

24. Gwich'in Council International, "Joint Statement on Arctic Council Cooperation Following Russia's Invasion of Ukraine," March 3, 2022.

25. Ibid.

26. Inuit Circumpolar Council (ICC), "Statement from the Inuit Circumpolar Council Concerning the Arctic Council," March 7, 2022.

27. Ibid.

28. Ibid.

29. Ibid.

30. Meral Jamal, "Arctic Council Resumes Work on Limited Scale, Without Russia," *Nunatsiaq News*, June 22, 2022.

31. Ibid.
32. Ibid.
33. Russian Association of Indigenous Peoples of the North (RAIPON), NGO in Special Consultative Status with the Economic and Social Council of the United Nations, Document No. 64, March 1, 2022.
34. Statement of the International Committee of Indigenous Peoples of Russia (ICIPR), March 11, 2022, https://polarconnection.org/international-committee-of-indigenous-peoples-of-russia.
35. Ibid.
36. Ibid.
37. Ibid.
38. Ibid.
39. Ibid.
40. Gloria Dickie, "Russian Officials Call Arctic Council Boycott 'Regrettable,'" *SaltWire*, March 4, 2022.
41. Ibid.
42. Aidan Chamandy, "Russian Invasion Puts Arctic Council's Future on Ice," *iPolitics*, March 4, 2022.
43. Ibid.
44. Ibid.
45. Ibid.
46. Arctic Council, "Interview with Nikolay Korchunov, Russia's Senior Arctic Official,"
March 30, 2020.
47. Ibid.
48. Ibid.
49. Ibid.
50. Ibid.
51. Tom O'Connor, "Russia Says Ukraine Crisis Must Not Reach Arctic, NATO Says It Already Has," *Newsweek*, March 31, 2022.
52. Ibid.
53. Ibid.
54. Ibid.
55. Ibid.
56. Ibid.
57. Ibid.
58. Ibid.
59. Ibid.
60. Ibid.
61. Ibid.
62. See Barry Scott Zellen, "As NATO Expands Across the North, Russia Faces a Strengthening Unity Among the Arctic 7," *Juneau Empire*, October 12, 2022, and *UConn Daily Campus*, October 26, 2022; "Finland's Accession to NATO Alliance Achieved at Last: Interview with University of Lapland Professor Julian Reid," *UConn Daily Campus*, April 10, 2023 (part 1); April 11, 2020 (part 2); April 12, 2023 (part 3); April 13, 2023 (part 4); and April 14, 2023 (part 5); and "Calculated Risk? Barry Scott Zellen Examines, Following Turkey's and Hungary's Consent, the Approval by All NATO Member States of Finland's NATO Accession," *Intersec*, April 2023, 12–14.
63. I first examined this flurry of Arctic development in several brief news commentaries starting in October 2022. See Zellen, "Russian Aggression"; "All Busy on

the Arctic Front: America Turns Its Attention North with Renewed Energy," *UConn Daily Campus*, November 1, 2022; "As War in Ukraine Continues," 16–18.

64. Jake Sullivan, @JakeSullivan46, Twitter, October 7, 2022, https://twitter.com/JakeSullivan46/status/1578386396724842498.

65. Elizabeth Buchanan, "The Punishment Paradox: Understanding the Unintended Consequences of Suspending Arctic Cooperation with Russia," *Modern War Institute Blog*, October 24, 2022.

66. For instance, see Simpson, "Rise and Sudden Fall of the Arctic Council"; and Al Jazeera, "Fears Mount for the Arctic as Cooperation with Russia Stalls: Is a Polar Body's Viability at Risk If It Cannot Cooperate with the Country That Controls Most of the Arctic Coast?" May 9, 2023.

67. John Last, "Future of Arctic Council in Doubt After End of Russian Chairship," CBC News, May 11, 2023.

68. Ibid.

69. Ibid.

70. Ibid.

71. Ibid.

72. Ibid.

73. Ibid.

74. Patsy Widakuswara, "US Wants Russia Iced Out Everywhere, Except the Arctic," *Voice of America*, August 4, 2023.

75. Ibid.

76. Ibid.

77. Heather Exner-Pirot, "Arctic Exceptionalism Is Over. Who Will Tell the Diplomats?," *Eye on the Arctic,* May 11, 2023.

78. Ibid.

79. Ibid.

80. Gwladys Fouche and Gloria Dickie, "Arctic Council Under Pressure as Norway Readies for Russian Handoff," Reuters, March 28, 2023.

81. Ibid.

82. Ibid.

83. Ibid.

84. Jan M. Olsen, "Norway Takes Over Presidency of Arctic Council from Russia," AP News, May 11, 2023.

85. Ibid.

86. Ibid.

87. Trine Jonassen, "Arctic Council After Russia's Handover: 'We Are Still Here,'" *High North News*, October 19, 2023.

88. Ibid.

89. Svein Vigeland Rottem, "The Arctic Council in Danger," *High North News*, March 28, 2023.

90. Ibid.

91. Ibid.

92. "Norway Takes Control of the Arctic Council at an Especially Troubled Time," *Bellona*, May 30, 2023.

93. Ibid.

94. Ed Struzik, "The Ukraine War Is Affecting Vital Arctic Research," *Bulletin of the Atomic Scientists*, February 17, 2023.

95. Zellen, "Arctic on Fire."

96. See Thomas Nilsen, "Border Trouble Not on Agenda When FSB Boss Visited Norway," *The Barents Observer*, October 8, 2023; and Hilde-Gunn Bye, "Nor-

wegian-Russian Summit in the Arctic," *High North News*, October 6, 2023. The benefits of Norway's continued engagement with Russia are discussed in Devyatkin, "Can Russian-US Scientific Cooperation Be Restored as Arctic Warming and the Ukraine War Intensify?"

97. Arctic Council Secretariat, "Arctic Youth High on the Agenda During Norwegian Chairship Meeting with Indigenous Permanent Participant Organizations."

98. Ibid.

99. Ibid.

100. Arctic Council Secretariat, "Arctic Council Advances Resumption of Project-Level Work."

101. Ibid.

102. Ibid.

103. Ibid.

104. Ibid.

105. Ibid.

106. Ibid.

107. Astri Edvardsen, "The Arctic Council with New Decisive Step Forward," *High North News*, February 29, 2024.

108. There were, in order of their publication and all authored by Barry Scott Zellen: "The World Needs the Arctic Council Now More Than Ever" (*The Barents Observer*, April 19, 2022; *The Whitehorse Star*, April 20, 2022; *PolarConnection .org*, April 20, 2022; and *The Inuvik Drum*, April 21, 2022); "As World Celebrates Earth Day, Arctic Council 'Pause' Endangers United Stand Against Climate Change" (*UConn Daily Campus*, April 26, 2022; My Turn, *Juneau Empire*, April 27, 2022; *Peninsula Clarion*, April 28, 2022); "Why a Freeze on the Arctic Council is Shortsighted," (*The Globe and Mail*, April 27, 2022); "Arctic Council 'Pause' Endangers Humanity's United Stand Against Climate Change" (*Nunatsiaq News*, May 7, 2022); "Cold Shoulder," 8–10; "Arctic Council Pause"; "Path to Peace," 16–18; "Sacred Promise Broken?"; "As Ongoing War in Ukraine Precipitates a Crisis in Circumpolar Unity, a Full Restoration of the Arctic Council Is Still Needed—Now More Than Ever" (Presentation for the 15th Symposium on Polar Law, University of Iceland, Reykjavik, Iceland, October 12, 2022); and "As War in Ukraine Upends."

109. Government of Greenland Ministry for Statehood and Foreign Affairs, *Greenland in the World,* Section 2: The Arctic Council, 13.

110. Ibid., Section 2: The Arctic Council, 13.

111. Ibid., Section 8: Connectivity, 27.

Selected Bibliography

Anderson, Alun. *After the Ice: Life, Death, and Geopolitics in the New Arctic*. Washington, DC: Smithsonian, 2009.

Anido, David. "Greenland Is an Important Neighbour." *Ottawa Citizen,* May 14, 2011.

Antrim, Caitlyn L. "The Next Geographical Pivot: The Russian Arctic in the Twenty-first Century." *Naval War College Review* 63, no. 3 (Summer 2010).

Arctic Council Secretariat. "Arctic Council Advances Resumption of Project-Level Work," February 28, 2024, https://arctic-council.org/news/arctic-council-advances-resumption-of-project-level-work/.

Arctic Council Secretariat. "Arctic Youth High on the Agenda During Norwegian Chairship Meeting with Indigenous Permanent Participant Organizations," January 29, 2024, https://arctic-council.org/news/norwegian-chairship-meeting-with-indigenous-permanent-participant-organizations/.

Arctic Governance Project. "The Canada-US Arctic Cooperation Agreement." http://arcticgovernance.custompublish.com/the-canada-us-arctic-cooperation-agreement.4668242-142904.html.

Armitage, Derek, Fikret Berkes, Aaron Dale, Erik Kocho-Schellenberg, and Eva Patton. "Co-management and the Co-production of Knowledge: Learning to Adapt in Canada's Arctic." *Global Environmental Change* 21 (2011): 995-996.

Armitage, Derek, Fikret Berkes, and Nancy Doubleday, eds. *Adaptive Co-management: Collaboration, Learning and Multi-Level Governance.* Vancouver: UBC Press, 2008.

Åtland, Kristian. "Mikhail Gorbachev, the Murmansk Initiative, and the Desecuritization of Interstate Relations in the Arctic." *Cooperation and Conflict* 43, no. 3 (September 2008): 289-311.

Barber, Benjamin R. *Jihad vs. McWorld: Terrorism's Challenge to Democracy*, New York: Time Books, 1995.

———. "Jihad vs. McWorld: The Two Axial Principles of Our Age—Tribalism and Globalism—Clash at Every Point Except One: They May Both Be Threatening to Democracy." *The Atlantic*, March 1992.

Bennett, Mia M. "Singapore: The 'Global City' in a Globalizing Arctic." *Journal of Borderlands Studies* 33, no. 2 (2018): 289–310.

Berger, Thomas R. "Conciliator's Final Report." Nunavut Land Claims Agreement Implementation Contract Negotiations for the Second Planning Period 2003-2013, March 1, 2006, https://www.tunngavik.com/documents/publications/2006-03-01%20Thomas%20Berger%20Final%20Report%20ENG.pdf.

———. *Northern Frontier, Northern Homeland: The Report of the Mackenzie Valley Pipeline Inquiry.* Ottawa: Minister of Supply and Services Canada, 1977.

———. *Village Journey: The Report of the Alaska Native Review Commission.* New York: Hill and Wang, 1985.

Berry, Mary Clay. *The Alaska Pipeline: The Politics of Oil and Native Land Claims.* Bloomington: Indiana University Press, 1975.

Beyerchen, Alan. "Clausewitz, Nonlinearity, and the Unpredictability of War." *International Security* 17, no. 3 (1992): 59–90.

Bockstoce, John R. *Furs and Frontiers in the Far North: The Contest Among Native and Foreign Nations for the Bering Strait Fur Trade*. New Haven: Yale University Press, 2009.

———. *White Fox and Icy Seas in the Western Arctic: The Fur Trade, Transportation, and Change in the Early Twentieth Century.* New Haven: Yale University Press, 2019.

Bogoyavlenskiy, Dimitry, and Andy Siggner, "Arctic Demography." Chapter 2 in *Arctic Human Development Report*, edited by Níels Einarsson, Joan Nymand Larsen, Annika Nilsson, and Oran R. Young. Akureyri: Stefansson Arctic Institute, 2004, pp. 27–41.

Boraas, Alan, and Aaron Leggett, "Dena'ina Resistance to Russian Hegemony, Late Eighteenth and Nineteenth Centuries: Cook Inlet, Alaska." *Ethnohistory* 60, no. 3 (July 2013): 485–504.

Borgerson, Scott G. "Arctic Meltdown: The Economic and Security Implications of Global Warming." *Foreign Affairs* 87, no. 2 (March/April 2008): 63–77.

Brigham, Lawson W. "The Northern Sea Route, 1997." *Polar Record* 34 (1998), 219–224.

———. "Russia Opens Its Maritime Arctic." *Proceedings Magazine* 137, no. 5 (issue 1299 (May 2011).

———. "Think Again: The Arctic—Everyone Wants A Piece of the Thawing Far North. But That Doesn't Mean Anarchy Will Reign at the Top of the World." *Foreign Policy* 181 (September/October 2010): 70–74..

Burke, Danita Catherine. *Diplomacy and the Arctic Council*: Montreal/Kingston: McGill-Queen's University Press, 2019.

Cavanagh, Edward. "A Company with Sovereignty and Subjects of Its Own? The Case of the Hudson's Bay Company, 1670–1763." *Canadian Journal of Law and Society* 26, no. 1 (2011): 25–50.

Chaffin, Tom. *Sea of Gray: The Around-the-World Odyssey of the Confederate Raider Shenandoah*. New York: Hill and Wang, 2007.

Chandonnet, Fern, ed. *Alaska at War: The Forgotten War Remembered.* Alaska at War Committee, 1995; reprinted by University of Alaska Press, 2007.

Cloe, John Haile. *Attu: The Forgotten Battle*. Anchorage: National Park Service, 2017.

Cloe, John Haile and Michael F. Monaghan. *Top Cover for America: The Air Force in Alaska, 1920–1983*. Anchorage: Anchorage Chapter-Air Force Association, 1984.

Coates, Kenneth S. *Canada's Colonies: A History of the Yukon and Northwest Territories.* Toronto: Lorimer, 1985.

Cohen, Stan. *The Forgotten War: A Pictorial History of World War II in Alaska and Northwestern Canada*. Missoula: Pictorial Histories Publishing Co., 1981.

———. *The Great Alaska Pipeline*. Missoula: Pictorial Histories Publishing Co., 1988.

Curry, Angus. *The Officers of the CSS Shenandoah.* Gainesville: University Press of Florida, 2006.

Dale, Aaron, and Derek Armitage. "Marine Mammal Co-management in Canada's Arctic: Knowledge Co-production for Learning and Adaptive Capacity." *Marine Policy* 35 (2011): 440.

Davis, Richard Clarke, ed. *Rupert's Land: A Cultural Tapestry.* Waterloo: Wilfrid Laurier University Press, 1988.

Declaration on the Establishment of the Arctic Council (also known as the Ottawa Declaration). Preamble, 1. Arctic Council Archives, 1996. https://oaarchive.arctic-ouncil.org/handle/11374/85.

Deep, Alex. "Hybrid War: Old Concept, New Techniques." *Small Wars Journal,* March 2, 2015.

Devyatkin, Pavel. "Can Russian-US Scientific Cooperation Be Restored as Arctic Warming and the Ukraine War Intensify?" *The Nation,* April 16, 2024. https://www.thenation.com/article/world/russia-global-warming-science-diplomacy/.

Dmytryshyn, Basil. "Russian Expansion to the Pacific, 1580–1700: A Historiographical Review." *Slavic Studies* (*Surabu Kenkyu*) 25 (Sapporo: Hokkaido University, 1980): 1–25.

Dodds, Klaus. "Flag Planting and Finger Pointing: The Law of the Sea, the Arctic, and the Political Geographies of the Outer Continental Shelf." *Political Geography* 30 (2010): 1–11.

Dodds, Klaus, and Mark Nuttall. *The Scramble for the Poles: The Geopolitics of the Arctic and Antarctic.* Cambridge: Polity, 2015.

Donahue, Katherine C., and David C. Switzer. *Steaming to the North: The First Summer Cruise of the US Revenue Cutter Bear, Alaska and Chukotka, Siberia, 1886*, Anchorage: University of Alaska Press, 2014.

Edwards, Wayne, and Tara Nataraja. "ANCSA and ANILCA: Capabilities Failure?" *Native Studies Review* 17, no. 2 (2008), 69–97.

English, John. *Ice and Water: Politics, Peoples and the Arctic Council.* Toronto: Allen Lane, 2013.

Exner-Pirot, Heather, and Robert W. Murray. "Regional Order in the Arctic: Negotiated Exceptionalism." *Politik* 20, no. 3 (2017).

Farrar, Victor J. *The Annexation of Russian America to the United States.* Washington, DC: W. F. Roberts, 1937).

Feinberg, Leonard. *Where the Williwaw Blows: The Aleutian Islands-World War II.* Boulder, CO: Pilgrims' Process, 2003).

Fenge, Terry. "Inuit and the Nunavut Land Claims Agreement: Supporting Canada's Arctic Sovereignty." *Policy Options* (December 2007–January 2008).

Flynn, Matthew J. *First Strike: Preemptive War in Modern History.* New York: Routledge, 2008.

Freeman, Milton M. R. "Looking Back—and Looking Ahead—35 Years After Inuit Land Use and Occupancy Survey." *Canadian Geographer* 55, no. 1 (2011).

———. "Persistence and Change: The Cultural Dimension." In *A Century of Canada's Arctic Islands, 1880–1980*, edited by M. Zaslow. Ottawa: Royal Society of Canada, 1981, pp. 257–266

Garfield, Brian. *The Thousand Mile War: World War II in Alaska and the Aleutians.* New York: Doubleday, 1969; and Fairbanks: University of Alaska Press, 1995.

Gerhardt, Hannes. "The Inuit and Sovereignty: The Case of the Inuit Circumpolar Conference and Greenland." *Politik* (2011): 6–14.

Golder, Frank A. "The Purchase of Alaska." *American Historical Review* 25 (1920): 411–425.

Goldstein, Daniel M., and Katherine V. Dillon. *Williwaw War: The Arkansas National Guard in the Aleutians in World War II.* Fairbanks: University of Arkansas Press, 1992.

Gray, Colin S. *Maritime Strategy, Geopolitics, and the Defense of the West.* New York: National Strategy Information Center, 1986.

Griffiths, Franklyn, Rob Huebert, and P. Whitney Lackenbauer. *Canada and the Changing Arctic: Sovereignty, Security and Stewardship.* Waterloo: Wilfrid Laurier University Press, 2011.

Hall, Sam. *The Fourth World: Heritage of the Arctic and Its Destruction.* London: The Bodley Head. 1987.

Hartwig, Georg. *The Polar and Tropical Worlds.* Springfield, MA: C. A. Nichols, 1874.

Hattendorf, John B. *The Evolution of the U.S. Navy's Maritime Strategy, 1977–1986.* Newport, RI: Naval War College Press, 1989.

Heaton, John W. "Athabascan Village Stores: Subsistence Shopping in Interior Alaska, 1850–1950." *Western Historical Quarterly* 43, no. 2 (2012): 133–155.

Heininen, Lassi. "Special Features of Arctic Geopolitics: A Potential Asset for World Politics." In *The Global Arctic Handbook*, edited by Matthias Finger and Lassi Heininen. Cham: Springer, 2019.

Heininen, Lassi, and Chris Southcott, eds. *Globalization and the Circumpolar North.* Fairbanks: University of Alaska Press, 2010.

Hertell, Hans H. "Arctic Melt: The Tipping Point for an Arctic Treaty." *Georgetown International Environmental Law Review* 21 (2009): 565–591.

Holmes, James R., and Toshi Yoshihara. *Red Star over the Pacific: China's Rise and the Challenge to U.S. Maritime Strategy.* Annapolis, MD: Naval Institute Press, 2010.

Hossain, Kamrul, and Hiroshi Maruyama. "Japan's Admission to the Arctic Council and Commitment to the Rights of its Indigenous Ainu People." *The Polar Journal* 6, no. 1 (2016): 169–187.

Huebert, Rob. "The Need for an Arctic Treaty: Growing from the United Nations Convention on the Law of the Sea." *Ocean Yearbook* 23 (2009).

Huebert, Rob, and Brooks Yeager. *A New Sea: The Need for a Regional Agreement on Management and Conservation of the Arctic Marine Environment.* Oslo: World Wildlife Fund, 2008.

Hughes, Dwight Sturtevant. *A Confederate Biography: The Cruise of the CSS Shenandoah.* Annapolis, MD: Naval Institute Press, 2015.

Jones, Dorothy M. *A Century of Servitude: Pribilof Aleuts Under U.S. Rule.* Lanham, MD: University Press of America, 1981.

Jones, Gordon. "The Role of Research Associations in Frontier Exploration Arctic Petroleum Operators' Association (APOA) and Eastcoast Petroleum Operators' Association (EPOA)." *Journal of Canadian Petroleum Technology* 20, no. 4 (October 1981).

Jonsson, Hannes, *Friends in Conflict: The Anglo-Icelandic Cod Wars and the Law of the Sea.* London: Hurst, 1982.

Keil, Kathrin, and Sebastian Knecht, eds. *Governing Arctic Change: Global Perspectives.* London: Palgr,ave Macmillan, 2010.

Kenney, Gerard I. *Arctic Smoke and Mirror.* Ottawa: Voyageur Publications, 1994.

Keskitalo, E. Carina H., and Mark Nuttall. "Globalization of the Arctic." Chapter 13 in *The New Arctic*, edited by B. Evengård, J. N. Larsen, and Ø. Paasche. New York: Springer, 2015.

Keskitalo, E. Carina H., and Chris Southcott, "Globalization." Chapter 10 in *Arctic Human Development Report: Regional Processes and Global Linkages*, edited by J. N. Larsen and G.Fondahl. Copenhagen: Norden, 2014.

Khrushcheva, Olga, and Marianna Poberezhskaya, "The Arctic in the Political Discourse of Russian Leaders: The National Pride and Economic Ambitions." *East European Politics* 32, no. 4 (2016): 547–566.

Kikkert, Peter, and P. Whitney Lackenbauer. "Canada's Arctic and Northern Policy Framework: A Roadmap for the Future?" Briefing Note, *2019 Arctic Yearbook*, https://arcticyearbook.com/arctic-yearbook/2019/2019-briefing-notes/332-canada-s-arctic-and-northern-policy-framework-a-roadmap-for-the-future.

Klimenko, Ekaterina, Annika Nilsson, and Miyase Christensen. *Narratives in the Russian Media of Conflict and Cooperation in the Arctic.* Stockholm: Stockholm International Peace Research Institute, August 2019.

Koivurova, Timo, and Erik J. Molenaar. *International Governance and Regulation of the Marine Arctic: Overview and Gap Analysis*. Oslo: World Wildlife Foundation, 2009.

Lackenbauer, P. Whitney, and Ryan Dean. "Arctic Exceptionalisms." Chapter 14 in *The Arctic and World Order*, edited by Kristina Spohr and Daniel S. Hamilton. Washington, DC: Foreign Policy Institute/Henry A. Kissing Center for Global Affairs, Johns Hopkins University, 2020.

Lackenbauer, P. Whitney, Adam Lajeunesse, James Manicom, and Frédéric Lasserre. *China's Arctic Ambitions and What They Mean for Canada.* Calgary: University of Calgary Press, 2018.

Laird, Robbin, Edward Timperlake, and Richard Weitz. *Rebuilding American Military Power in the Pacific: A 21st Century Strategy*. Santa Barbara: ABC-Clio / Praeger Security International, 2013.

Landriault, Mathieu, Andrew Chater, Elana Wilson Rowe, and P. Whitney Lackenbauer. *Governing Complexity in the Arctic Region.* New York: Routledge, 2019.

Larsen, J. N., and L. Huskey. "The Arctic Economy in a Global Context." Chapter 12 in *The New Arctic*, edited by B. Evengård, J. N. Larsen, and Ø. Paasche. New York: Springer, 2015.

Lazarus, Arthur, and W. Richard West. "The Alaska Native Claims Settlement Act: A Flawed Victory." *Law and Contemporary Problems* 40, no. 1 (Winter 1976): 132–165.

Mackinder, Halford John. *Democratic Ideals and Reality: A Study in the Politics of Reconstruction.* Washington, DC: NDU Press, 1996.

———. "The Round World and the Winning of the Peace." *Foreign Affairs* 21, no. 4 (July 1943): 595–605.

Marine Corps Historical Reference Pamphlet. *The United States Marines in Iceland, 1941–1942.* Washington, DC: U.S. Marine Corps Historical Division, 1970.

Mason, Rachel H. "You Can't Go Home Again: Process of Displacement and Emplacement in the 'Lost Villages' of the Aleutian Islands." *Alaska Journal of Anthropology* 8, no. 2 (2010), 17–29.

Mazour, Anatole G. "The Prelude to Russia's Departure from America." *Pacific Historical Review* 10, no. 3 (September 1941), 311–319.

McGhee, Robert. "The Nineteenth Century Mackenzie Delta Inuit." Pp. 141–152 of *Supporting Studies*, Vol. 2 of *Inuit Land Use and Occupancy Project Report*, edited by M.R.F. Milton. Ottawa: Supply and Services Canada, 1976.

McGrath, Melanie. *The Long Exile: A Tale of Inuit Betrayal and Survival in the High Arctic.* New York: Vintage Books, 2006.

Mearsheimer, John J. "A Strategic Misstep: The Maritime Strategy and Deterrence in Europe." *International Security* 11, no. 2 (Autumn 1986): 3–57.

Ministry for Statehood and Foreign Affairs, Government of Greenland, *Greenland in the World: Nothing about us without us – Greenland's Foreign, Security and*

Defense Policy 2024-2033—an Arctic Strategy. Nuuk, Greenland: Naalakkersuisut, February 21, 2024.

Mitchell, Martin D. "The South China Sea: A Geopolitical Analysis." *Journal of Geography and Geology* 8, no. 3 (2016).

Morison, Samuel Eliot. *History of United States Naval Operations in World War II.* Boston: Little, Brown, 1947.

Naske, Claus-M, and Herman E. Slotnick. *Alaska: A History of the 49th State.* Norman: University of Oklahoma Press, 1994.

Native American Rights Fund. "1988 Amendments Provide Stop-Gap Protection for Native Land and Corporations." *NARF Review* 13, no. 2 (1988): 1–5.

———. "'The Fifth Disaster': The Colonization of the North Slope of Alaska." Native American Rights Fund Announcements 3, no. 1 (1975): 1–14, 47.

Newman, Peter C. *Caesars of the Wilderness.* Toronto: Penguin Books Canada, 1988.

———. *Company of Adventurers: The Story of Hudson's Bay Company.* Toronto: Penguin Books Canada, 1985.

———. *Empire of the Bay.* Toronto: Penguin Books Canada, 1998.

Noble, Dennis L., and Truman R. Strobridge. *Captain "Hell Roaring" Mike Healy: From American Slave to Arctic Hero.* Gainesville: University Press of Florida, 2017.

Nord, Douglas C. *The Changing Arctic: Creating a Framework for Consensus Building and Governance Within the Arctic Council.* New York: Palgrave Macmillan, 2016.

Notzke, Claudia. "A New Perspective in Aboriginal Natural Resource Management: Co-management." *Geoforum* 26, no. 2 (May 1995): 187–209.

Nuttall, Mark. "Aboriginal Participation, Consultation, and Canada's Mackenzie Gas Project." *Energy & Environment* 19, no. 5 (2008): 617–634.

O'Connor, Tom. "Russia Says Ukraine Crisis Must Not Reach Arctic, NATO Says It Already Has." *Newsweek,* March 31, 2022.

O'Neill, Dan. *The Firecracker Boys.* New York: St. Martin's Press, 1994.

———. *The Firecracker Boys: H-Bombs, Inupiat Eskimos, and the Roots of the Environmental Movement.* New York: Basic Books, 2007.

O'Rourke, H. Kevin, and J.G. Williamson. "When Did Globalization Begin?" *European Review of Economic History* 6 (2002): 23–50.

Parlato, Nicholas, Gail Fondahl, Viktoriya Filippova, and Antonina Savvinova. "The Evolution of Forming 'Territories of Traditional Nature Use' in the Sakha Republic (Iakutiia)." *Sibirica* 20, no. 1 (Spring 2021): 1–27.

Parlee, Brenda L., and Ken Caine, eds. *When the Caribou Do Not Come: Indigenous Knowledge and Adaptive Management in the Western Arctic.* Vancouver: University of British Columbia Press, 2018.

Penikett, Tony. "An Unfinished Journey: Arctic Indigenous Rights, Lands, and Jurisdiction?" *Seattle University Law Review* 37, no. 4 (2014): 1127–1156.

Perras, Galen Roger. *Stepping Stones to Nowhere: The Aleutian Islands, Alaska, and American Military Strategy.* University of British Columbia Press, 2003.

Pincus, Rebecca. "Three-Way Power Dynamics in the Arctic." *Strategic Studies Quarterly* 14, no. 1 (Spring 2020): 40–63.

Pope, Dudley. *73 North: The Battle of the Barents Sea.* London: Weidenfeld and Nicolson, 1958.

Rearden, Jim. *Cracking the Zero Mystery: How the U.S. Learned To Beat Japan's Vaunted WWII Fighter Plane.* Mechanicsburg, PA: Stackpole Books, 1990.

Ricks, Thomas E. *Fiasco: The American Military Adventure in Iraq.* New York: Penguin, 2006.

Rodon, Thierry, and Aude Therrien. "Resource Development and Land Claims Settlement in the Canadian Arctic: Multilevel Governance, Subsidiarity and Streamlining." In 2015 *Arctic Yearbook: Arctic Governance and Governing*, edited by Lassi Heininen, Heather Exner-Pirot, and Joel Plouffe. Akureyri: Northern Research Forum, 2015.

Ross, W. Gillies. *Arctic Whalers, Icy Seas: Narratives of the Davis Strait Whale Fishery*. Toronto: Irwin, 1985.

———. "Whaling, Eskimos and the Arctic Islands." In *A Century of Canada's Arctic Islands, 1880–1980,* edited by Morris Zaslow, pp. 33–50. Ottawa: Royal Society of Canada, 1981.

Sale, Richard, and Eugene Potapov. *The Scramble for the Arctic: Ownership, Exploitation and Conflict in the Far North*. London: Frances Lincoln, 2009.

Schooler, Lynn. *The Last Shot: The Incredible Story of the CSS Shenandoah and the True Conclusion of the American Civil War*. Waterville, ME: Thorndike Press, 2005.

Schøyen, Halvor, and Svein Bråthen. "The Northern Sea Route Versus the Suez Canal: Cases from Bulk Shipping." *Journal of Transport Geography* 19, no. 4: 977–983.

Sergunin, Alexander. "Russia and Arctic Security: Inward-Looking Realities." In *Understanding Arctic Sovereignty and Security*, edited by Wilfrid Greaves and P. Whitney Lackenbauer. Toronto: University of Toronto Press, 2021.

Shadian, Jessica. "From States to Polities: Reconceptualizing Sovereignty Through Inuit Governance." *European Journal of International Relations*, May 2010: 1–26.

———. "Remaking Arctic Governance: The Construction of an Arctic Inuit Polity." *Polar Record* 42, no. 222 (2006): 249–259.

Shibata, Akiho, Leilei Zou, Nikolas Sellheim, and Marzia Scopelliti, eds. *Emerging Legal Orders in the Arctic: The Role of Non-Arctic Actors*. New York: Routledge, 2019.

Simpson, Brett. "The Rise and Sudden Fall of the Arctic Council: With Russia No Longer Involved, It's Hard To See What Arctic Politics Can Still Accomplish." *Foreign Policy*, May 31, 2023.

Singer, J. David. "The Level-of-Analysis Problem in International Relations." *World Politics* 14, no. 1 (October 1961): 77–92.

Smith-Peter, Susan. "Creating a Creole Estate in Early Nineteenth-Century Russian America." *Cahiers du Monde Russe* 51, nos. 2–3 (2010): 441–459.

Snook, Jamie, Ashlee Cunsolo, and Aaron Dale. "A Conceptual Representation of the 'Shared Space' of Co-management Research, Recommendations, Actions, and Decision-making." *Northern Public Affairs*, July 2018: 52–56.

Storey, Ian. "The Arctic Novice: Singapore and the High North." Roundtable: Polar Pursuits: Asia Engages the Arctic. *Asia Policy* 18 (July 2014): 66–72.

Takahashi, Minori, ed. *The Influence of Sub-state Actors on National Security Using Military Bases to Forge Autonomy*. Cham: Springer 2019.

Thomas, Clive S. "Intergovernmental Relations in Alaska: Development, Dynamics and Lessons." *Northern Review* 23 (2001).

Thomas, M. E. "Alaska Native Claims Settlement Act: An Update." *Polar Record* 24, no. 151 (1988): 328–329.

Torrey, Barbara Boyle, and Agafon Krukoff. *Slaves of the Harvest*. Anchorage: Tanadgusix (TDX) Corporation, 1980.

Vidal, Gore. *Williwaw*. New York: E. P. Dutton, 1946.

Waddell, James I. *C.S.S. Shenandoah: The Memoirs of Lieutenant Commanding James I. Waddell*. Annapolis, MD: Naval Institute Press, 1960.

Wegge, Njord. "The Political Order in the Arctic: Power Structures, Regimes and Influence." *Polar Record* 47, no. 2 (2011): 165–176.

Whittle, William C. *The Voyage of the CSS Shenandoah: A Memorable Cruise*, edited by D. Alan Harris and Anne B. Harris. Tuscaloosa: University of Alabama Press, 2014.

Williams, Andrew. *The Battle of the Atlantic: Hitler's Gray Wolves of the Sea and the Allies Desperate Struggle To Defeat Them*. New York: Basic Books, 2003.

Wilson, Gary N., Christopher Alcantara, and Thierry Rodon. "Inuit Diplomacy in the Circumpolar North." *Canadian Foreign Policy* 13, no. 3 (2007).

———. *Nested Federalism and Inuit Governance in the Canadian Arctic*. Vancouver: University of British Columbia Press, 2010.

Wintermute, Ira F. "War in the Fog." *American Magazine*, August 1943: 40–42, 100–101.

Wonders, William C. "The Dene/Inuit Interface in Canada's Western Arctic, NWT." In *For Purposes of Dominion: Essays in Honor of Morris Zaslow*, edited by Kenneth S. Coates and William R. Morrison. Toronto: University of Toronto Press, 1989.

Wood, Margaret Mary, "The Russian Creoles of Alaska as a Marginal Group." *Social Forces* 22, no. 2 (December 1943): 204–208.

Woodman, Richard. *Arctic Convoys: 1941–1945*. Barnsley, UK: Pen and Sword Books, 2007.

Worl, Riccardo. "The 1991 Time Bomb Defused." *Alaska Native Magazine*, June 1988.

Worl, Rosita. "Models of Sovereignty and Survival in Alaska." *Cultural Survival Quarterly* 27, no. 3 (September 15, 2003).

Yoshihara, Toshi. "China's Vision of Its Seascape: The First Island Chain and Chinese Seapower." *Asian Politics and Policy* 4 (July 2012): 293–314.

Yoshihara, Toshi and James R. Holmes. "Responding to China's Rising Sea Power." *Orbis* 61, no. 1 (2017), 91–100.

Young, Oran R. "The Age of the Arctic." *Foreign Policy* 61 (1985): 160–179.

———. *Arctic Politics: Conflict and Cooperation in the Circumpolar North*. Hanover, NH: University Press of New England, 1992.

———. "Governing the Arctic: From Cold War Theater to Mosaic of Cooperation." *Global Governance* 11, no. 1 (Jan.–Mar. 2005): 9–15.

Zellen, Barry Scott. "The Arctic Council Pause: The Importance of Indigenous Participation and the Ottawa Declaration." *Arctic Circle Journal*, June 14, 2022.

———. *Arctic Doom, Arctic Boom: The Geopolitics of Climate Change in the Arctic*. Santa Barbara: Praeger, 2009.

———. "The Arctic on Fire: Intensifying Wildfires and Climate Change as a Top Security Threat in the North." *Arctic Circle Journal*, November 29, 2023.

———. *The Art of War in an Asymmetric World: Strategy for the Post–Cold War Era*. New York: Bloomsbury Academic, 2012.

———. "As War in Ukraine Continues, America Refocuses Its Attention on Arctic Front." *The Polar Times: The Journal of the American Polar Society* 5, no. 1 (Spring–Summer 2023): 16–18.

———. "As War in Ukraine Upends a Quarter Century of Enduring Arctic Cooperation, the World Needs the Whole Arctic Council Now More Than Ever." *The Northern Review*, no. 54 (2023).

———. *Breaking the Ice: From Land Claims to Tribal Sovereignty in the Arctic*. Lanham, MD: Lexington Books, 2008.

———. "Calculated Risk? Barry Scott Zellen Examines, Following Turkey's and Hungary's Consent, the Approval by all NATO Member States of Finland's NATO Accession." *Intersec: The Journal of International Security*, April 2023: 12–14.

———. "China and the 'Near-Arctic': An Opportunity Lost Over 150 Years Ago." *Georgetown Journal of International Affairs*, September 5, 2019.

———. "Clan, the State, and War: Lessons from the Far North." *Joint Force Quarterly*, no. 58 (2010): 20–21.

———. "Cold Front: Hillary, Ottawa, and the Inuit: A Year after the Inuit Re-Assert Their Sovereignty, Washington Takes Their Side." *Journal of Military and Strategic Studies* 12, no. 3 (Spring 2010).

———. "Cold Shoulder: Barry Scott Zellen Examines Why the Arctic Council 'Pause' Puts Arctic Cooperation into a Deep Freeze." *Intersec: The Journal of International Security*, May 2022: 8–10.

———. "Geopolitics, Indigenous Peoples and the Polar Thaw: Sub- and Transnational Fault Lines in the Coming Arctic Cold War." *Nordicum Medterraneum* 16, no. 2 (July 2021).

———. "High Stakes in the High North: Alternative Models for Greenland's Ongoing Constitutional and Political Transformation." *Nordicum Mediterraneum* 16, no. 2 (July 2021).

———. "A Missed Opportunity: How China Ceded Its Claims to What Is Now the Russian Far East, Leaving Japan as Asia's Pre-eminent 'Near-Arctic state'." *Intersec: The Journal of International Security*, October 2019: 26–28.

———. "Multinational Corporations in the Arctic: From Colonial-Era Chartered Companies to Contemporary Co-management and Collaborative Governance." In *The Palgrave Handbook of Arctic Policy and Politics*, edited by Ken Coates and Carin Holroyd. New York and London: Palgrave Macmillan, 2020.

———. *On Thin Ice: The Inuit, the State and the Challenge of Arctic Sovereignty.* Lanham, MD: Lexington Books, 2009.

———. "Path to Peace: Barry Scott Zellen Explains Why as the Arctic Thaw Continues, Lifting the Arctic Council Pause Presents Russia a Viable Route to Resolution." *Intersec: The Journal of International Security*, June 2022: 16–18.

———. *The Realist Tradition in International Relations: The Foundations of Western Order.* Westport, CT: Praeger Security International, 2011.

———. "Rethinking the Subcomponents of World Order." *Small Wars Journal,* August 15, 2012.

———. "Russian Aggression Gives US Excuse to Focus Military, and More, on Arctic," *Responsible Statecraft*, October 29, 2022.

———. "A Sacred Promise Broken? Barry Scott Zellen Explains Why the Decline in Indigenous Consultation at the Arctic Council Signals a Retreat from the Spirit of the 1996 Ottawa Declaration." *Intersec: The Journal of International Security*, July/August 2022: 22–24.

———. "Stability and Security in a Post-Arctic World." In *The Fast-Changing Arctic: Rethinking Arctic Security for a Warmer World*, edited by Barry Scott Zellen. Calgary: University of Calgary Press, 2013.

———. "Stepping Stones to a Secure Arctic: Strategic Importance of Arctic Island Chains Grows in Lockstep with Rise in Polar Temperatures, State Rivalry." *Proceedings* 78, no. 1 (Spring 2021): 26–20.

———. "U.S. Defense Policy and the North: The Emergent Arctic Power." In *The Fast Changing Arctic: Rethinking Arctic Security for a Warmer World*, edited by Barry Scott Zellen. Calgary: University of Calgary Press, 2013.

———. "Where East and West Converge: The U.S. Embrace of Collaborative Security for the Arctic." In the *Handbook of the Politics of the Arctic*, edited by Leif Christian Jensen and Geir Hønneland. Cheltenham, UK: Edward Elgar Publishing, 2015.

Index

About the Book

For some three centuries, the Arctic region has been a zone of collaborative governance. The interests of diverse sovereign states, indigenous peoples, NGOs, and other stakeholders have been aligned—even during periods of global conflict. Now, however, these consensus-based foundations are being tested.

In *Arctic Exceptionalism*, Barry Scott Zellen considers: What explains the enduring cooperation in the region? Will new international dynamics upend the consensual approach? Could the intensifying nationalism across the Inuit homeland likewise endanger it? Zellen traces the region's long diplomatic history to show how competing interests have managed to establish an enduring stable order, and how escalating state rivalries and renewed nationalism are likely to affect it.

Barry Scott Zellen is research scholar in the Department of Geography at the University of Connecticut and an Arctic security fellow at the Institute of the North.